Virtual Prototyping

IFIP – The International Federation for Information Processing

IFIP was founded in 1960 under the auspices of UNESCO, following the First World Computer Congress held in Paris the previous year. An umbrella organization for societies working in information processing, IFIP's aim is two-fold: to support information processing within its member countries and to encourage technology transfer to developing nations. As its mission statement clearly states,

> IFIP's mission is to be the leading, truly international, apolitical organization which encourages and assists in the development, exploitation and application of information technology for the benefit of all people.

IFIP is a non-profitmaking organization, run almost solely by 2500 volunteers. It operates through a number of technical committees, which organize events and publications. IFIP's events range from an international congress to local seminars, but the most important are:

- the IFIP World Computer Congress, held every second year;
- open conferences;
- working conferences.

The flagship event is the IFIP World Computer Congress, at which both invited and contributed papers are presented. Contributed papers are rigorously refereed and the rejection rate is high.

As with the Congress, participation in the open conferences is open to all and papers may be invited or submitted. Again, submitted papers are stringently refereed.

The working conferences are structured differently. They are usually run by a working group and attendance is small and by invitation only. Their purpose is to create an atmosphere conducive to innovation and development. Refereeing is less rigorous and papers are subjected to extensive group discussion.

Publications arising from IFIP events vary. The papers presented at the IFIP World Computer Congress and at open conferences are published as conference proceedings, while the results of the working conferences are often published as collections of selected and edited papers.

Any national society whose primary activity is in information may apply to become a full member of IFIP, although full membership is restricted to one society per country. Full members are entitled to vote at the annual General Assembly, National societies preferring a less committed involvement may apply for associate or corresponding membership. Associate members enjoy the same benefits as full members, but without voting rights. Corresponding members are not represented in IFIP bodies. Affiliated membership is open to non-national societies, and individual and honorary membership schemes are also offered.

Virtual Prototyping
Virtual environments and the product design process

Proceedings of the IFIP WG 5.10 workshops on virtual environments and their applications and virtual prototyping, 1994

Edited by

Joachim Rix
Fraunhofer Institut für Graphische Datenverarbeitung
Darmstadt
Germany

Stefan Haas
Fraunhofer Center for Research in Computer Graphics
Providence
USA

and

José Teixeira
Centro de Computacao Grafica
Coimbra
Portugal

Published by Chapman and Hall on behalf of the
International Federation for Information Processing (IFIP)

 **CHAPMAN & HALL**
London · Glasgow · Weinheim · New York · Tokyo · Melbourne · Madras

Published by Chapman & Hall, 2–6 Boundary Row, London SE1 8HN, UK

Chapman & Hall, 2–6 Boundary Row, London SE1 8HN, UK

Blackie Academic & Professional, Wester Cleddens Road, Bishopbriggs, Glasgow G64 2NZ, UK

Chapman & Hall GmbH, Pappelallee 3, 69469 Weinheim, Germany

Chapman & Hall USA, 115 Fifth Avenue, New York, NY 10003, USA

Chapman & Hall Japan, ITP-Japan, Kyowa Building, 3F, 2-2-1 Hirakawacho, Chiyoda-ku, Tokyo 102, Japan

Chapman & Hall Australia, 102 Dodds Street, South Melbourne, Victoria 3205, Australia

Chapman & Hall India, R. Seshadri, 32 Second Main Road, CIT East, Madras 600 035, India

First edition 1995

© 1995 IFIP

Printed in Great Britain by TJ Press, Padstow, Cornwall

ISBN 0 412 72160 0

A catalogue record for this book is available from the British Library

∞ Printed on permanent acid-free text paper, manufactured in accordance with ANSI/NISO Z39.48-1992 and ANSI/NISO Z39.48-1984 (Permanence of Paper).

CONTENTS

Preface

Virtual Prototyping - a new keyword from the applications point of view!

Based on the enormous developments in the areas of CAD, CAM or CIM as well as in visualization and interaction techniques, a development towards virtual environments is on its way. The computer technology more and more provides the possibilities to support the needs of industrial applications. Design support, planning and simulation tools, presentation and interaction mechanisms, and group work support are building the environment for product design and manufacturing. The major goal is the integration of the product development process with the support of the CA- techniques.

A Virtual Prototype is a major interim step towards the final product. Based on the design information, like geometry and topology, simulation results, like FEM or kinematic calculations, combined with material, tolerances and other information, it will be possible to generate a prototyp with the computer ready for realistic presentations as well as interaction with the product even in an early stage of the development.

Developing this technology and the integration in the product development process promises major advantages for the industrial process:

- Reduction of time - the time parameter is somehow the most important factor today, as time-to-market is a key marketing issue related to the competitors.

- Saving cost - virtual prototyps will reduce the number of physical prototyps needed, which reduces not only the development time, but also the manpower as well as the tools and material. Also early prototyping results can feedback to the design stage in time before the production costs are already fixed.

- Increase of quality - the evaluation of different alternatives of the design can be realised in the virtual environment much faster and cheaper. This allows for a better evaluation of the appropriate solution to serve the given requirements best.

To support these developments the Working Group 5.10 of IFIP TC 5 (Computer Applications in Technology) organised two workshops in the fall of 1994.

The first one addressed the more general issue of "Virtual Environments and its Applications". Topics like simulation, interaction, concepts and tools, as well as support services were addressed besides the application perspectives of Virtual Reality. This workshop was organized by the Centro de Computacao Grafica (CCG) and took place in Coimbra, Portugal on October 24 and 25, 1994.

The second workshop was more focussed on the application of the "Virtual Prototyping". From this point of view the conceptual design and framework aspects of an environment, new modelling and data management needs for an integrated product modelling, and the basic technologies for rapid prototyping, interaction and virtual environments were presented. These presentation sessions were followed by small group discussion sessions on specific topics. This workshop was organized in cooperation by the Fraunhofer-Institut für Graphische Datenverarbeitung (IGD) in Darmstadt, Germany and the Fraunhofer Center for Research in Computer Graphics (CRCG) in Providence, RI, USA. It took place in Providence, RI on September 21 - 23, 1994.

As the two workshops addressed related areas and covered common topics of interest, after the events it was decided to put the two results together in one volume of an IFIP publication, as it is presented in here.

The book is set up in six separate sections:

- Perspectives for Virtual Prototyping

  This section describes the developments towards the use of Virtual Prototyping. A historical view towards virtual reality, information technology as an enabler for process optimization and going beyond manufacturing technology are basic aspects leading to new architectures for cooperative environments supporting virtual prototyping.

- Advanced Product Modelling Techniques

  Two major aspects are addressed in this section. First, the importance of an integrated product model, as under development with the STEP standard, was stated to support a virtual prototyping system on the data integration level. Second, new modelling techniques for semantic modelling were presented from different perspectives. The further needs for geometric representations and the integration of features in the object specification, as well as the needs to integrate in a concurrent engineering and virtual prototyping environment were discussed.

- Architectures for Distributed Systems

  With the perspective to allow the use of virtual environments also for distributed applications, the support of cooperative work among the users was discussed. This section presents different architectures based on the application requirements for cooperation, real-time needs and network support.

- Advances in Virtual Reality Technologies

  Interaction with and presentation of the information in real-time are the major goals to dive into your immersive environment. This section addresses these issues and describes some advances in interaction techniques in a 3D environment and in using simulation results in the virtual world. Also the task performance using 3D displays and the enhancement of computer vision using virtual reality and vice versa are covered by two papers.

- Industrial Applications

  This section demonstrates the potential application scenarios for Virtual Prototyping.

  In addition, the issues of system and data integration of different applications in the virtual environment are addressed in these papers. The scenarios are showing a variety of application areas of virtual prototyping, from visual evaluation, design and product development to robot or kinematic simulation, vehicle motion planning and assembly.

- Workshop Summaries

  This section is summarizing the results of the two workshops and is reflecting the discussions during the workshops.

Both of the workshops had a good international participation from 8 different countries and 56 respectively 35 participants. Based on the review of the submitted papers, the programme committees selected 15 and 17 papers for presentation at the workshop, as a mix of scientific and application oriented contributions. The papers as well as the discussions offered a great opportunity for this field of development.

Virtual Prototyping, using virtual environments for the product development process, seems to have an enormous potential for future applications in the industrial areas. Because of the promising results of the workshop the participants agreed to plan for a follow-up workshop, which is currently planned for late 1995, to take place in Arlington, Texas in the US.

The editors would like to express their gratitude to all those, who supported the two workshops and helped to get the book ready for publication. First of all, those are the speakers, participants, and the authors.We also would like to thank the members of the programme committees, the sponsors of the workshops, and Chapman&Hall as the publisher, namely Susan Hodgson. Our gratitude also apply to IFIP and especially to Prof. Encarnação, the chairman of the working group 5.10, for initiating the two events. Finally, many thanks to Ms. Podlich and Mr. Quester for their effort to put the final document together.

Darmstadt, Providence, Coimbra   May 1995

J. Rix, S. Haas, and J. Teixeira

IFIP TC5 (Computer Applications in Technology) Working Group 5.10

Workshop on Virtual Environments, Providence, RI (USA)
Programme Committee

P. Bono, Fraunhofer CRCG, Providence, RI (USA)
P. Brunet, University P. Catalunya (Spain)
S. Bryson, NASA (USA)
R. Earnshaw, University of Leeds (UK)
J. Encarnação, GRIS (Germany)
B. Falcidieno, IMA (Italy)
A. Figueiredo, University of Coimbra (Portugal)
M. Gigante, RMIT (Australia)
M. Göbel, Fraunhofer IGD (Germany)
G. Grinstein, University of Massachusetts (USA)
M. Gross, ZGDV (Germany)
R. Hubbold, University of Manchester (UK),
T. Kunii, University of Aizu (Japan)
L. Magalhães, Unicamp, FEE-DCA (Brasil)
J. Rix, Fraunhofer IGD (Germany)
J. Teixeira, CCG, Coimbra (Portugal)

Workshop on Virtual Prototyping, Coimbra (Portugal)
Programme Committee

R. Anderl, Technical University of Darmstadt (Germany)
G. Barth, Daimler-Benz (Germany)
P. Bono, Fraunhofer CRCG, Providence, RI (USA)
R. DeLosh, Ford Motor Company, Dearborn, MI (USA)
B. Falcidieno, IMA, Genova (Italy)
H. Fuchs, UNC-CH, Chapel Hill, NC (USA)
J. Hooker, Caterpillar, Inc., Peoria, IL (USA)
F. Jouy, Dassault Systèmes, Paris (France)
E. Jungmann, Siemens, Munich (Germany)
H. Kagerer, Siemens Nixdorf Informationssysteme, Burlington, MA (USA)
F.-L. Krause, Fraunhofer IPK, Berlin (Germany)
S. Lu, UIUC, Urbana-Champaign, IL (USA)
G. Olling, Chrysler Corporation, Detroit (USA)
A. Requicha, USC, Los Angeles, CA (USA)
U. Rethfeld, Siemens Nixdorf Informationssysteme, Munich (Germany)
J. Rix, Fraunhofer IGD, Darmstadt (Germany)
W. Steger, Fraunhofer IPA, Stuttgart (Germany)
J. Teixeira, CCG, Coimbra (Portugal)
T. Tomiyama, University of Tokyo (Japan)
M. Wozny, NIST (USA)

PART ONE

Perspectives for Virtual Prototyping

1

About "Plato's Cave"

J. C. Teixeira
CCG/ZGDV - Centro de Computação Gráfica
Rua de Moçambique nº 17, R/C Esq., 3030 COIMBRA, PORTUGAL
email: teixeira@ccg.uc.pt.

V. B. Murtinho
Departamento de Arquitectura - Univ. Coimbra
Largo D. Dinis, 3000 COIMBRA, PORTUGAL.

> " Imitation is, therefore, away from truth, and if it moulds up all the objects it is as it seems, only because it respects a minimum portion of each one of them, which is nothing more than a shadow."
>
> Plato, Book X, *The Republic*

Abstract

Every investigation is vivified through the constant search for sublime. The development of a mimetic illusion goes until the limit of the virtuosity allowed by each techniques related with its own time. Man wants to feel himself as being the *Nature's Pencil*[1]. However, in the application of new working methods, what could be expected from each one remains on a lower level comparing with the discoveries based on themselves.

In this paper, based upon an historical evolution of the methods, means and representational spaces, we would like to stress that *Virtual Reality* is a decisive step to the conquest of the perceptive space. Man, throughout his life, operates the reality in different ways, feeling always the need of immersing in the representation of the surrounding reality. Therefore, it is

---

[1] *The pencil of Nature (1844-46)* Is the title of a book by Fox Talbot, one of the photograph's pioneers.

important to settle down the effective conscience of the trajectory, described by the human geniality during the conquest of representational processes to create illusions of the reality.

1 INTRODUCTION

When the utensil became instrument[2] and the plain surface has offered itself as a support, the spirit, the eye and the hand could finally converge to the translation of the already seen into the new visible. It was always man's intention to produce images which he intended to be close to the visible ones. He always wanted to cause the effect of impression of the reality. Nevertheless, the man soon realised that the technical means available were quite limited. In other words, the image as an equivalent is determined by its environment in which it is expressed. Therefore, any system or method that represents the real is, certainly, another step into ingenious artifice of the illusion of the senses; visual, tactile or audible. Such an advance in the illusive process was always due to a convergence between human geniality and scientific techniques.

We believe that the cavern men's amazement before parietal paintings was certainly the same that was caused by the paintings of the Greek Parrhasius, the panels of Brunelleschi, the daguerreotypes of Mister Daguerre, the first film projection by the Lumière Brothers, the first television live program or the first virtual environments. Let it be the écran, a wall, a *vetro tralucente*, a projection panel, a monitor or the retina itself one thing is for sure: they all support images of great amazement.

2 THE REPRESENTATION AND SPACE APPROPRIATION

Since Lascaux, man make an effort to control the real. Initially, because he could reproduce it, he had a gift of magic and repetition. The magician-artist can draw certain objects many times as he wishes. He could reinvent the real, but he has always been restricted in participating on the created environment. The drawing materialises the retina image and gives physical existence to mental image, convergence point between the sensitive and the intelligible. Parietal images, the first inventory of the world, are remarkable testimonies of the changes and the new knowledge of Palaeolithic man. As Bergson decidedly claimed, it was in prehistory and due to "intelligence, a step into the original", that "the capacity to manufacture artificial objects, in particularly tools to make tools, varying indefinitely the manufacturing" (Bergson, 1911) was propitiated.

---

[2] One must clarify the distinction thar it is made in our work between **utensil** and **instrument**. Utensil must contribute to external actions, while instruments are sued to adequate sensorial perception to phenomenons, allowing a better observation off things.

Figure 1 Lascaux caverns - France (about 15.000 a.C.) (Gombrich, 1989).

It was in Greece culture, as the cradle of the western culture and also a culture of the image, that the representation's aporia had an enormous liveliness. All art in ancient time is founded on a *model*, or *idea*, as Plato says, which can be placed in front of the artist or simply, the artist can have it in his mind (Seneca, 1991). Every time the artist copies the model he evidences its truth, that means, he certifies its consistence and allows a model's better knowledge. In representation, once we are limited to the plan, it becomes impossible to "reproduce" or "copy" the things. We can only execute an *imitation by simulacrum*. One example of the reproduction of "equivalent" signs or simulacrum is the painting of grapes' bunches from the Greek Parrhasius: This painting has such a skilfulness that it is said that real birds would attack it, thinking those bunches were real.

In his book, *The Sophist,* Plato defines two forms of **imitation**: a copy as reproduction of the real in its proportions of length, width, depth and suitable colours and, **simulacrum** towards something resembling the thing when it is seen from a unfavourable position. Maybe this critic was not only addressed towards painting, as one clearly understands by reading the mentioned excerpt. It certainly was also addressed to the famous Phidias statue's question: the inferior part of the sculpture was too short when seen at close distance but its dimensions seemed correct when it was placed on the level of the sight[3]. This episode shows the importance of the artifice, in what concerns the subjectivity of vision. In reproducing the external shape, the model is built not only by the relation with the real object, but mainly through the way in which the object will be perceived.

From Roman-Hellenistic culture came the idea of representation as a consequence of the object's real image seen by the artist. The real image itself is no longer considered. Plotinus

---

[3] According to Pliny, Alcamenes and Phidias have on a building of a statue of Athena. This statue was to be placed on a column: Phidias made an elongated and apparently deformed figure, that was judged as inferior to the one of Alcamenes but, when Phidias statu was placed on the pedestal, it was the one that gare the better seeing thanks to counter-perspective effect.

Neoplatonism[4] reassures this idea which would influence all representation in most Middle Ages and Renaissance through Alberti's philosophic reflections. Nevertheless, in Seneca's *epistulae morales* it is still indifferent the fact that imitation can born from an "idealised" and interior image or from a natural object.

Christianity brought a generalised movement in western world, a movement of symbolic representations which were absolutely esoteric and which evoked a superior and super-sensible reality. Generalised disinterest towards sensible was settled. Consequently representation broke with the process of life — abjection for the visible — and stopped working with materiality and "physical supports", becoming an aprioristic and privileged way of subjective experience, a support of diffusion of the massesë feelings.[5] If in Ancient time the Gods participated in daily life, becoming promiscuity between mythology and humanity a commonplace, in the subsequent historical period an hierarchical superior world was elected and any event of the under-world was attributable to divine will. Every representation on this period is "more a symbol than an image, more a convention than a vision" (Murtinho, 1993).

As a result, any biblical image has an implacable, mesmerising effect. It is a kind of open window below Heaven and it maintains an exclusive relationship with the divine, through spiritual and doctrinal contents. Such is the reality creatures conceive and try to explore.

When Giotto, a Florentine painter used a natural scenery on the life of S. Francisco, he proposed a sensitisation of the visible in what concerned representation. Maybe because that whole scene was somewhat contemporaneous — it is said that Giotto actually met S. Francisco — it produced great astonishment which resulted from a big similitude between figures and real persons. Images are not allusive: they are solid, tangible and vigorous in expression and gesture. S. Francisco was "in fact" S. Francisco (Battisti, 1990). The naturalistic quality of Giotto is attested in post-classic art, on introducing "the insect that deceives in the eyes"[6].

---

[4] About Neoplatonism and Plotinus, we sugest the reading of *"Le Neoplatonisme"* from Jean Brun PUF, France, 1988.

[5] In Arras Council (1025 b.C.) the aims of painting are: the catechetical, the historical and the aesthetic ones. From here a maxim is born, the same one that would justify christian themes: *Quod legentibus scritura, hoc idiotis pictura* (Painting is a way of bringing God's word to the illiterate people).

[6] According to Vasari on his *Vites*, Giotto would have painted on a nose of a recent made figure by his master Cimabue — a fly, which was tried to be repeatedly performed by his master Cimabue. This quality of a notable performer is certainly the one which is exalted on Dante's (*Divine*) *Comedy*.

Figure 2 S.Francisco - Giotto (1296-1297) (Battisti, 1990).

A century later, Filippo Brunelleschi, while elaborating a pair of panels representing *Baptistery* and the *Palace of the Landlord*, achieved an unequalled deep, through a precise and attentive study of reality and through rigorous geometry: with the aid of a fixed mirror and the look on a certain determined point, the image of the real reflected on that mirror would be the image of the painting itself when reflected on the mirror. Certainly, these were images of amazement.

The mirror will acquire a strong demonstrative slope, giving, in practice, a comparison between the real shape in space and the seen shape. *In principio erat speculum...*

Through images, pictorial or specular, life and the world become duplicated. The mirror is the judge. Only with "the judgement of the mirror" (Hollanda, 1549) can the validity of the relationship be determined: "*if you want to see if your painting faithfully reproduces the real, you take up a mirror and try to reflect the living thing on it, and then you can compare this image with your painting. You must pay attention if both are in compliance with each other.*" (Vinci, 1943)

Renaissance brings the attraction by reflection on nature and the observation method as a real way leading to science. Representation offers direct relations between the objective real and its registry seen from a particularly point of view. Man sees the world depending on his knowledge, and it is with his own knowledge that man represents it.

Objective illusionism is the underlining form of affective communion — *empathy* — and it is an accomplice between man and the world. The artist shares the experience he had with the visible and awakes the energy of seeing. The Albertian *vetro tralucente* allows the representation of things and perspective is the exact method leading works to *perfezzione*; that is the universal judgement. Representation covers the reproduction of reality or the projection, as an anticipation, of a reality to come. Painting rectifies life and the world and it is a means of triumph over nature. Therefore, "the second way of the Renaissance which would represent the art's decisive moment in a process of conquest of perfect composition, carries itself the seeds of distrust over the value of objective rules. These rules would generate the interest on pretended representation of the space. Now it is the *art of pretending*. That is when Leonardo da Vinci begins the process of discovering the psychology of shapes" (Tavares, 1994). The *trompe-l'œil* effect and its simulated architectures propitiate a transparency illusion of the screen and the ephemeral stupefaction of look.

Figure 3 Albrecht Dürer (1525) (Wright, 1983).

Inexorably, the insatiable appetite of vision leads to certain limits, such as laboriously perspectives founded on the most vigorous geometric constructions of, for instance, Paollo Ucello. Another example is met on Albrecht Dürer. He tried to build entangled projection mechanisms of body figures on a plain surface; mathematization of representative constructions or the elaboration of artifices which materialise the visible experience. However, as Panofsky showed in a well-known essay (Panofsky, 1927), perspective is a *symbolic* — or conventional — *form* of representation of a retinical (spherical) image on a plain surface (Panofsky, 1927). Consequently imitation is a "cultural codified" process which presupposes "a relationship of equivalence or *similitude* between image and imitated object" (Modica, 1992).

Figure 4 San Romano Battle (1450) (Gombrich, 1989).

Karl Popper's pertinent critic towards David Hume's belief in science infallibility recognises the provisional value of conclusion. Science is not an absolute truth, but it is the absolutely relative truth. Available technology constitutes inevitably a formulative limitation. Interpretation or representation are conditioned by the previous knowledge of the thing and the world. That is why one can understand Kepler's definition: "vision is produced by the painting of the visible thing which builds up itself on the white and concave retinic wall."[7] Certainly three centuries later Kepler would compare vision to photograph but in his time painting was the art drawing near retinic image.

On seeing, mixed with the will to observe, visibility must take its place, *"painting itself on the receptive surface of the retina though the action of the most insignificant, colourful light brushes"* (Frade, 1992). Kepler, taking his *camarae obscurae* as a model of vision, observes the most fundamental act of vision and his *mechanic eye* enables the tracing of the *imago rerum visibilis* captured on the inside of the equipment.

When Niepce and Daguerre succeeded in a chemical fixing on a sensible emulsion, that means when they invent the photograph film, they gave viability to the promise of infinitive duplication of world's appearances through the simple action of a shaft of light reflected on a support. Photographic images soon become a certain type of objects virtually omnipresent. Photograph, almost obligatory in modern society, becomes the most confortable way giving

[7] In the original, "Visio igitur per picturam rei visibilis ad album retinae et cauun parietem". (Kepler, *Ad Vitellionem Paralipomena quibus astronomiae pars optica traditur*, 1604; Extracted from *Figuras de Espanto*, Pedro Miguel Frade, Asa Edições, Oporto, Portugal, 1992, p. 37).

access to the world; photograph is the proof which satisfies visibility, the certificate of existence and the guarantor of authenticity. It is a miniature and a fragment of the world. The camera imposes, according to Moholy-Nary in *Photography Film* (1925), *the hygiene of the optical* and it eventually *limits our standard of pictorial and imaginative association which was recorded in our vision by great painters* (Sontag, 1986).

Each image propitiates a revelation that gives us back the lost vision[8]; the vision is the omnipresent witness. Fixed images work as captured experiences, where the camera is the privileged instrument on the acquisitive act. Image establishes an obvious relationship of similitude with the referent, and it is, therefore always a presential symbolic form. Humanity, *hopelessly attached to Plato's Cave* — as Susan Sontag declared — *begins to delight itself with images of truth.*

However, the discoveries in the field of the technologies of visible unveil the possibility of motorization of the snapshot[9] — 17 images per second and later 24 images per second allowed an optical illusion on film projection. *The art of the engine* is the possibility of reproduction in its real time. The effect is born when the rhythm of the succession of images makes an image appear while a precedent image is still recorded on the sensorial impression of the retina. Still, in the translation of the illusion of movement, a steady dimension is always necessary. Movement is a relationship between phenomena: movement exists because fixed points also exist.

The first projection of the history of film making, *L'entrée du train en gare de Ciotat* by the Lumiére brothers, projected in the Salon Indien in 1895 frightened the audience before the optical illusion, nowadays risible, that a locomotive was precipitating towards them. Certainly the audience didn't feel outside the action when confronted with a mechanical transposition of the external world to the screen . Initially, it were the places of the images with the images of the places which amplified and rectified our vision of the world.

Also Vladimir Zworyki's *Iconoscope[10]* was, according to him, a means to raise vision, but it soon became a precise means of communication of the masses. With a sender, a receiver, broadcasting by Hertzian waves to a long distance, image can reach even the most recondite place.

---

[8] More than twenty years over the mission of Apollo 11, Edwin Aldrin confessed that he remembers more clearly his mission to the Moon through the pictures that immortalized the event than what he saw with his own eyes. The adventure belonged him more intimately because it had been transformed into a patrimony of Humanity than a patrimony of his own.

[9] Long before cinema appeared, experiences on the persistence of the retinic image had already made — the first experiences were made in the end of the 10th century. From this experience, Joseph Plateau, a Belgian physicist, invented the *phenaquistiscope*, in 1832, which allowed to vision the illusion of movement through a minimum of ten images per second (in *Histoire du Cinéma*, Gérard Betton, Presses Universitaires de France, 1984).

[10] Iconoscope is the first designation to electronic television.It was invented by Vladimir Zworykin and presented thr first time in 1933.

The television set is the *witness lamp* which one can switch on and off in order to see the "transparency" of the world. The speed-limit of electromagnetic waves guarantees a transmission in real time; video is no longer the most modernised representation of the fact, but a live presence of the place (Virilio, 1990).

Side by side with sensible reality, there is from now on the presence of a *telereality*. In the *box that changed the world*, the real is a reference of a non-thinking reality. From a epistemological point of view, truth was guaranteed when the things had its presence before the eye. *Video* gives visibility and the potential presence and also the abolition of the distance. In the *cone of the visibility of appearences* one has a vertigo of the revelation of the extensive space.

The *video*, as Paul Virilio said (Virilio, 1990), is the active participation in the constitution of an instantaneous and interactive localisation of a new concept of "space-time", which has nothing in common with the topography or Euclidian geometry, but with implications on the vision and perception of the world.

The live tele-vision on planetary scale depends on the performances of the satellites. The speed of the events and the fact that they reach us in real time — remember, for instances, the war of the Gulf — place doubts in the concept of reality. For a long time, the idea of reality was a slow process of progressive assimilation. The idea of the real is mostly a presential genesis. One must still see before he believes. With the acceleration of time and mutations, and the dematerialization of informative supports, the concept of static reality does no longer functions. There is a crisis in the concept of reality and a loss towards the value of experience. The image is more and more the opium of the world. Solid world is more and more a fluid world.

3 VIRTUAL REPRESENTATION OF REALITY

It is clear that modernity brought the rupture with the idea of reality legated from the past: the idea of a space and homogeneous and isotropic time in which lays down a reality well implanted in the solidity of the materials of which objects are made (Manzini). Contemporaneity consecrate vulgarisation of instruments such as the photograph or television. Systematic intimacy with a simulated reality, daily apprehended alters the idea of things. The idea of things is constructed not upon a physical reality but, most of times, upon images.

Since the primordium, the real thing representation was always subordinated towards its own image — see, for instances, the unfruitful efforts on the Plato's model. Never in history was the thesis of Feuerbach on its 2nd edition (1843), *The Essence of Christianity* so actual. This thesis makes a critic of the society of the 19th century, because it *"preferred image than the thing, copy than the original, representation than reality, the looks than the being"* (Feuerbach, 1992).

Nowadays, images are considered substitutes or interpretations of things. In this sense, images are of two types: one, considers image as an emanation (of light waves or shafts) or as a material vestige (kind of trace); the other sees image as a representation or a simulacrum.

One example of the first type are photographic or *video* images; objective painting or computer environments are examples of the second type. Through images such as photographic ones, one looks for a substitute for the world, while through artificial images, such as cybernetic images one tries to build a substitute world.

Image and thing are two different forms of existence. Therefore, or the thing remains itself because coincides with the object, or the image is the object, despite referring to referential object. The presence of image appeals for a memory effort this means that every time we are confronted with a certain image, its nature-sign is directly proportional to our subjacent or pre-existent culture. The richer our information remains, richer will be its decodification. An image simply by itself doesn't mean anything.

Our experience in inhabit the world tells us that the truth is transmitted by our senses. The act of seeing, listening or feel converge to a redefinition of the visible and the sensible. Our coupling to the real world is made through sensorial organs. It is through the senses, mainly through vision, that a way of access to the world is propitiated. "In experience things are no longer subjects of vision; they become *seen things*. What we call vision belongs, now to the potency of thinking which attests that this appearance has answered to our eye's movements" (Merleau-Ponty, 1992). In vision, man is not aware of the eye; things come straight into his brain. Therefore, our vision is not wide, what we see are merely perspectives, and *our reality*, as Descartes pointed out, *is the thought*.

Greek culture's imagetic axiality, as Umberto Eco said on an interview to the *Nouvel Observateur*, made intelligence go through an intuition of the image. Things and the world are the *object* of thought. Also, Knowledge is based on a visual process, even on a visuality of abstraction: reasoning. The great divulgation of painting in the Midde Ages was a way of bringing God's word to illiterate people. One must make a critic at this point because the spiritual must not necessarily be translated into images, or internal images materialised in external images. "The power to create needs a point of support the crutch of reality" (Pessoa, 1986).

In virtual systems, visual sensations and physical stimulus converge to the effect of reality. The new Virtual Reality Systems problematize concepts such as: the exterior and interior, mental and material. Cybernetic simulacrum cancel the epistemological difference between the concepts of exterior and interior. They interiorize any form of exteriority.

Traditionally the sensible and the intelligible are opposed and any sensation belongs to a model: there is, on one side, the model, and, on the other side, the idea. In virtual environments, every time the image is projected on the retina, those two spheres are connected in an operative way. "From now on, *sensible images* are potentially able to modify models, the same ones that had generate them. Images and models are, in fact, of the same nature: they are two faces of the same reality, a sensible and a intelligible face"[11].

As Ovid suggests, Pygmalion a legendary sculptor of Cyprus, has fallen in love with a statue from his chisel and asked Aphrodite to give a woman made from that statue's image. The goddess promptly satisfied Pygmalion's wish, giving life to that statue. In this episode,

---

[11] in an interview of Phillippe Quéau, special dossier of the magazine *La Recherche*, May 1994.

Aphrodite as a transcendental potency becomes a potency that give life and provokes a metamorphosis of the figure in the reality of her model. The work is usually conceived on the image of the being, but, with Galathea, the being exists as the work's image. As a threshold phenomenon, it is the copy that originate the being on a efficient excess of the image. It is the formulation of a "vow which can not subsistute the representation of a being, even an *ideal* one, for a double's presence in the real" (Marin, 1993). It is *Pygmalion's Power*, the hyper simulacrum power, that humanity aspire since Ancient time.

Photograph achieved the perspective in real time. It was only surpassed by the *video*, because *video* allowed a real perspective in time. Technological developments in post-modernity, together with cybernetic models, are expected to bring a certain expectation on the elaboration of models close to those of Pygmalion. But, here, model and its graphical representation is not a double of the object; the goal is that the graphical representation could be seen as something that has all the characteristics of objects and could operate as stimulant field which could be accessed. Simulation is the capacity to pretend existing what doesn't exist, creating a situation of the real perception being replaced by the real of the assisted perception — Virtual Reality. "The mirror of the beings and the appearances of real and its concept no longer exists. Imaginary coextensivity doesn't exist no longer anymore: genetic miniaturisation is the dimension of the simulation. The real is produced from miniatures of cells, from matrix and from memories, from command models and, from here, it can be reproduced an indefinite number of times. It doesn't have to be rational anymore, because it can't be compared with any ideal or negative instance. It is merely operational. Indeed, it is no longer the real because it is no more entangled in an imaginary. It is a hyper-real, a synthesis' product that irradiate combinatory models in a hyper-space without atmosphere" (Baudrillard, 1981).

In these representations there is an artificial resurrection of the bodies. These bodies acquire a kind of ductile condition that through geometric combinations of lines and surfaces have the capacity to transform themselves in objects of experimentation, conditioned by the environment around them. Through protocol, the truth-record (computer) enables the realistic immersion in a synthetically and interactive world. The "visitor" immerges on the screen and inside it an unlimited and non-dimensional world appears.

The computer is a substitute of the world, which can be *neutral, extensive* (as the lens) or *intrusive* (as the Periscope). The first case deals with the translation of things as they are seen before one's eyes, for instance the virtual visits to domestic spaces. The second case intends to lengthen the reach of sight, as when one simulates certain molecular reactions, and the third case deals with the possibility of penetration in areas where one cannot physically enter, such as the interior of human arteries.

As an illusive phenomenon, virtual environments resemble a dream. Only the dream is not observable. The dream is the illusion to see what we don't see, *video* is the illusion of seeing and not having what we see. As Merleau-Ponty said, "the illusion of the illusions is to believe" (Merleau-Ponty, 1992).

Through synthetic images one can build a world of possible or non-plausible situations which can suggest various perceptions and emotions in a reserved environment: that is the deed of *Alice*. One is transported to that world in Lewis Carrol[12] brilliant tale, *Alice's Adventures in Wonderland*. By Alice's hand, dream gives place to the perception of multiple appearances of the world or to the experiences of truth. There, likelihood is the more believable as experiences of truth are the more better. Cyberspace is the "materialisation" of Alice's deed, that is mirror's crossing.

In his *Essay on Mirrors*, Umberto Eco refers to the mirror as being a *threshold-phenomenom that settles the limits between the imaginary and the symbolic*. Any specular experience belongs to the imaginary domain, because it is an illusory phenomenon. As an illusory phenomenon it can be integrated in the symbolic system, in order to be potentially operative. It is the mirror that profiles the passage of the image to the thing (Eco, 1989).

If we look into the mirror, cybernetic device confine us in "representative and symbolic illusions that give it consistence and specular nature on its appearance" (Rodrigues, 1990). Cybernetic and specular images are real because they can be accepted as a physical and consistent object and they can be fixed on a surface (one can take their photograph). They are also virtual images because they are apprehended as if they existed inside the screen, despite the screen — a kind of Odin's record (Borges, 1975). — having only one face. That is virtual reality's paradox.

Available technical means soon propitiate proliferation of generalised simulacrum production inside reality. Cybernetic electronic devices are an example of what has been said: they enables the figure of simulacrum which with a plain and reticular form, is at the same time irradiate and impulsive. But, beyond simulacrums of the real, new models of closed significance to themselves start to emerge. They are not open to any referential sense. The computer of *War Games* does not make a distinction on the game of reality, "between total war simulation and its effective beginning, because the real and the simulacrum confound themselves on the discursive order of the programme". The figure of simulacrum provokes an "impossibility of distinction between the figure and the real and a deadly dissuasion of the representation itself" (Rodrigues, 1990).

In Escher's lithography, *Magic Mirror*, figures cross a mirror and freely emerge on a surface supposedly rigid. A real dog crosses a mirror, and while he moves away from the mirror, his specular image withdraws to the opposite side. On the other hand, when a reflected image reaches the limit of the mirror, it acquires autonomously reality contours. The real is duplicated throughout specular image and this same image gets a new real. At the end, the real and image bear a phenomenon of objective duplication of the thing.

---

[12] Lewis Carroll is a pseudonym of Charles Lutwidge Dogson. It's interesting the distinction between mentioned and real authory.

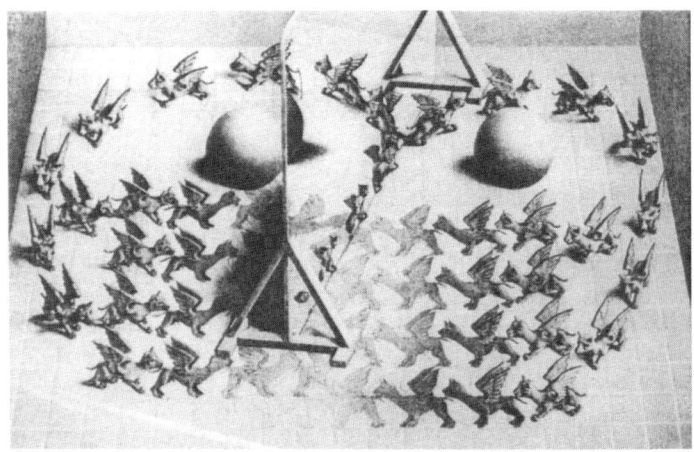

Figure 5 Magic Mirror - M. Escher (1946) (Ernst, 1991).

One of the problems that is subjacent to cybernetics and to the real is the definition of reciprocal frontiers. In the mirror world of *Alice*, this obstacle is dissolved provoking a conscious fusion between the two realities, one appearing following the other (*Alice* enters wonderland throughout a burrow). In a Virtual Environment that frontier is a physical conscious barrier, because when the user immerses in the environment, he cuts the stimulus of the outer world. Only through illusion one can reach the virtual environment. In the conscience of illusion, imagination is operative.

The objective of virtual systems is the relaxed co-operation between the visitor and the scene. The scene should be mediated by an easy handling machine less and conscious to the user. One might talk of transparency of the system which supports an illusive effect. In other words, interesting technology is the one that vanishes itself. Therefore, the more invisible technologies are and the more they embody themselves in the processes and in navigation practices, a better resolution one can take from illusive suggestion. When an artist draws, we doesn't feel his hand or the drawing object, the spirit must be directly attached to the drawing surface. Such must be the sensation of a user of any system: he doesn't have to feel the machine or the technology. That technology must be a passport that cancels the frontier.

An interesting point is the way one proceeds towards the knowledge of the object trough an interacting relationship between subject and object. The subject can act on the object and formulates, by this way, his own knowledge or, possessing already a certain knowledge of the object, it is the object itself that acts on the subject, causing a new knowledge of the object. It is why this relationship is interactive. In Cyberspace knowledge is provisional. It can be, however mould in the relationship of virtual subject/object — and vice versa — and adjusted in the relationship of real subject/object.

Related to this fact is also the question of knowing and of recognition. Or the thing is recognisable and in virtual system one needs external experience of the subject to validate action/reaction with the object, or it will be the thing itself that later will validate the interior experience one had on the virtual environment. As pure experience, virtual system only validates the virtually of the "experience" itself. Nowadays, sometimes is the simulacrum itself that validates the thing; it is familiar the usage of the expression "it almost looked like a movie" to narrate a situation and to make it true. Other times this expression appears in the movie itself and has the objective of transporting ourselves into reality.

Virtual Environments are like means of transport: they withdraws us from the environments where we physically are and takes us to other "spaces". Virtual Reality as a *medium*, symbolically allows to accede to physical object, despite it dispenses it. In real, the bodies are substituted by an illusory but operative artifice, functioning as a dissuasion process of that real: image becomes real and the real is imaginary. Through the suppression of truth, the systems of virtual reality create an illusion of supplementary truth by attesting what is credible. The virtual is an intervention work which seeks in reality the control over perception, replacing world's materiality by the dematerialization of "our reality". In cybernetic devices, apparently permeable, one operates with immaterial material and one moves in a non-dimensional space. This is a form of absent presence.

Virtual environments confine themselves a world to which one cannot demand more than to what the system has on its programme. In other words, this discovery will be always "controlled". Virtual environment are made with a key to experiment things. These environment presupposes always the total immersion of the subject by voluntary contract and always denunceable. All interactions between subject and the machine must be simulations in real time. It is through emotion and not through reactive interactivity that one are transported into spaces.

In this symbiosis man/machine/virtual environment a double presential genesis is established: physical presence on the real and virtual presence in Cyberspace. This is the gift of ubiquity.

If with the sentence *Johanes de eych fuit hic* written under the wall mirror of the nuptial camera which reflects the visible and invisible space (situated beyond the painting which canvas is "transparent mirror" and the painter stands still) the author of *Giovanni Arnolfi and his Wife*, signalises the ubiquity of the producer subject and of the represented subject (?) — eye and witness. In *Las Meninas* of Velásquez such an artifice is suppressed when the artist contemplates the model and he focuses his look on an external point outside the scene, coinciding with the eye of the spectator. There is, objectively, an inversion of roles: the spectator is the painter's model and the model on the linen is the audience represented on the scene. The artist is portrayed on the painting and the canvas made by the painter is not seen. In the moment that the painter places the spectator in his sight, he projects his image on the canvas is has turned is over and which has a visible reverse. But, surreptitiously, in the painter's *atelier* back window, among a series of suspended pictures, a mirror is found which offers the real image of the canvas, the one canvas marginalizes; the mirror gives back the invisibility of the painting (Foucault, 1966).

Figure 6 Giovanni Arnolfini and his Wife - Jan van Eyck (1434) (Gombrich, 1989).

The two examples we present (Figures 6 and 7) attest the problem of ubiquity in what concerns psychological and representative issues. They are, therefore worthy witnesses of the artist's skills. They also point out another important question: the symbolism of self-re-presentation and the abolition of the frontier that the painting constitutes between the world represented and the physical world.

Ubiquity is subjacent to virtual world. Cyberspace is an ubiquitous world of physical and mental immersion on image, where we can sail through virtual universes created by communication. But the most surprising thing is that one can choose between look at virtual world through his own eyes or with somebody else's in extra-corporeal experiences — *impartiality principle* — and one can, without any mirror, re-examine oneself. Through stereoscopic laser helmets capable of stimulating the retina, one creates a sensation of immersion and navigation on a controlled environment, capable of interacting and handling images.

Figure 7 Las Meninas - Velázquez (1656) (Gombrich, 1989).

The effort of seeing is no longer needed, the image meets the eye and settles down on the inside of the retina. Mutual attraction between the eye anxious to see and the object anxious to be seen, always mediated by the image, becomes a state, as Paul Virilio would say, of *Polar Inertia* where active image dispenses the object of which it is referential and gives attention to a passive observant. Traditionally it is the look that reveals the identity of the figure, in Cyberspace it is the image that reveals itself over the look. The image of synthesis assumes itself as an objectual latent potency.

As Paul Valery wrote:

"The desire of *realism* search more and more for the powerful means of reproduction. Reproduction leads to technique. Technique leads to classification and order. Order leads to the systematic, to the complete exploration of the most generous employment of all the resources, to his largest freedom, above anything else to fulfil. And taking exact reproduction of a concrete fact as a basis one will reach a kind of gymnastic which includes *false* and *true*."(Valery, 1994)

It is this duplicity that the painter René Magritte caricatures:

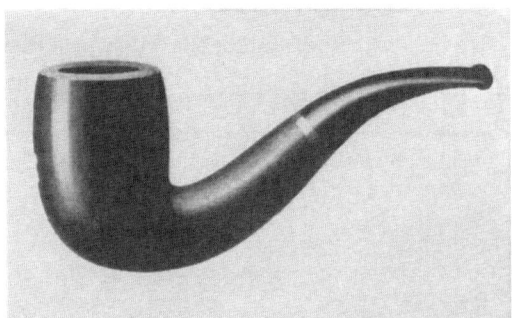

Figure 8 Ceci n'est pas une pipe - Magritte (1928/1929).

REFERENCES

Battisti, E. (1990) *Giotto*. Editions Albert Skira, Genève.
Baudrillard, J. (1981) *Simulacres et Simulation*. Éditions Galillée, France.
Bergson, H. (1911) *L'Évolution Créatrice*. PUF, Paris.
Borges, J.L. (1975) *El Libro de Arena*. Emecê Editores, Buenos Aires.
Eco, U. (1989) *Sugli Specchi e Altri Saggi*. Portugueses version of Difel, Lisbon.
Ernst, B. (1991) *Espelho Mágico de M.C. Escher*. Benedikt Taschen Verlag, Berlin.
Feuerbach, L. (1992) *L'Essence du Christianisme*. Editions Gallimard, France.
Foucault, M. (1966) *Les Mots et les Choses*. Editions Gallimard, France.
Frade, Pedro Miguel (1992) *Figuras de Espanto*. Edições ASA, Porto.
Gombrich, E.H. (1989) *The story of the art*. Phaidon Press Limited, England.
Hollanda, F. (1549) *Do Tirar Polo Natural*. Livros Horizonte, Lisboa.
Manzini, E. (1993) .Mutamenti Percettivi. *Lotus* , **75**, Milan.
Marin, L. (1993) *Des Pouvoirs de l'Image, Gloses*. Seuil, Paris.
Merleau-Ponty, M. (1964) *Le Visible et l'Invisible*. Editions Gallimard, France.
Modica, M. (1992) *Imitation*, vol. 25 of Einaudi Encyclopedia, port. version by INCM.
Murtinho, V. (1993) *Perspectivas: O Espelho Maior ou o Espaço do Espanto*. FCTUC, Coimbra.
Panofsky, E. (1927) *Die Perpektive als "symbolische form"*. Vortrage der Blibliothek Warburg, Leipzig-Berlin.

Pessoa, F. (1986) *O Livro do Desassossego*, from the heteronymous Bernado Soares, Publicações Europa-América, Portugal.

Rodrigues, A.T. (1990) *Estratégias da Comunicação*. Editorial Presença, Lisbon.

Seneca, L.A. (1991) *Ad Lucilium Epistulae Morales*. Portuguese version, Fundação Calouste Gulbenkian.

Sontag, S. (1986) *On photography*. Translation of José Afonso Furtado, Publicações Dom Quixote, Lisboa.

Tavares, D. (1994) *História da Arquitectura*. FAUP, Porto.

Valery, P. (1994) *Tel Quel I. Choses tues. Moralités. Les Principes dán-archie pure et appliqués*. Editions Gallimard, France.

da Vinci, L. (1943) *Trattato della Pittura*. Castelian version by Mario Pittalluga, Losada Editorial, Buenos Aires.

Virilio, P. (1990) *L'Inertie Polaire*. Christian Bourgois Éditeur, France.

Wright, L. (1983) *Perspective in Perspective*. Routledge & Kegan Paul, London.

BIOGRAPHY

José Carlos Teixeira is an Auxiliary Professor at the University of Coimbra in the areas of Computer Graphics and Geometric Modelling, head of its Computer Graphics Research Group and President of the CCG/ZGDV Executive Board. His main research interests are in Geometric Modelling, Virtual Environments, CSCW and new Interaction Techniques.
Responsible and involved in different European and Portuguese Projects, is member of the Editorial Board of the journal "Computer & Graphics" (Pergamon Press), founding member of the WG 5.10 on Computer Graphics of IFIP TC 5, President of the EUROGRAPHICS Portuguese Chapter and head of the Portuguese Technical Committee for Standardisation on Computer Graphics - CT 109. He is a member of EUROGRAPHICS, ACM, ACM-Sigraph and IEEE.

Vitor Murtinho is an Assistant since 1988 in the Architectural Department at University of Coimbra. He received his degree in the Architecture Faculty from the Technical University of Lisbon in 1988.

2

Information Technology, a Catalyst for Process Optimization

K. Fliess
Siemens Corporation
186 Wood Avenue South, Iselin, New Jersey 08830 USA
Tel: 908-321-3058 fax: 908-603-7213 email:
internet:kevin.fliess@sc.siemens.com

A. Jain
Siemens Nixdorf Information Systems
300 Wheeler Road, Burlington, MA 01803 USA

Abstract

Rapid product development and virtual prototyping are fast becoming commodities of world class companies. In order to maximize their effectiveness, one must be able to measure and change business processes.

Efforts like Business Process Reengineering (BPR) and Concurrent Engineering (CE) that help define process improvement methodologies are being adopted by more and more organizations. This paper will cover a small array of information technology (IT) tools that aid the paradigm shift from traditional, functionally-oriented operations to process-oriented ventures. Software tools are plentiful but cannot be used as stand-alone resources. Successful utilization is dependent upon a holistic approach to change management with adequate planning, personnel and financial resources, know-how and information technology.

Analysis of organizations in terms of processes and process flows increases our understanding of core competencies and weakness enabling business success. This paper addresses issues raised when implementing IT as a catalyst for finding process-oriented solutions. BPR and CE are initiatives that bear similar attributes and are equated to be different sides of the same coin. The two have been freely intermixed in the following paper.

Keywords

Process optimization, business process reengineering, process management, process analysis, PEAT, SIFRAME

1   INTRODUCTION

The demise of protected market structure due to recent political revolution, and multinational trade agreements (i.e., GATT) have not only opened gates of once restricted markets, but also created new challenges for organizations on a global scale. Technology has provided and continues to provide solutions to a series of information related problems. The cost of technology (Figure 1) is decreasing at a phenomenal rate.

The rapid change in technology has also created an environment where the market life of a product has decreased due to the introduction of new and more advanced products on a regular basis. Companies that cannot meet the challenges of decreasing the development time cycles to a level comparable with product life without losing quality, are bound to end up like dinosaurs.

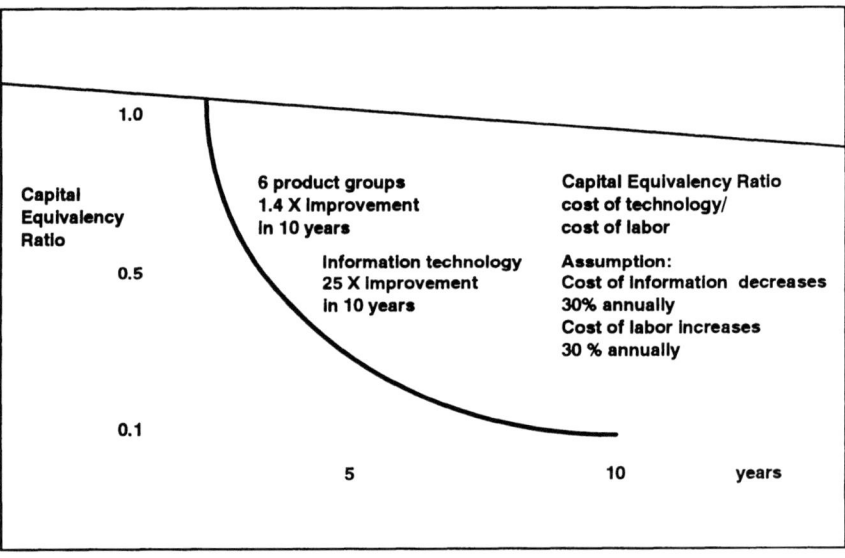

Figure 1 Cost of Information Technology (Benjamin and Yates)

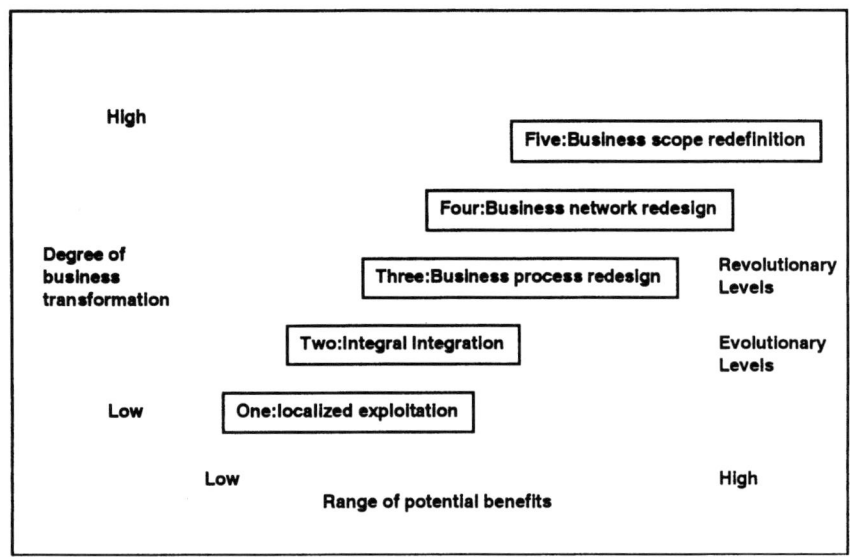

Figure 2 Five levels of IT induced reconfiguration (Venkatraman)

2 BPR AND CE

The extent of change an organization attempts to induce reflects the extent of failure or benefit that may be realized. (Figure 2) BPR and CE are inter-related. "At the heart of business reengineering lies the notion of discontinuous thinking -- identifying and abandoning the outdated rules and fundamental assumptions that underlie current business operations."(Hammer and Champy) CE deals with cross-disciplinary groups working together with the aim of achieving the goal of rapid product development. The key to both CE and BPR is the understanding of "What is a Process?". "In definitional terms, a process is simply a structured, measured set of activities designed to produce a specified output for a particular customer or market."(Davenport)

2.1 Process awareness and understanding

Before undertaking an improvement initiative a company must understand its current situation. The area of process analysis is becoming more scientific -- focusing on hard data and metrics to identify the weaknesses and strengths of an organization. Not only is IT a catalyst for CE, but it also allows for effective change management associated with CE and BPR initiatives. Effective change management enhances functional agility, translating into rapid new product introduction, quicker customer response time, prompt service and other factors that characterize successful organizations.

2.2 Reengineering demystistified

Reengineering the Corporation, by Hammer and Champy, alerts industry to the need for a change from task based thinking to a process-oriented focus. Task based thinking -- the fragmentation of work into its simplest components and their assignment to specialist workers has influenced the organizational design of companies for the last two hundred years. The shift to process based thinking is already under way, and is illustrated in the radical changes that mainstream companies have made. Not all BPR and CE engagements succeed. It is estimated that 70% of all so-called efforts fail. (Datamation) Failure can be attributed to socioeconomic, sociotechnological and humanistic reasons.

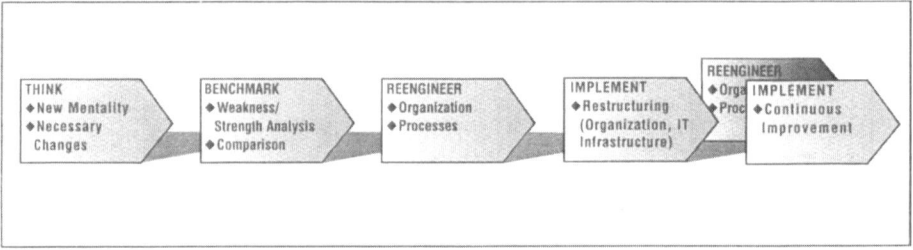

Figure 3 A simplified view of steps that an organization follows to analyze its processes and perform CE-based reengineering.

The first three steps require a series of brainstorming sessions to define how the organization should conduct its business. These steps, though driven by human input, must be aided by process optimization software and process modellers (e.g. Siemens PMG PEAT methodology and SNI GRAPES tool). These tools use predefined matrices to identify the cost and attributes of the processes which make up an organization. These tools do not eliminate the thinking involved, but merely make it easier to play the "what if" game. The results of the analysis are operational benchmarks for processes and process flows, as well as a series of guidelines on how business should be conducted.

The latter steps identify the need for an IT-supported structure that allows the recommendations of the first three steps to be implemented, incorporating also the mechanism for capturing execution matrices, thus aiding in the continuos improvement process. It is at this stage that most engagements fail due to the difficulty in merging methodology with technology.

3 IT REQUIREMENTS

The first three steps call for the analysis and simplification of processes. IT is the gateway to successful process-oriented change management, and is an catalyst for fast process analysis, simplification, and optimization. This requires three distinct steps which must be performed sequentially.

3.1 Process mapping

The technique of flowcharting processes using IT. Mapping exercises begin with an interview of the process owner -- the individual with the most intimate understanding of the flow of the business. It is critical that issues of time, cost, staffing and value be addressed accurately for each step in the process. A well constructed map should illustrate all the steps in a specific process regardless of the number of functions involved or their geographic locations; as well as, salient subprocesses, interactions, hand-offs, loops, decision trees and case statements. The map is the prerequisite for all future improvement steps. The success of the reengineering endeavor is directly linked to the accuracy of the diagram. (Figure 4)

3.2 Process evaluation and analysis

The second step is a metrical diagnosis of data in the process map, cost and time being the key criteria. The diagnostic outputs provide the organization with an empirical assessment of its current operational status. The data allows the organization to examine how the process flows [normal vs. best and worst case scenarios] which leads to internal benchmarking and/or the articulation of quantifiable goals for process simplification and improvement. Information extracted during the analysis stage becomes the basis for process simplification and reengineering. The IT structure must support the realization of the goals and recommendations identified at this stage.

3.3 Process simplification, optimization and reengineering

Process simplification, true to its name, is the examination and potential elimination of non-value-added steps in a process. An "optimized" process falls out of simplification. Non-value added activity can be identified by asking the question: *Is the customer willing to pay for this?* Note that not all non-value added work should be automatically eliminated and should not be confused with cost added. It may be critical to the process. If the task in question does not add value, then it must be ruthlessly examined and if possible changed or eliminated. Simply cutting and pasting steps in a process is detrimental to the initiative. Optimization is a prerequisite to reengineering, and mandates the obliteration of superfluous, cost/time adding steps and the adoption of a progressive way of thinking. Perhaps the greatest obstacle to this paradigm shift is "unlearning" bad practices. IT is a catalyst for identifying *why are we doing what we are doing* and making recommendations for improvement.

Given the removal of excessive non-value added steps, the initiative ascends to the next level, reengineering. This is the culmination of all diagnostic work and the beginning of a rearchictecture of key business elements: methods, metrics, processes, and organization.

Without adequate IT support, reengineering is an impossible endeavor, requiring elaborate calculations and exorbitant time and personnel costs. Effective software catalysts have become essential, and therefore must be examined. SIEMENS PEAT package, tabulates relational data rapidly and generates summary outputs of process maps. Built in flexibility allows the user to analyze probable reengineering strategies without jeopardizing the financial or operational well being of the organization.

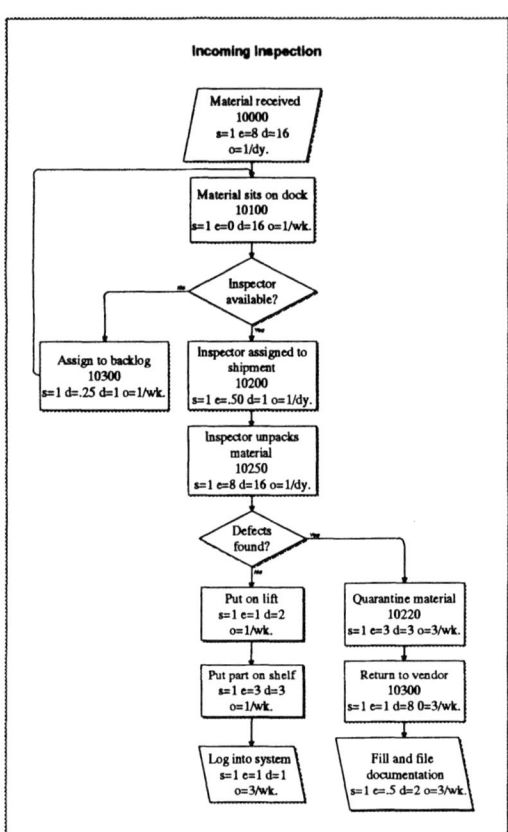

Figure 4 Process map example

3.4 Process mapping criteria

Figure 4 represents a snapshot of a much larger process map. Each box in the diagram represents a separate task and includes:

effort	(e)	or the focused, "hands-on" work time
duration	(d)	or the total time for the step from start to finish
staffing	(s)	or the number of persons involved
and occurrences	(o)	the frequency of the step

3.5 Data aggregation and analysis

Data from the map, as well as additional statistics gathered during the interview are tabulated by PEAT and include the following summary outputs for each discrete process. (An aggregate process analysis can also be done with the same results): Total Duration, effort, occurrences, staffing, value-added activity, non-value added activity, functional interactions, process owners, hand-offs, major task category (i.e. shipping, receiving, installing, manufacturing, engineering, etc.), minor task category (i.e. communicating, validating, preparing, core action, reformatting, etc.)

The reengineering team or cross functional CE task force examines the process flow and cross references it with summary PEAT outputs. Tasks are scrutinized discretely and collectively. Finally, points of entry into the process are identified, so that the change initiative can begin in earnest. Decisions to change should be based on indisputable evidence not on emotional ties. This scientific approach to process simplification and reengineering provides an objective portrayal of an organization with little attention paid to functional blocks. Furthermore, the "why me?" victim mentality common to CE and BPR initiatives can be confronted with honest, empirical feedback. The realization must also be flexible enough to allow for ongoing changes, as neither the market place nor the technology used are stagnant in nature. CE and BPR initiatives by their nature induce drastic, holistic change but continue to evolve after goals are met.

4   CRITICAL SUCCESS FACTORS

Any IT realization of the recommendations of process analysis will need to address four major issues in a very flexible yet disciplined manner:

-   Process
-   Teamwork
-   Tools
-   Data

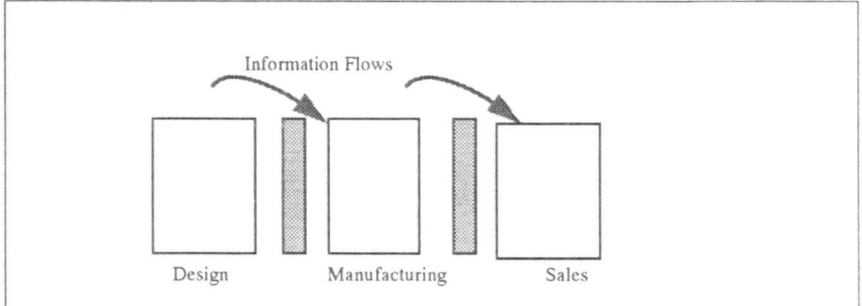

Figure 6 Functional Blocks

4.1 Process

In an organization based on Tayloristic philosophies, the functional blocks, manned by specialists are responsible for performing independent tasks without much outside influence. The designers in the design block perform their function without the influence of either the marketing or manufacturing blocks. This structure leads to higher development costs and delayed product release. In the past, the issue of quality and customer satisfaction was not dependent upon the organizational structure, rather on the abundance of resources.

Today, in order to obtain maximum return on investment, product development must be organized in terms of processes, identifying the best possible techniques for achieving organizational development goals. The processes chosen to achieve the organizational goals must be both flexible and cross functional.

The flexibility of the processes defines the organization's capability to react to changing market needs. The dynamic and diverse state of the market demands that organizations with long presence potential must be both proactive and reactive.

In a functionally based architecture, borders between the functional blocks as well as dissemination of information among hierarchical levels leads to time and resource waste. By only addressing functional processes, the above stated problem is not removed. To realize improvement, processes must be cross functional.

4.2 Teamwork

The main challenges to cross-functional teams is communication. Often times, individuals from different functional areas develop independent languages and systems. IT provides a catalyst for finding a solution to the communication issues. The IT structure supporting cross functional teams must allow free communication on a common plane.

IT systems supporting team behavior must provide concurrent access to global information.

4.3 Tools

Tools are instruments used by people to perform their tasks more effectively and achieve organizational goals. This can range from simple word processors to complex artificial intelligence applications. Investment in IT tools can only be justified when tools are used to optimum capacity. IT system should help reduce the learning curve as new tools are introduced.

4.4 Data

Data generated during the reengineering of an operation reveals information about performance and for process improvement. In the product development process, data generated pertains to innovation and technology used in design, manufacturability, service, etc. The aerospace industry requires the storage of airplane design and construction data for decades. IT structures must provide the capability and flexibility for storing information and allow easy retrieval and access.

Use of relational databases (RDBMS) has simplified storage aspects. Nevertheless, lack of distributed IT architecture has not allowed for their extensive use in an enterprise wide solution. Siemens PEAT package uses a simple relational database to store a multioperational company's process data in one repository. This allows simplified and continuously updatable benchmarking information.

The lack of distributed databases, however, results in islands of independent, non-connected repositories, resulting in inconsistency and maintenance overheads. Only recently, with the introduction of distributed databases is this problem being addressed. The introduction of Object Oriented Database Management System (OODMS), allows for a more realistic, flexible and abstract repository design. Companies are still reluctant to use the new database technique in their IT solution due to the lack of international standards. The emergence of new standards (e.g. CORBA ...) will have a profound effect on this.

5   SYSTEMS

The previous sections covered the Siemens PEAT package used to analyze and simplify an organizations processes. The subsequent sections covered the four factors to be considered in the implementation of an IT structure. The following sections cover two types of tools used to provide an IT solution.

5.1 Product data management systems (PDMs)

PDMs are in essence data vaults based either on a RDBMS or OODBMS technology. PDMs offer powerful functionality for classification, management, access and control of data stored in the databases. This allows users to manipulate and manage data in a very flexible manner. PDMs use a non distributed database as the underlying repository -- creating database islands, and rendering the information exchange difficult.

Most PDMs encourage teamwork via the use of locking mechanism, be it on an individual or cooperative level. This form of interaction is not sufficient for it does not provide project clarity for all participants.

Some PDMs are emerging with adhoc workflow management systems. A workflow in general refers to the execution of jobs according to particular rules. The workflow incorporated within PDMs allows data to be grouped into work packets. This work packet is transmitted electronically to users per rules identified during the think, benchmark and reengineer phase of BPR/CE.

PDMs allow for a collection of independent workflows, offering only a partial view of the project. Furthermore, it is not possible to interconnect these partial views into a matrix of processes that define the project in its entirety. As well, workflow systems are sequential in nature and cannot guide users in determining the downstream effect of process execution. Workflows do not provide an effective mechanism to capture process metrics nor aid in the continuos improvement process

5.2 Process management systems

Systems are available which use processes as the basis for modeling, unlike PDMs which use data. PEAT allows organizations to analyze their business in terms of processes and identify a simplified model, a network of interconnected, interrelated business flows. Systems are appearing where the network process model can be addressed and implemented directly. Siemens Nixdorf's SIFRAME is one such system.

SIFRAME's emphasis on processes allows work to be broken down into its discrete parts. Each work element is assigned to a team, associated with a process flow, that defines work procedure. Each process is further associated with tool(s), its environment and related data set. In this manner, four aspects of processes, team, tool and data are covered in a unified domain.

6   CONCLUSION

This paper addresses an array of issues concerning the relationship between IT and process improvement. It is designed to give the reader a broad overview of the subject. Suffice it to say that the key to successful BPR, CE, process simplification, and other related topics, is an awareness, understanding, and utilization of the four critical factors: process, teamwork, tools, and data.

The above sections have only covered a single IT package aiding in process analysis and simplification and two IT packages used to implement recommendations from the process analysis phase.

There are other tools available that also provide IT solutions to the above mentioned problems but were not addressed in this paper.

7   REFERENCES

R. I. Benjamin and J. Yates (1991) The Past and Present as a Window on the Future in *The Corporation of the 1990s* (ed. M. S. Scott Morton), Oxford University Press.

Datamation (Aug. 1993) Does Reengineering Really Work?.

T. H. Davenport (1993) *Process Innovation*, Harvard Business School Press.

M. Hammer and J. Champy (1993) *Reengineering the Corporation*, Harper Collins.

N. Venkatraman (1991) IT-Induced Business Reconfiguration in *The Corporation of the 1990s* (ed. M.S. Scott Morton), Oxford University Press.

8 BIOGRAPHY

Kevin Fliess graduated from Washington and Lee University in Lexington, VA in 1993. He holds a BA in Politics and German. As a student, Kevin interned with Siemens at their world headquarters in Munich, Germany in corporate purchasing and corporate logistics. He is currently working as an associate consultant for Siemens Corporation USA's Process Management Group: A highly focused, internal consulting organization dedicated to the enterprise success of the North American operating companies. He is responsible for on-site client management, technology development, and the continuing advancement of the process optimization service. Kevin would like to acknowledge Robert C. Daniell, Marc A. Kind, and Frank Wilhelm for their support and devlopment of the PEAT methodology and tools.

Adidev Jain earned a BE from Middlesex University, London, U.K. in 1987. His area of concentrations were Microwave and Telecommunications theory. He has since worked on topics ranging from digital signal processing, business process reengineering, data warehousing, process and object oriented design. His business and personal interests have taken him to various parts of the world, including Asia, Europe and the US This paper was written during his tenure as a senior systems engineer for the electronics giant Siemens Nixdorf Information Systems. Currently, Adidev is employed as a senior consultant, implementing object oriented designs with TRECOM Business Systems, a full service systems integration and consulting firm on the east coast.

3

Rapid Prototyping
An Approach Beyond Manufacturing Technology

Dipl.-Inform. Jürgen Wagner[1]
Dr.-Ing. Wilhelm Steger[2]

Abstract

This paper defines Rapid Prototyping as a new approach to reducing cycle times in product development, thus going beyond the view of Rapid Prototyping as a collective term only for the class of new, generative manufacturing technologies. In particular, the use of different categories of prototypes for the development of new products, and the organizational impact of Rapid Prototyping are shown to form a holistic framework for evolutionary product development. Depending on the extent to which the organizational principle of Rapid Prototyping is implemented, and depending on the individual utilization of specific prototypes, there may be a tremendous benefit in terms of the factors time, quality, and costs of the entire process. Furthermore, these benefits are not only due to the availability of physical prototypes. In all areas where virtual prototypes can provide answers to the core questions of a development process in the same manner as their physical counterparts, virtual prototypes perform at least as well (e.g., through reduced building time or cost reductions).

Before the influence of prototypes on the development process is discussed, there will be some general remarks on classes of prototypes and purposes they are used for. The need for a new organizational form is then derived from the requirements of designing, manufacturing, and evaluating the described prototypes, with cooperation and the respective enabling technologies turning out to be the essentials to the success of Rapid Prototyping in product development.

---

1. Fraunhofer Institute for Industrial Engineering and Organization (FhG IAO)
   Nobelstr. 12, D-70569 Stuttgart, Germany.
   E-Mail: wagner@iao.fhg.de. WorldWideWeb: http://www.iao.fhg.de.

2. Fraunhofer Institute for Manufacturing Engineering and Automation (FhG IPA)
   Nobelstr. 12, D-70569 Stuttgart, Germany.
   E-Mail: steger@qt.ipa.fhg.de.

1 Current Situation and Motivation

International competition is characterized by an increasing dynamics of innovation. The gradually decreasing time of product marketing - in single cases, it is even shorter than the time of product development - requires enterprises to establish a growing number of new product developments and faster prototype development cycles in order to compete on the international market successfully. Thus, customer-oriented products which are tailored to the needs of particular target groups, as well as the early advertising of products gain more importance. The strategy of taylorization of work processes and the resulting, highly-specialized work distribution lead to deeply structured, hierarchical forms of organization, inhibiting not only fast reactions to changing customer requirements and short iteration cycles in product development, but also disabling innovation in enterprises. Consequently, such enterprises are not flexible enough to adapt to today's market's dynamics [Bull92].

Consequently, individual enterprises have to specialize their product lines while shortening iteration cycles during product development and maintaining a high innovation rate. Shorter product life cycles reduce monetary gains and lead to manufacturing sites moved off to countries offering cheap labour. As a direct result, the mutual interdependencies between and within companies become more and more significant and require the underlying organization to provide for a well-coordinated collaborative development process. As traditional tayloristic approaches quickly reach their limits due to the increasing complexity of the resulting processes, new methods for organizing the product development process are called for ([Bull93], [Bull94]).

However, not only the organization form has to be blamed for the decreasing success of many companies. There is also an insufficient support of the development process by information and production technology. Prototypes are still manufactured manually because of a missing CAD description, or worked out in a time-consuming milling or casting process, whereas a stereolithography apparatus could do it much quicker. Especially with regard to the increasing complexity of new and innovative products, prototypes play a key role for sharing ideas in multi-disciplinary teams. If this does not work, people fail to identify themselves with the product and become demotivated, which has tremendous influence on the quality not only of the work process, but also of the products.

All these problems mark simultaneously the goals and guidelines for a successful company. There is no doubt about the competitive edge of a faster time-to-market, but in order to fulfill all individual customer needs, increasing quality and decreasing costs of the products have to be considered as well.

Approaches like CIM (Computer-Integrated Manufacturing) or CSE (Concurrent/Simultaneous Engineering) deal with some of the described problems by introducing a high degree of information integration and well-defined information flows and processes. While such approaches are suitable for at least partially well-known tasks like the redesign of an existing product in order to improve functionality or quality, complex engineering tasks in the development of innovative products do not follow rules which can be modelled in the required, strict ways.

Innovation is closely related to creativity, i.e., a highly flexible work environment is needed to facilitate innovation and to eliminate constraining systems and rules, especially in the early phases of conception, where a formal product as such does not yet exist. It can be concluded that a integrative approach granting high flexibility is needed for product development [Bull94].

2   Goals of Rapid Prototyping

The Rapid Prototyping approach attempts to solve these problems by introducing into the product development process from the earliest stages

- an organizational form of distributed, autonomously responsible expert teams,

- techniques for the holistic integration of all knowledge along the product development process,

- methods for providing an integrated development and testing environment with both virtual and physical prototype manufacturing technologies, and

- techniques for networking all involved team members by means of intelligent cooperation and communication systems.

The key issue of Rapid Prototyping is the tight organizational and information-technological integration of the development team and all knowledge required to fulfill the development task, despite spatial separation, in order to shorten development iteration cycles, to decrease development cost by employing virtual as well as physical prototyping technologies, and to increase the utility of individual cycle steps through collaboration.

As a prerequisit for the right definition and design of a development process we first want to point out the use of prototypes, and how they can benefit in general the critical success factors of a company (i.e., time, costs, quality, flexibility).

3   Prototypes in a Product Development Process

3.1   Classes of Prototypes

A prototype is the result of the design and generation of one or more product characteristics which help the design team to test them against user requirements. According to this definition almost everthing can be a prototype. A classification is therefore recommended [Steg94] (see figure 1):

- Design prototypes: They serve first and foremost for a design review under the consideration of optical, esthetical and ergonomical requirements, whereas mechanical aspects or accuracy are normally neglected.

- Geometrical prototypes: They are employed for testing accuracy, form and fit of the later series parts. Therefore the focus is on geometry and not on material aspects.

- Functional prototypes: They represent a set of features which allow the test of some funtional aspects. A functional prototype is usually a subsystem of a product.

- Technical prototypes: They cover all functional aspects of the part and can be used as such, but the manufacturing process is usually different from the one which will be later used in series production. The technical prototype may also consist of different material.

Design prototypes		Design review under the consideration of optical, esthetical and ergonomical requirements
Geometrical prototypes		Employed for testing accuracy, form and fit of the later series parts. The focus is on geometry and not on material aspects.
Functional prototypes		Functional aspects which are represented as a set of features are reviewed (subsystem of a product)
Technical prototypes		All functional aspects of a part but the manufacturing process is different from the one which will be used in the series production. The material may be different.

Figure 1:   Classes of prototypes

However, not all prototypes have to be available in a classical sense as physical parts. For an example, take a CAD system which allows to visualize first design drafts, FEM software for the mathematical analysis of part structures, or a virtual reality machine that is envisaged to become an important design, evaluation and co-operation tool in the future. All these systems provide a non-physical, virtual prototyping, for all above mentioned classes of prototypes.

3.2   Use of prototypes

The benefit of prototypes emerges out of their use (see [Poll94], [UIri94]); four different cases can be identified (see figure 2).

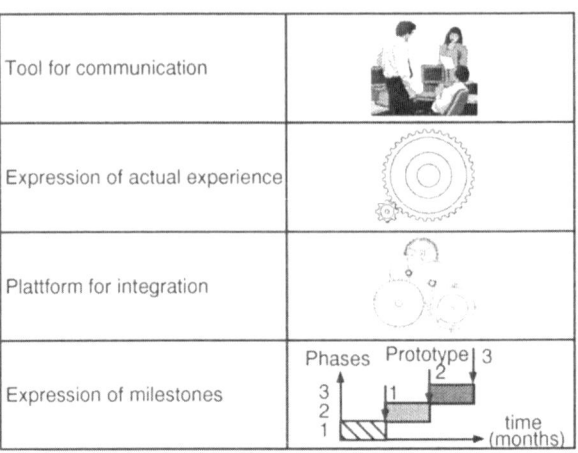

Tool for communication	
Expression of actual experience	
Plattform for integration	
Expression of milestones	

Figure 2:   Use of prototypes

- Tool for communication:
  Design is a process where a lot of people with different skills and views have to work together on the same product. For example a member of the top management has to explain his requirements to a designer, a mould maker and an expert for assembly tasks and vice versa. A prototype acts as a catalyst for such a discussion process.

- Expression of actual experience:
  Maybe that the customer's requirements are very well defined at the beginning of a development process. Some uncertainty, however, may exist on how each requirement is fullfilled by a set of individual product features and how they can be put into reality. A prototype is therefore a tool for validation and verification, it expresses in each stage of the process the consolidated experiences of the customer and supplier.

- Platform for integration:
  As we remember a functional prototype represents a subsystem of the final product. All subsystems have to be integrated and tested where constraints of assembly and co-operation have to be considered as vital. Such prototypes are well known as alpha- or beta-prototypes.

- Expression of mllestones:
  The management and/or customers normally want to evaluate the progress of the development process at some stages. Only in case of a positive evaluation the project may go on.

3.3   Benefits of Prototypes for a Development Process

Regarding the classes and the use of prototypes there can be indentified again four situations how a prototype can positively influence the development process (see [UIri94]). An interesting point is that only in the first case the speed of the prototyping process impacts directly the development process; in all other cases the mere availability of a prototype (instead of having none) is responsible for a shorter and - due to less iteration cycles - cheaper development.

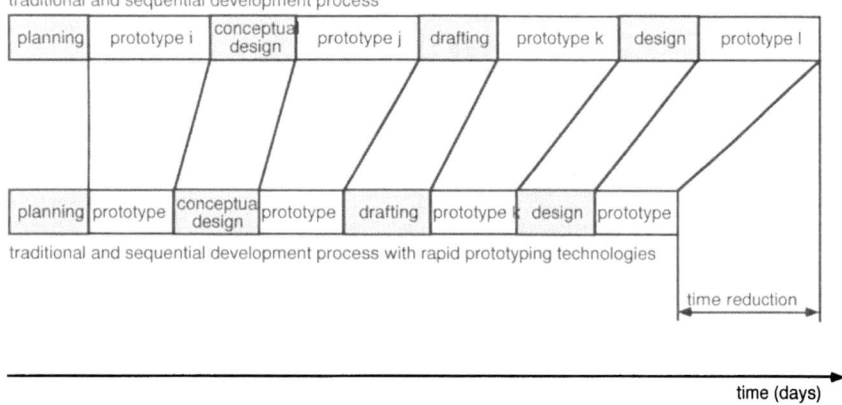

Figure 3:   Acceleration of the prototyping process

3.3.1   Acceleration of the prototyping process

By virtual prototyping prototypes can be made more quickly then with traditional physical methods (see figure 3). Yet the benefit depends on the complexity of the prototype and the set of requirements which should be evaluated. Furthermore there is a large number of prototypes in industry which encorporate electrical or hybrid (both mechanical and electrical) issues. Virtual prototyping, however, concentrates more on mechanical characteristics (geometry, esthetics, stress, fatigue, etc.) and can - at least within a short-term view - only apply in a phase of the development process which addresses these issues.

3.3.2   A prototype influences later development phases

In this case the prototype acts as a communication tool within the team and improves the decision finding. For example tool design can be done more quickly if a 3d description of the part - instead of a set of 2d drawings - is available (see figure 4) or a complex physical prototype may be no longer necessary because a computer simulation provides the desired results (e.g., crash behaviour simulation).

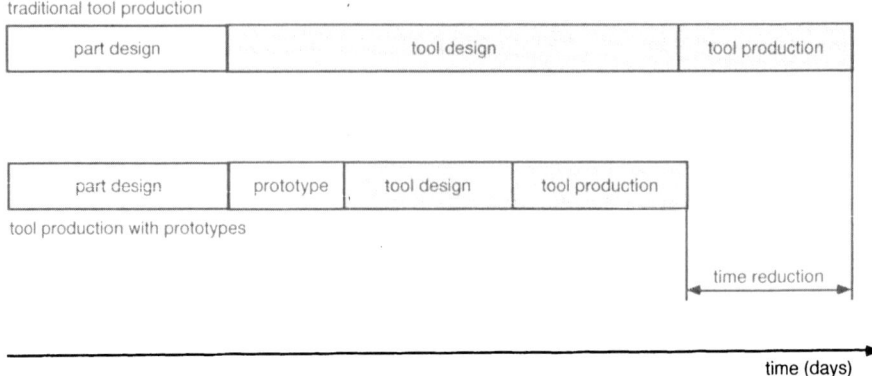

Figure 4:   Influence of a prototype on later phases

3.3.3   A prototype improves the success rate of a development process

Especially an early prototype allows a quick verification of the assumptions of the product development process. The reliabilty of these information, which serves as input to the next development phase, increases and also the probability of cost consuming changes in a later stage decreases (see figure 5). The costs of a prototype, however, have to be compared with those of a product change. Therefore prototyping is only recommended for risky and expensive products.

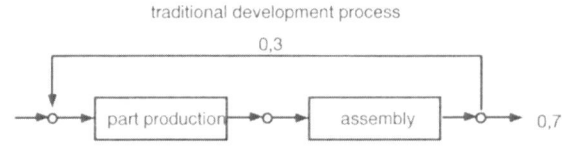

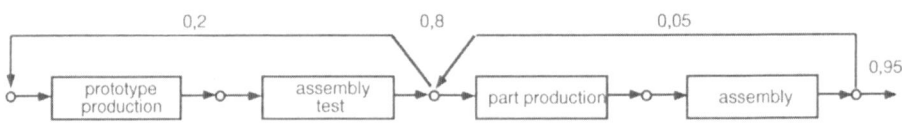

Figure 5:   Increase of the success rate of a development process

3.3.4 Prototypes influence the sequence of the development phases

The conventional sequence of the design of an injection mould is explained in figure 6. Tests of the mould can be done only after it has been almost finished. Final assembly and test are influenced in line by a sequential process. A prototype (physical or virtual) helps to test the tool at an early stage and reduces both development time and uncertainty.

traditional development process

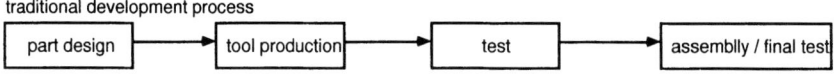

re-organised development process with prototypes

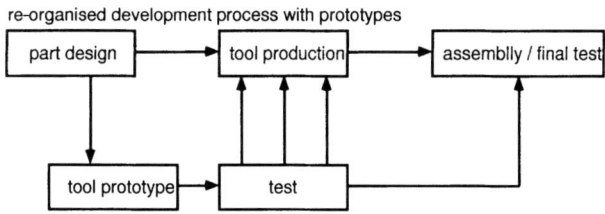

Figure 6:   Reorganisation of the development process by prototypes

The following section will focus on the organizational framework required to gain the optimal utility from each of the described prototypes by applying Rapid Prototyping requirements to cooperation structures. In particular, a sketch of a future information-technological background directly facilitating this form of expert cooperation is described.

4   Cooperation in Rapid Prototyping

The inherently contradictory requirements of ideally maximizing quality while minimizing time and costs for both, development and production of prototypes or products, necessitate trade-offs to be made in most decisions regarding which prototypes may be designed and built in a particular stage of product development, in order to yield the desired results.

• Often, limitations on the development process are imposed by the availability of resources, e.g., supercomputing facilities. In such cases, simple prototypes may be used to investigate simple cases, while complex cases need to be studied in more detail, which requires more or a different type of resources.

- In virtual prototyping, some features of a prototype design may not be demonstrable using entirely virtual prototypes, so a hybrid combination with a partially physical system may be required in order to combine the advantages of both approaches.

- In physical prototyping, the need to plan and to determine the utility of combinations of manufacturing technologies for a certain prototype arises as individual technologies implicitly define different physical properties and characteristics. For example, stereolithography may not be feasible if the thickness of separators lies below the manufacturable limits.

- Last, but not least, the managerial coordination of a development project requires trade-off decisions if several feasible alternatives exist but resources determine that only one or a few alternatives can be carried through. A similar situation can also be found in the evaluation phase of prototypes, where each aspect which was planned to investigate on a particular prototype, is examined. The gained results in the respective interation cycle step then need to be evaluated in a following exploitation step.

There is one common property in all of the above examples: two or more of the involved experts have to collaborate in order to obtain the desired results or decisions. The predominant reason for this is that whenever a decision involving even only partly opposing alternatives needs to be made, any single person will very likely be unable to make the best decision possible if the domain of alternatives is characterized by a significant amount of complexity. Therefore, the domain and decision complexity has to be reduced by letting individuals with specific know-how arrive at a partial interpretation of results or at a partial decision, and have these discussed in a team to check their mutual consistency or feasibility.

4.1 Types of Complexity in Product Development

Four distinct types of complexity can be identified in product development.

4.1.1 Product complexity

New products in mechanical and electronic engineering (e.g., trucks, cars, aircrafts, spacecrafts, ships, electronic devices supporting mechanical or thermodynamical systems) may involve thousands of parts. Even with a much smaller number of individual parts, fully understanding the interdependencies of functions of different components or assemblies usually requires thorough knowledge from a number of different engineering domains. With innovative products, this problem is even more significant, as there may be no long-standing know-how as there is for conventional products.

4.1.2 Technological complexity

The development of innovative products in fields of engineering implies the use of innovative technologies or the use of well-known technologies in new applications. Either way, the well-defined behaviour of new technologies or new combinations of manufacturing processes may not be well researched. In that case, product development also implies the first-time or refined development of stable and consistent characterizations of the processes in question.

4.1.3 Knowledge complexity

Given product and technological complexity, it is clear that several experts have to collaborate by providing expert knowledge on particular aspects of the entire product or prototype. The task of only partially integrating these different types of knowledge into a common pool in order to arrive at a common understanding and basis for investigations of possible solutions is one of the most difficult endeavors in multi-disciplinary work. The failure of traditional departmentalism with a plethora of formal interfaces, information-passing protocols, and release procedures stems from the lack of the common basis for discussing results.

4.1.4 Organizational complexity

Hierarchical development project structures oriented on the departmental structure of a company may imply a separation of work tasks that is not beneficial to reaching common goals by collaboration. The departmental or organizational separation of experts whose joint knowledge would be even more beneficial to the development process than each one's by itself causes artifical barriers where information reduced to mere data has to be passed. Clearly, well-defined results of work units are a prerequisite for a manageable project in terms of progress monitoring and controlling. However, this does not necessarily imply that those have to be the pieces of information passed between departments. They may as well be other, jointly achieved results which are defined based on the goals of the development project, not on the particular departmental structure of an organization.

4.2 The Need for Cooperation

The examples given above clearly indicate that a close collaboration between the different experts involved in achieving a common goal is required in order to facilitate

- an efficient exchange of information,
- the consideration of as many facettes of a problem as possible, and
- the time-effective accomplishment of a task.

Due to their inherent complexity, advanced development projects present problems of unpredictability (i.e., the product specification and work plan cannot be stated definitively at the outset) and concurrency (i.e., many interdependent subtasks are performed in parallel and may need to continuously exchange information). As a result, modifications of the original work plan or product specifications may be required within or after each iteration cycle. This type of work organization clearly cannot be accomodated by traditional management hierarchies [Amad93], but call for ad-hoc group decisions in conferences among the involved experts. This style of organization has been called „the networked organization" [Malo91].

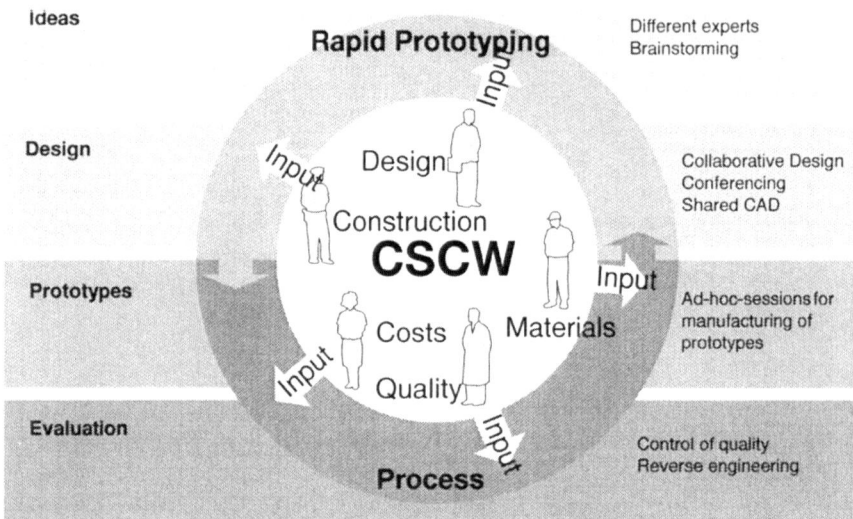

Figure 7:   Cooperative work in Rapid Prototyping as a group process

Collaborative work is defined as the process of two or more individuals working together towards a common goal (e.g., a design, a prototype, an evaluation result, a managerial decision), with each individual contributing to the ultimate result. This organizational form requires all participating individuals to share a common stock of knowledge aiding in communicating essential information between them, and to reach a common understanding of subjects (figure 7). Consequently, expert training has to take a „T model" into account: the horizontal bar symbolizes general knowledge needed to facilitate the understanding and participation in cross-domain, group decision processes, while the vertical bar represents detailed knowledge in an area of specialization. In order to contrast this type of experts to those found in tayloristic organizations, the term „generalist" or „generalistic expert" is used.

Computer-aided systems for various specialized fields of product development have been broadly employed during the last two decades. Despite the principal possibility of networking them together into an integrated environment, application programs (e.g., FEM, CAD) are single-user oriented, with only limited capabilities for data sharing and cooperation between users [Toye92]. As a result, they further solidify the work organization based on high specialization, and inhibit ad-hoc on-line cooperation. However, cooperation and communication are especially important in early phases of product design, where product data in the conventional sense only start to evolve as a result of team creativity and cooperation.

Rapid Prototyping is based on intensified collaboration. The most important prerequisites for effective cooperative engineering in distributed expert teams are

- the ability to utilize the flexibility gained from implementing the organizational form of teams of generalists, i.e., experts with a broad overview of not only their own field of specialization, but also that of others, and

- the support of the complex product engineering process by information and communication technology to achive an integration on the data, the application, and the team level.

The following section will elaborate the differences between the Rapid Prototyping approach and traditional development methods.

4.3   Rapid Prototyping vs. Traditional Development Methods

Rapid Prototyping can be viewed as an evolutionary process combining all technological, methodological, and organizational tasks from product conception to manufacturing into a single framework. In opposition to traditional approaches ([Bull92], [Köni93]) assigning specific prototypes to particular development phases, in Rapid Prototyping, the product of the development process evolves as a whole from early prototypes to the final product. This way, different dimensions of interest can be investigated in each iteration step, yielding information on particular aspects of the respective prototypes.

The rapidity of Rapid Prototying is achieved by the utilization of fast prototype manufacturing technologies (virtual prototypes, generative technologies), but also through the intensified cooperation of the development team. The constant interaction of team members guarantees dynamic, cross-functional links on demand, facilitating ad-hoc coordination and planning in direct response to questions and problems arising during the process of evolutionary product development. Clearly, this close interaction of developers also requires a seamless integration of tools.

Rapid Prototyping is orthogonal to sequential (conventional) or parallelizing (Concurrent Engineering) approaches suitable for development projects with mostly well-known structures. The strength of Rapid Prototyping shows in projects requiring the holistic, creative development of new, complex products, i.e., products where neither product nor project structures can be defined well in advance because they are subject to development within the project itself. However, a resonable combination of the three approaches proves to be the key to success: Rapid Prototyping may be used in the product definition and early design phases, including (in particular, virtual) prototype manufacturing and testing, whereas sequential and Concurrent Engineering methods may be employed in the subsequent refinement and optimization phases of the then well-defined product.

4.4   IT Support for Rapid Prototyping

In analogy to the different areas of complexity identified in a product development process, information and communication technology support must be granted to the development team.

4.4.1   Product Design and Evaluation

Major sources of product complexity are the number of components involved and the properties of functional or geometrical interdependencies between components. A Rapid Prototyping environment must accomodate this by integrating aids for appropriately organizing and intelligently retrieving these pieces of information. For retrieval, the individual needs of different experts have to be taken into account, i.e., it is essential that design and evaluation tasks may be performed cooperatively while allowing for individual views on commonly manipulated data. The acceptance and adequacy of a computer-based system significantly depends on its capability to provide each expert with the data representation or visualization he or she is used to work with.

4.4.2   Prototype Manufacturing Technologies

A correct supply of information is essential to all implementations of technology. The necessary information integration described for product design and evaluation also pertains to technological knowledge for manufacturing virtual and physical prototypes. A close cooperation of experts on candidate technologies for building a particular prototype is essential for a best-choice selection among alternatives.

4.4.3   Knowledge

The integration of active knowledge facilitates not only a simplified exchange of information between experts from different domains, it also allows for an evolution of the knowlegde base from a first manual cooperation model to a more and more automated environment monitoring and controlling responsibilities and roles of development team members. This way, violations of imposed knowledge constraints (e.g., the modification of a diameter which also unknown to the designer affects the dimensioning of certain other components) can be handled by an emergency exit to a CSCW (Computer-Supported Cooperative Work) component of the Rapid Prototyping environment, letting the affected developers solve the problem, rather than trying to automate every decision. In fact, some decisions may require changes or relaxations to previously defined constraints, so an automatic treatment of these cases would not have been possible. However, by incrementally automating certain propagations of knowledge and decision guidelines, developers may delegate trivial aspects of their work to the IT support, concentrating on the creative task of designing the product.

4.4.4 Organization

On the organizational level, there are three principal issues to be considered:

Project Management components need to accomodate the autonomous, spatially distributed work of team members and subteams. The project organization is likely to be flat, as opposed to deeply hierarchical structures in conventional organizations. Also, classical workflow approaches as provided in many Engineering Data Management Systems are not suitable to refine project tasks to a tool level, as they tend to impose a too rigid form on the product development process, incapable of accomocating rapid changes.

Computer-Supported Cooperative Work techniques need to be employed to support the synchronous or asynchronous interaction of humans through or with the aid of computer-based systems. The benefits of ad-hoc meetings for immediatly tackling problems, as opposed to filing engineering change requests or resorting to weekly meetings, have to be used in order to shorten communication paths between people. The „natural" character of synchronous interactions with shared applications and audio (optional video) will also create a subjective feeling of togetherness within the team, despite spatial and temporal separation. In particular, once a „team spirit" can be created, the readiness to accept the joint responsibility for the product development as a whole increases, thereby improving the motivation and quality of the work environment. An important role will play multi-media as an intuitively appealing, simple way of representing even complex information [Fran91].

Team Qualification is an important issue when an existing development team needs to be transitioned from the old departmental organization to a team organization. It may be necessary to offer special training courses in cooperation and team management, in order to facilitate effective team work. In contrast to group work in production, the social aspects and human factors involved in the collaboration of experts are widely unresearched.

5   Conclusion

This paper presented a view on Rapid Prototyping that includes all organizational, knowledge and technological aspects of product development.

Some key conclusions can be drawn:

- Rapid Prototyping as a process covering all stages of product development draws its rapidity in individual development cycles not only from technologies allowing engineers to quickly manufacture prototypes. It is vital for the success of a product development process to take as well Human Factors and organizational issues into account, and to specially encourage and support cooperative work. Cooperation is the key to effectively managing different types of complexity in product development.

- The benefit of prototypes for development processes comes merely from the fact that we use prototypes without regard whether they are physical or virtual, but rather considering the degree to which new knowledge can be obtained from their evaluation. The organisation of the development process (i.e., in particular which prototype has to be used in which stage) has to be identified therefore as a critical point. There is, however, no generic organisation of a development process. It depends on the individual product.

- A flexible, fast, cost- and quality-effective development process combines a series of prototyping technologies. Virtual methods (CAD, simulation, FEM, Virtual Reality, etc.) and physical techniques (stereolithography, laser sintering, conversion tools, etc.) should both be applied dependent on the specific tasks in the process.

- The integration and best practice of many technologies requires a set of enabling technologies (e.g. expert systems, CSCW, adaptive information systems, databases and knowledge bases, product and process modelling). Both, the development of the enabling technologies and the prototyping technologies themselves have to be taken into account.

Many of the open research questions presented in this paper will be investigated within the Special Research Field (Sonderforschungsbereich) SFB 374 – Development and Testing of Innovative Products (Rapid Prototyping) – at the University of Stuttgart, which is partially funded by the German National Research Foundation (Deutsche Forschungsgemeinschaft – DFG).

References

[Amad93] Amadio, P.; Fassina, I.: Case Studies. In: Power, R.J.D. (ed.): Cooperation aming organizations: the potential of Computer Supported Cooperative Work". Springer-Verlag, Heidelberg, 1993.

[Bull92] Bullinger, H.-J.: Triaden-Management – Produktentwicklung im Blickwinkel veränderter Strukturen. In: Bullinger, H.-J. (Hrsg.) 4. F&E-Forum: Offensivstrategien für die Produktentwicklung, gfmt-Verlags-KG, München 1992.

[Bull93] Bullinger, H.-J.; Fremerey, F.; Fuhrberg-Baumann, J.: Innovative Production Structures - Precondition for a Customer-Oriented Production Management. In: Orpana, V.; Lukka, A. (ed.): Production Research 1993. Proceedings of the 12th International Conference on Production Research,Lappeenranta, Finland, 16-20 August, 1993, Amsterdam, London, New York, Tokyo: Elsevier 1993, S. 15-39.

[Bull94] Bullinger, H.-J.: Rapid Prototyping - Schneller zu neuen Produkten. In: Proceedings Fertigungstechnisches Kolloquium (FTK) '94. Stuttgart, November 1994.

[Köni93]   König, W; Eversheim, W.; Celi, I; Nöken, S; Ullmann, C.: Rapid Prototyping –
            Bedarf und Potentiale. In: VDI-Z, 8/1993, pp. 92-97.

[Poll94]   Pollmann, W.: Prototyping at Daimler Benz: State of the Art and Future
            Requirements. In: Proceedings of the IMS International Conference on Rapid
            Product Development, Jan 31 - Feb 2,1994, Stuttgart, Fraunhofer-Institute for
            Manufacturing Engineering and Automation (IPA), Stuttgart, 1994, pp. 241 ff.

[Fran91]   Francik, E.; Ehrlich Rudman, S.; Cooper, D.; Levine, S.: Putting Innovation to
            Work: Adoption Strategies for Multimedia Communication Systems. In:
            Communications of the ACM, Vol 34, No. 12, pp 53-63.

[Malo91]   Malone, T.W.; Rockart, J.F.: Computers, networks, and the cooperation. In:
            Scientific American, 265(3), pp. 84-91.

[Steg94]   Steger, W.: Generative Fertigungsverfahren - Gewinn für den Prototypenbau und
            den Entwicklungsprozeß. In: Proceedings CAT/Quality '94,17. - 20. Mai 1994,
            Stuttgarter Messe und Kongreßgesellschaft, Stuttgart, 1994, pp. 156-164

[Toye92]   Toye, G. et al.: SHARE: A Methodology and Environment for Collaborative
            Product Development. In: Proceedings of the IEEE Infrastructure for Collaborative
            Enterprises (CDR-TR #19930507).

[Ulri94]   Ulrich, K.: The Role of (Rapid) Prototyping in (Rapid) Product Development. In:
            Proceedings of the IMS International Conference on Rapid Product Development,
            Jan 31 - Feb 2,1994, Stuttgart, Fraunhofer-Institute for Manufacturing Engineering
            and Automation (IPA), Stuttgart, 1994, pp. 195-205

4

Cooperative Working on Virtual Prototypes

Dr. Stefan Haas, Fraunhofer CRCG, Providence, RI 02906, USA.
email: shaas@crcg.edu
Uwe Jasnoch, Fraunhofer IGD, 64283 Darmstadt, GERMANY
email: jasnoch@igd.fhg.de

Abstract

Virtual Prototypes allow the product development process to concentrate on a fully digital basis. Unfortunatelly, most CAD environments do not support concurrent data access of several users. This makes it necessary to include refinement loops, an often time consuming process.

In this paper, we define an CAD framework which allows coopertaive working of multiple users using different tools. This enables a new working paradigm, we call COOPERATIVE ENGINEERING. Instead of time consuming refinement loop iterations, intercative multi-disciplinary work is introduced. The benefits of this new working paradigm are manifold. Most important are the shortening of the design cycles and quality improvements due to the enhanced information contexts build by the shared sessions.

Keywords

Virtual Prototyping, Reengineering, Concurrent Engineering, Cooperative Engineering, Cooperative Working, Multi-Disciplinary Work

1 INTRODUCTION

In the last decade, computer-aided (CAx) and computer-integrated (CIx) technologies had a strong influence on the product development process. Prototypes are needed for the feedback of evaluation experience back into the electronic design.

Physical prototypes need a refinement loop of design, preparation and evaluation, step-by-step. Fully digital prototypes overcome this refinement as discussion and modification can take place at the same time.

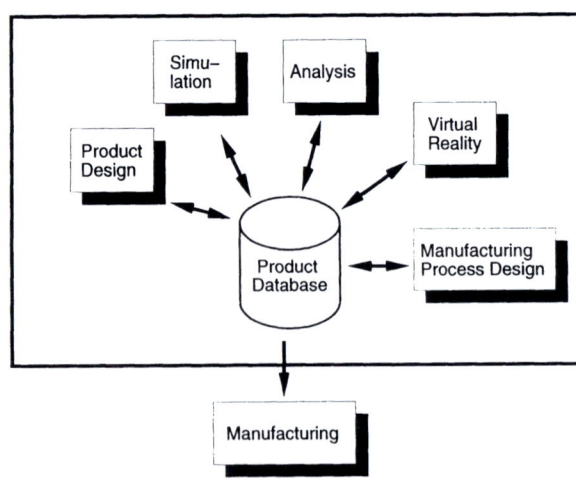

Figure 1 Environment considered for Virtual Prototyping

This cooperative working (CSCW) is the next step for making the product development even faster and efficient. In contrary to most basic CSCW applications, an open environment with heterogeneous application (see fig. 1) has to be considered.
Therefore, certain assumptions have to be made:

- each user may use a different (specialized) application
- virtual prototypes are more than a single digital files but rather a collection of data structures with overlapping contents
- coordinating design activities has to create a common u understanding about objects and activities
- frameworks have to glue existing applications by providing global services

Therefore, this paper focuses on the definition of these advanced global services and their usage within existing frameworks to show the benefits of cooperative engineering.

2 PROTOTYPING FRAMEWORK

The Virtual Prototyping Environment is based on a framework and integrated applications, as indicated in figure 2. The task of the framework is to offer libraries for the application development process and to provide a set of services for the applications during runtime.

The global services extending existing frameworks are:

- Cooperative communication and conferencing
- Multi-media documentation and mail
- 2-D/3-D user interface
- Distributed object oriented data management

Cooperative Communication and Conferencing

This service provides on the one hand an asynchronous broadcast and point to point message service for communication. On the other hand a synchronous communication service can be used for conferencing and handshake request. The asynchronous service is used by heterogeneous applications to exchange messages, thus informing each other about a special topic. Here it is important, that the applications are independent each application can select those messages individually which are considered important for the application and the kind of reaction.
The synchronous service is also used to support applications in the area of CSCW. Here, the basic services like group management in a conferencing session are provided.

Multi-Media Documentation and Mail

Integrated documentation becomes an important part of the product development process. The framework provides support for integrated documentation and enables the handling

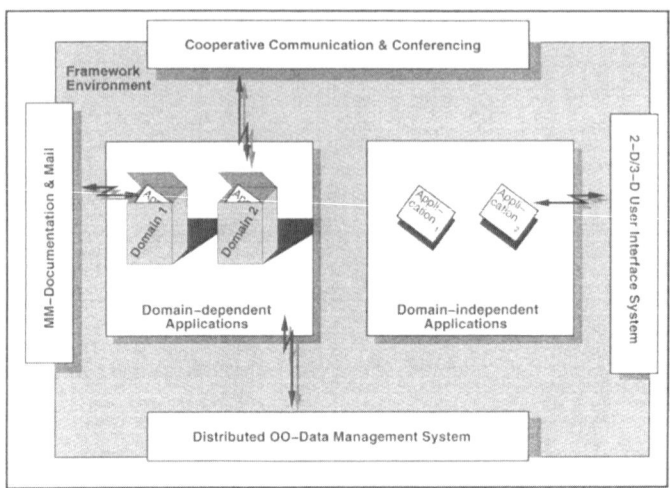

Figure 2 Extended Framework for Virtual Prototyping

and exchange of multi-media objects and the conversion and integration of different formats.

2-D/3-D User Interface

The user interface system provides services necessary for the development of homogeneous 2-D and 3-D user interfaces. Homogeneous user interfaces at the integrated applications raise the acceptance of the end users and help them working in the virtual prototyping environment.

Together with the communication and conferencing system, these services enable and simplify the development of wrappers to transfer single-user systems into conferencing application.

Distributed Object-Oriented Data Management

The data management component offers all services for handling with objects of the distributed environment. The basic goals for the component are to hide the location of an object for the application and to enable a uniform working with transient and persistent objects.

Advanced concepts, like working data bases for user session, enabling concurrent access to objects, and advanced check-in/check-out facilities are also topics of this service.

These services are necessary to improve the virtual prototyping environment for the end user in several directions. First off all, they are the basis for a common look and feel of the applications, necessary for the acceptance of the end-user.

With sharing the same data and sending information between, applications are not longer isolated, they are an integrated part of a homogeneous product development environment. Thus, the different steps of the product development are not longer separated and the problems of data exchange between the different phases could be solved.

3 APPLICATION DOMAINS

The integrated applications cover all aspects of virtual prototyping and are grouped in application domains. The virtual prototyping environment consists of the following domains:

• Product Design

This domain consists of the CAD-systems involved in the design of the product. These systems can either be geometry-based or feature-based. The purpose of this domains' applications is to create and modify the design entities of the product.

• Simulation

The simulations domain embraces applications for the physical simulation and also for the analysis of the product.

• Virtual Reality

The modeled products are put into the virtual reality domain to be presented and examined in an advanced virtual environment. The virtual reality domain provides also the environment for the compound visualization of physical simulations or analysis results together with the correlated shapes of the design parts.

• Manufacturing Process Design

The domain covers all applications for the computation of the manufacturing process. Here, based on the designed and simulated product data, the data for the real manufacturing process are computed. These domains will help to reduce the time for prototype development and will help to improve the quality of the last domain, Manufacturing.

4 COOPERATIVE WORKING TOOLS

The main task of the communication system is to deliver messages between homogeneous and heterogeneous application and/or framework components throughout the environment. Hereby, different scopes of the messages must be supported by the communication system, according to the purpose of the message.
One basic requirement is the communication between heterogeneous applications. In contrast to CSCW applications (computer supported cooperative work) where different instances of the same application exchange messages, communication between heterogeneous application means the communication between applications with different purposes.

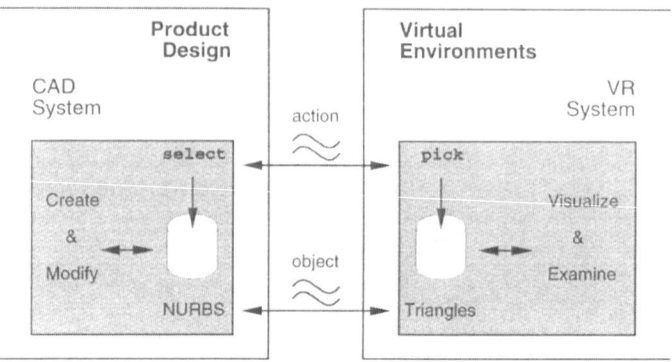

Figure 3 Similarities between domains

For example the message exchange between a presentation application ,e.g., a Virtual Reality System, and a constructive application ,e.g., a CAD-System. Here, picking an object in the presentation system could lead to an corresponding action, e.g., highlighting, in the constructing system (see fig. 3).

A basic requirement hereby is the common understanding of the objects and activities. This common understanding can be established by an application-independent message set, e.g., as used for desktop messages [Fra91b] or software development [Dig92].

ToolTalk Message Set

Tooltalk [Fra91a] has defined a message set for *Desktop services* which unifies many of the window and process based informations, such as

- `Created/Deleted`,
- `Iconified/DeIconified`,
- `Mapped/UnMapped`,
- `Modified/Reverted`,
- `Quit` and
- `Started/Stopped`.

These messages can be mapped to any kind of window system and window manager making it an ad-on to existing software components. These messages are mostly broadcasted to all other applications in the local runtime environment.

For applications performing *Document and Media Exchange* messages such as

- `Abstract`,
- `Display` and
- `Edit`

can be used to inform about multi-media objects and their usage.

Similar effort has been taken to unify message in *CASE* systems, based on the software development systems requirements.

The *CFI message set* [CFI92] is a first step towards unifying messages for CAD systems. This can be taken as a basis for a the common understanding spanning several other domains.

For virtual prototyping, these messages have to be generalized in a domain independent way. This can be done as many of these messages appear in similar forms also in other domains, e.g. select and pick in fig. 3.

This strategy also results in requirements for the data management component of the virtual prototyping framework which has to keep track of the corresponding object instances in the different applications. e.g., a particular patch in the CAD's NURBS model and a corresponding triangle mesh in the VR system.

Based upon a common topological structure, appropriate mappings have to be found to translate entity identificators, e.g, names or numbers, to other representation forms [JKSU94].

Considering the case, that the shape was made using NURBS, the simulation using FE rasters and the visualization using triangle meshes (see fig. 3), a mapping can be difficult. In any way, this mapping, based on the deltas to be applied, is much easier and far more effective than running the conversion several times and invoking a manual search from the scratch.

5 SCENARIO

Virtual prototyping in a distributed working environment is extending the basic meaning of having a prototype that no longer needs to be physically present to having a prototype that is not even represented as a single entity but rather as a collection of information that all together comprise the prototype.

This situation is already common in every existing prototype development environment as different applications deal with different entities and formats. Up to now, the underlying framework can keep track about the used data types and formats but hardly ever takes care about the heterogeneous nature of the prototype.

Basically, a virtual prototype is characterized as

- comprised by several and even redundant informations,
- partial information stored at different locations,
- accessed by different users and applications,
- consists of roughly the same topology but may have different structure,
- entities may be handled differently but applications,
- all of them represent the prototype, but through different views.

The virtual prototype frameworks offers global services to meet these advanced requirements of heterogeneous and open environments. They are globally available and no longer restricted to a single environment, but rather can link any application in every runtime environment of the virtual frame work that comprises the different sites involved (see fig. 4).

A major task to enable distributed interactive work based upon variable bandwidth net-

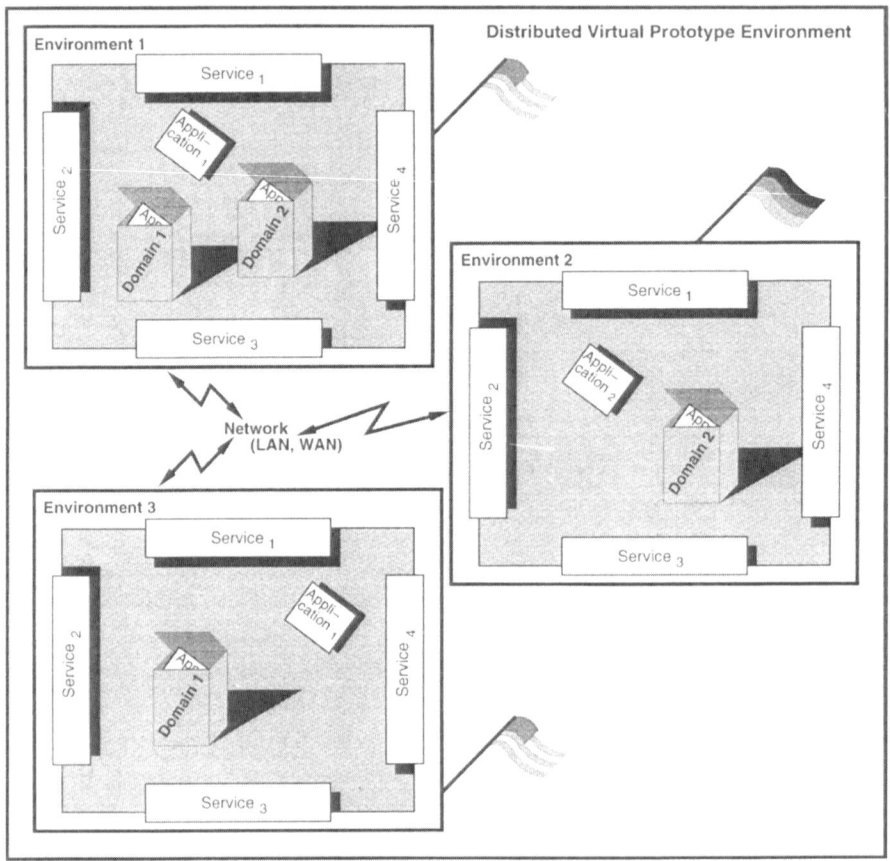

Figure 4 A Distributed Framework

work is a graceful message delivery, the main task of the communication server. Depending on the available network connections, an appropriate connection has to be established. Depending on the kind of message, e.g, frequency, message size (memos, images, video etc.), appropriate basic network services, have to be used.

The network services can be

- socket programs for specific applications
- customized RPC calls
- smnp (e-mail)
- ftp
- X11 client messages
- ToolTalk

- other framework specific communication components

The global communication service will select an appropriate line to the applications in the current session. Depending on the kind of message, e.g., broadcast messages or point-to-point request exchange, asynchronous or synchronous delivery, each connection may use a different transport mechanism (see fig. 5).

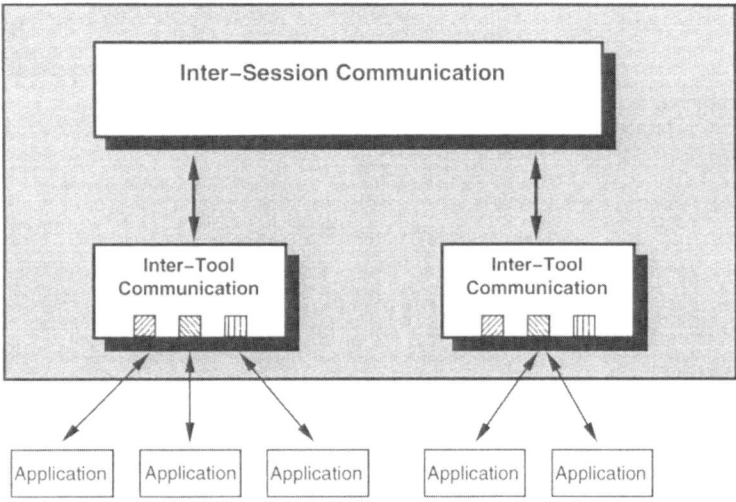

Figure 5 Server architecture

Also the messages itself have to take care of this delivery situation. In a situation where a remote partner can only reached via e-mail, only very few messages should be send to inform about the final state. It would not make much sense, if every interaction is transmitted that way, except for an off-line session replay. A similar services can be used to obtain initial information for new participating members to an already started session. Besides the elementary inter-tool-communication (ITC) service, an inter-session-communication service has to be provided. Its task is to monitor all sessions and the usage of appropriate ITC modules (see fig. 5).

The heterogenouity thus also reflects on the network structure of the participating applications. In fig. 6, a request message from application 1 in environment 1 is transmitted to its recipient, application 5 in environment 2, and vice versa the reply to it. Other applications may be passive listeners at either or other sites, each having it's own connection to the local communication server of the local runtime environment.

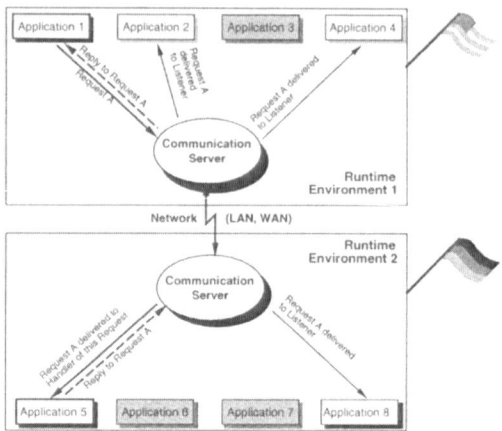

Figure 6 Distributed message delivery

6 CONCLUSION

The virtual prototyping framework offers significant extensions to existing product development frameworks in the areas of

- concurrent and cooperative engineering,
- open and distributed environments and
- heterogeneous data activity and user environments.

This advantages help decreasing the product development times (time-to-market) as well as customization times (time-in-market) for plain prototypes and mass production preparation.

In times of rapidly increasing requirements for faster and cheaper development these factors can cut development costs before they arise, e.g. due to refinement cycles.

The versatility of digital communication and data exchange help to make better use of machine and human capabilities in extending networkability towards cooperative working. Information translation is also essential in such system as it can consume serious amount of valuable human capital especially in open environments with multitudes of information types and formats.

Graceful handling of network services and their bandwidths enables interactive cooperative work even with application which have never been designed for these goals.

This shows the capabilities of an integrated framework which can be used with existing applications by add-on global services.

REFERENCES

[CFI92] Framework Architecture Reference, 1992.

[Dig92] Sunsoft Digital, SGI. The case interoperatability message sets: Version 1.0. Digital, SGI and Sunsoft, September 1992.

[Fra91a] R. Frankel. Solaris OpenWindows: Introduction to the ToolTalk service. *A White Paper*, 1991.

[Fra91b] R. Frankel. Solaris OpenWindows: ToolTalk in electronic design automation. *A White Paper*, 1991.

[JKSU94] Jasnoch, Kress, Schroeder, and Ungerer. Coconut: Computersupport for concurrent design using step. *IEEE 3rd Workshop on enabling technologies: Infrastructure for Collaborative Enterprises (WET ICE)*, April 17 - 19 1994.

5

Integration of Virtual Prototyping Applications with a Distributed Conferencing System

Norbert Schiffner
Fraunhofer-Institut für graphische Datenverarbeitung
Wilhelminenstraße 7
64283 Darmstadt
Germany
EMail: schiffne@igd.fhg.de

Abstract

Conference system support communication and collaboration among geographically dispersed user groups. In our contribution we introduce an environment that allows users to cooperate with the help of an electronic conferencing system. During a conference, the users can start all kinds of applications. I describe different approaches to offer a solution to integrate these applications in a conferencing system. At last I will introduce our new conferencing system WIDE (Workstation Integrated Distributed Environment). In this system we have tried to implement some of the integration approaches.

With these applications, the engineer can perform his interactive cooperative work in the conference scenario. Some of these applications are truly distributed, some were especially developed for industrial designer, others are standard tools.

In the future, most product prototyping work will be done on computers with all-digital data bases, interfaces, and output. The emphasis on concurrent engineering, total quality management, manufacturability, just-in-time manufacturing, and life-cycle costing makes prototyping a group collaborative effort. Similarly, the emphasis on virtual enterprises that span the world with their customers and suppliers make prototyping a truly distributed activity.

Keywords

CSCW, Conferencing System, Teleconference, Communication

1 Motivation

The increasing development of local and wide area networks and multimedia workstations allow the realization of interactive communication tools for conferencing environments. These distributed software systems use improved hardware and network features to enable synchronous communication.

An electronic conferencing environment enables the replacement of face-to-face meetings by computer supported conferences. This can be two people meeting, as well as organized conferences. Interactive communication and information exchange between geographically dispersed users is of great interest in different application areas like medicine, office organization, education and teaching, or the product development process. We want to describe the product development process as an example for the necessity of synchronous interactive communication tools provided by a conferencing environment.

One of the main goals in the product development process is to shorten the product development cycle by reducing the designing and manufacturing time period. To achieve this goal, hierarchical organization structure within the product development process must change to parallel organization and collaboration between different departments to improve parallel product development. One restriction for cooperation, communication and teamwork is that especially complex products are developed and manufactured by different companies at different locations. The realization of teamwork and parallel development needs sophisticated technologies to enable the involved persons to communicate and cooperate in a flexible and time efficient way. Short response time, feedback, early problem detection are extremely important. Technological support for the teamwork required by concurrent engineering is highly relevant to the study of computer supported cooperative work [FM 91]. Teamwork can be extended to the notion of virtual teamwork if the members of a team are geographically divided and communicate via electronic communication tools. Tools for asynchronous information exchange like fax, e-mail, file transfer etc. are widely used in industry. There is still a lack of tools for synchronous information exchange.

Designing, engineering or manufacturing specialists have to discuss many technical problems during the product development cycle, from, e.g. esthetical aspects of a design to manufacturability. Besides technical discussions, a great amount of commercial problems will be discussed by marketing, purchasing or managing specialists. The different locations where specialists work are often bottlenecks for early problem detection and decision making. A problem of today's work is that too much information is channeled through a team leader who is overloaded with traveling and work.

Tools of interactive communication integrated in a conferencing environment enable not co-located team members to communicate and collaborate whiteout time consuming and expensive meetings

The conferencing environment we want to introduce allows synchronous information exchange over wide area networks. It is comprised of multimedia technologies, communication and network infrastructure and complex user interactions. We divided the conferencing environment in to two major part, the conference management system and the conferencing applications.

2 The Conference System

The conference system is responsible for the administration of the conferences. It includes a distributed application server with functionality for organizing the interactions between users, conferences and applications. Because of general problems in distributing any kind of data we offer a special appropriate strategy.

Managing the flow of data is one of the most critical tasks in distributed systems. Because there is a large variety of data with different requirements in size and time, the organization of the communication channels is fundamental for the efficiency of the platform. It should work in high speed networks (i.e. Ethernet or ATM) and also in wide area nets like ISDN. The system should be highly stable against external influences and should rebuild automatically as much as possible. Even in case of errors there must be mechanisms to guarantee the reliability of data transmission.These requirements are solved using a client-server communication model.

In Figure 1 the data connections in the system are shown by an example. The Conference Server provides the distribution of the inner conference requests and messages. The only responsibility of the server that concerns the applications is starting and finishing.

The server has a connection to the conference database. where the 'public' data is stored. All information which needs to be distributed can be read or written via this database. Therefore it is guarantied to get all necessary informations when joining or leaving the conference.

The most important data is information about the users, conferences and applications. There are also other sets of data stored which are for internal usage, like hosts or screens for the X-Window system.

The application server is the basic tool for managing the conference. After building a connection to the conference server by starting the system, the user is able to begin his work in the given context. The system provides read and write access to the database. The block diagram in Figure 2 shows the realization.

A user-interface allows interactions via mouse and keyboard. At the beginning it will show a list of all connected users and actual running conferences. It is possible to join an existing conference by selecting a menu button. A conference member can leave the conference whenever he wants. It is also possible to create a new conference which others can join.

A very important task of the Application Server is providing functions for invoking and removing applications. Normally these applications are independent of the connection between manager and server and can be even used as stand-alone versions. So the stability of data transmission is as safe as possible against internal and external influences.

In conferences with multiple users or applications, there is a special input control system available which organizes the flow of user actions. In this mode only a single conference member can interact with the applications, like talking to the others or using the application. This is necessary because the optical and acoustical feedback is much lower than in face-to-faces conferences. The input control works like a fifo-queue and after having done his actions the user can pass the input rights to the next one in the queue.

3 Integration of Applications

The handling of applications from an integrated conference environment is one important feature in every conference system. The user should be able to start and stop applications during a

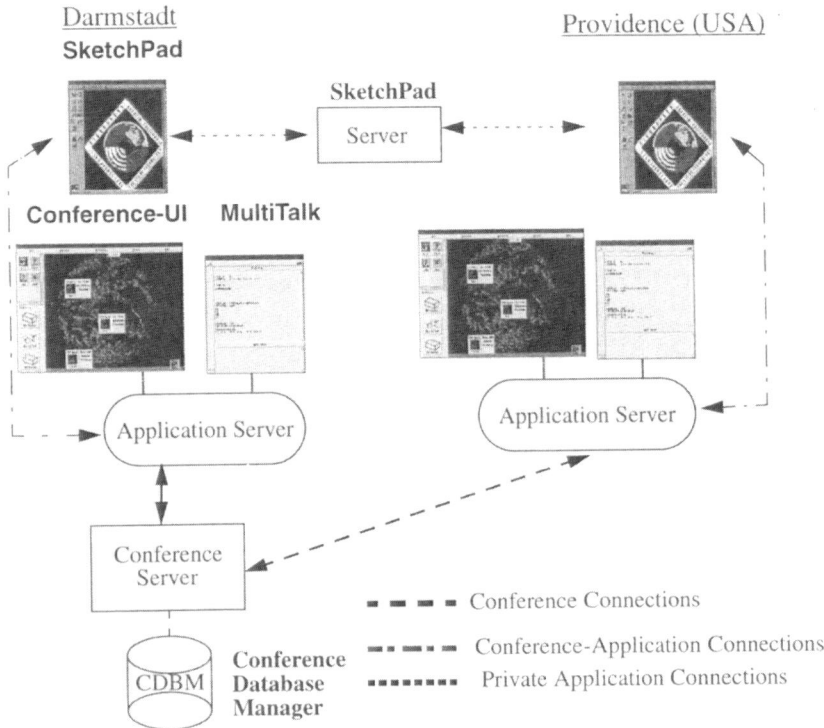

FIGURE 1 Connections in the Conference System

conference.

There exist three different types of applications:

1.Applications written for a special conference system.

2.Stand-alone CSCW Applications.

3.Non CSCW Applications.

A conference system needs different methods to be able to start those applications. We describe these methods in the following chapter.

3.1 Integration of Special CSCW Applications

Some conference systems offer a protocol to communicate with the external applications. With

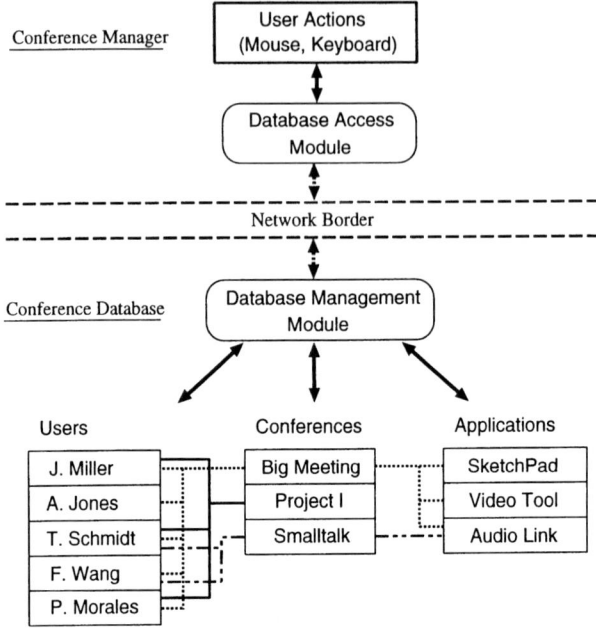

FIGURE 2 **Database access of the Conference Management System**

this protocol the applications can receive all information about the user and conference status. They are also able to control some parts of the conference system.

This way of integration is the best way because the application can use the same information. This is very important for the user interface, e.g. to use the same user name or colors to identify the user. This method can offer a seamless connection between the conference and the application user interface.

It is also important to control the applications. This can be done only with this method. The external applications can receive the status of the floor control and can control the own input mechanisms.

To spread this method, it will be helpful to create a standard for this protocol.

3.2 Integration of Stand-alone CSCW Applications

For existent stand-alone CSCW applications the conference system offers a shell script call. The conference system passes information to the shell script, e.g. host, user name. After the execution of the shell script the conference system has no chance to control the application. The feasibility of controlling is poor.

Unfortunately, most stand-alone CSCW Applications are not able to switch their own user interfaces off. The user must insert all the required information in the application user interface. Sometimes the external application needs some internal information from the conference system. If this happens, it is necessary to train the user.

3.3 Integration of Existing non CSCW Applications

In most cases it is necessary to integrate an already existing application into a conferencing environment. The reasons might be one of the following:

•There is no distributed version of the application available.

•It is to difficult to design the application for a conference environment in a short time. An example here is a CAD system. Such an application has been developed in hundreds of person/years and could not be extended to multi-user support in just a few months.

•The users won´t use a new unknown application just to present some information in conference environment.

In these cases traditional applications can be integrated into the conferencing environment using CSCW tools. First the user interface of the application must be shared between the conference members. This will be done by duplicating the output of the application to all users. All conference members have the possibility to see the same information.

Input capability to the application should be switched to only one of the conference members at a time. The floor holder mechanism of the conferencing system manages input switching. During conferencing time, one user ---the floor holder --- loads the application through that CSCW tool. The applications output will be distributed and hereby the floor holder presents his information.

It is necessary that the inactive partners have a possibility to follow the floor holders actions, especially if the conference goal is that conference members learn something about the usage of an application This will be possible only if the inactive users see all prompting and echoes. Therefore the mouse pointer movements etc. of the floor holder must be distributed as well.

As another feature of the CSCW tool sharing applications a sketching functionality should be integrated onto the distributed applications output. The floor holder can chose if he wishes to add input to the application or if he will draw into an overlay positioned on top of the applications output window. These sketches are distributed via a CSCW tool to the other conference members. They have --- being inactive users --- only the possibility to make sketches, too. Only when becoming an active user is there a need to distinguish between application input and sketching in the layer.

The scenarios components are illustrated in Figure 4. The CSCW tool for sharing applications between more that one user is a tool like shared X [Alt90]. Functionality for pointer movement distribution must be added to that tool. The sketching functionality on the other side could be added as well. A separate tool organized as a collection of transparent windows on top of the applications output windows containing the sketches is another possibility for a realization.

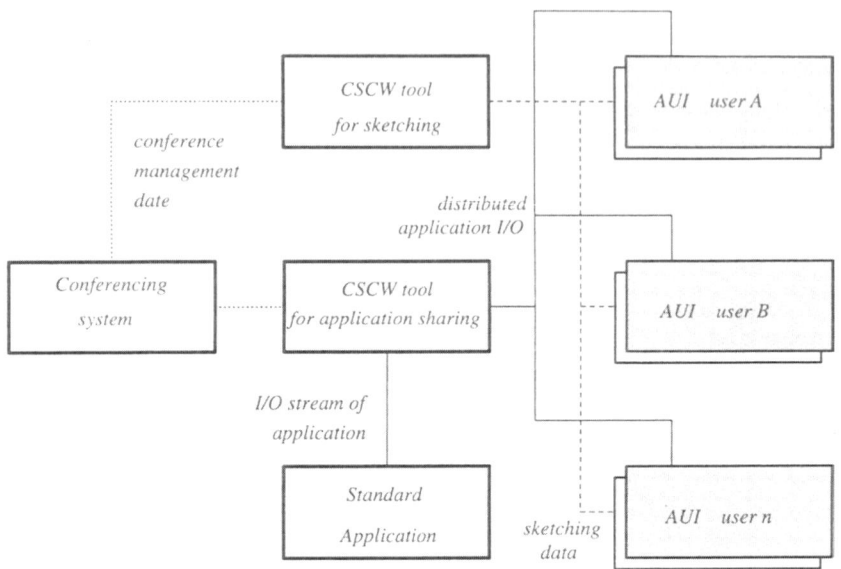

FIGURE 3   Integration of an application into conference environment

4   Conference System WIDE

4.1   An Integrated Environment for Synchronous Information Sharing

The conferencing system is a communication platform integrating applications, components and services for running a multi-party conference.

Geographically divided conference members can use audio, video, specific conferencing tools and standard application software together in an integrated environment. The conference system provides the user interface for the interactions, communication services for sharing of applications and administration services.

4.2   Functional Description

The conference system consist of two major parts, the conference server and the application server.

The conference server is a framework for administration of the conference, the conference members and the conference applications.

It includes the following functionalities:

- enable the user to join and leave a conference
- administrate information about the conference and the conference participants
- invoke and close conference and standard applications
- providing floor handling mechanisms

The application server are software systems running in a conference environment to support communication and collaboration. Conference applications include systems, which are particularly designed for conferencing like the SketchPad and the shared 3D Viewer as well as standard applications.

Standard software will be distributed with a shared window system to all conference members. The floor control is handled by the conference manager, it exist different kind of control mechanisms.

The server has a connection to the conference database, where the 'public' data is stored. All informations which need to be distributed can be read or written via this database. So it is guarantied to get all necessary informations when joining or leaving the conference.

4.3 Conference

In future versions it is planned to have more than one conference, where different groups of users are able to collaborate.

These conferences will have special security mechanisms for having private conference or group related permissions for using special conference applications.

5 References

[Alt90] Michael P. Altenhofer. Shared X. Technical report, NESTOR Project, Digital Equipment Corporation, CEC, Karlsruhe, 1990.

[BNRS91] M. Bever, S. Noll, J. Rix, and C. Schottmueller. Kooperative graphische Anwendungen in Hochgeschwindigkeitsnetzwerken. In J. Encarnacao, editor, *Telekommunikation und multimediale Anwendungen der Informatik, Informatik–Fachberichte (293)*, pages 313 – 324. Springer–Verlag, 1991. Proceedings.

[FM91] Susan Frontczak and Kathy Miner. Distributed computing and organisational change enable concurrent engineering. In *Proceedings of the Second European Conference on Computer–Supported Cooperative Work, Amsterdam*, pages 131 – 146. Kluwer Academic Publishers, London, 1991.

[Gav91] William Gaver. Sound support for collaboration. In *Proceedings of the Second European Conference on Computer–Supported Cooperative Work, Amsterdam*, pages 293 – 308. Kluwer Academic Publishers, London, 1991.

[GRWB92] S. Greenberg, M. Rosan, D. Webster, and R. Bohnet. Human and technical factors of distributed group drawing tools. In *Interactions with Computers: Special issue on CSCW*. December 1992. In Press.

[LM92] I.M. Lu and M.M. Mantei. Idea management in a shared drawing tool. In *Proceedings of the ACM Conference on Human Factors in Computing Systems (CHI92), Monterey*. The Association for Computing Machinery, New York, May 1992.

[MLB90] Morgan, Lach, and Bushnell. ISDN as an enabler for enterprise. *IEEE Communications Magazine*, pages 23 – 27, April 1990.

[NS91] S. Noll and M.G. Schendel. Cooperative sketching in a network environment for the automotive industry in europe. In *Proceedings of the Eurographics 1991, Technical Report Series, Vienna*, 1991.

[San92] A. Santos. Using a cooperative hypermedia editing tool to enhance group communication and productivity. *The Visual Computer: An Internal Journal of Computer Graphics*, 1992. In Press.

[SK93] K. Schroeder and H. Kress. Distributed conferencing tools for product design. In *Proceedings of the IFIP Workshop on Interfaces in Industrial Systems for Production and Engineering, Darmstadt*, March 1993. In Press.

[SNR91] M.G. Schendel, S. Noll, and J. Rix. Distributed SketchPad System: A tool for cooperative sketching in a network environment. In *Contribution to COMICS– Workshop, Toulouse*, June 1991.

6 BIOGRAPHY

Norbert Schiffner was born in 1961 in Frankfurt/Main in Germany and is working as a research assistant at the department of „Computer Supported Cooperative Work (CSCW)"since June 1992. He was studying Computer Science with specialization in Computer Graphics at the Technical University of Darmstadt from 1987 to 1991. From January 1992 to June 1992 he was working at the external department in Providence, USA. He holds a diploma (Dipl.-Inform) in Computer Science from the Technical University of Darmstadt.

PART TWO

Advanced Product Modeling Techniques

6

Product and Shape Representation for Virtual Prototyping

M. S. Bloor, A. McKay, Department of Mechanical Engineering and
M. I. G. Bloor, M. J. Wilson, Department of Applied Mathematical Studies
University of Leeds, Leeds, LS2 9JT, UK

Abstract

Instead of being an exciting step towards more rapid product development, virtual prototyping could become a new term to cover the existing computer tools collected under the terms CAD, CAE, CAx. The computer technology to enable realistic "handling" and distributed design consultation is now demonstrable. However, effective management and distribution of the information to be processed and, hence, use of the results could remain constrained by the many ad hoc data conversions imposed by our current inability to most effectively represent that information for shared use and reuse.

The first issue is to provide a framework for the information that supports product development and that is developed during product development. In order to create the integrated information system that virtual prototyping implies, the Open Distributed Processing Reference Model (ODP-RM) [deMeer 1992] for distributed information systems is an apt basis and a useful route map.

The role of STEP in such an information system needs explanation. It is encouraging that the ISO Standard for Exchange of Product Model Data (STEP) [Owen 1992] development methodology is using a similar map to that provided by the ODP-RM. The STEP work, particularly the tools used developing Application Protocols (AP) and the integrated resource models themselves, is an integral part of our way ahead. The learning from that community is already exploitable. As discussed below, however, their difficulty in defining the requirements of STEP, and the general lack of understanding of what STEP can provide, needs urgent resolution.

Another issue is the constraints imposed by representational techniques for geometry used in today's CAx systems, particularly parts with sculptured surfaces. Because shape is so fundamental to mechanical design and visualisation on computers, it is important that shape information can be related properly to other information and altered simply. This is not currently so.

Finally, one must understand the relative value of real and virtual prototypes and more importantly the need for physical models to test in order that confidence in computer simulations is well founded. This is particularly necessary where the virtual prototyping uses complex flow analyses, namely Computer Fluid Dynamics (CFD). The advent of layered manufacturing (additive manufacture, rapid prototyping) potentially expedites the making of real prototypes with minimal manufacturing planning.   "Computer Aided Rapid Prototyping" and the "Integration of CAD, CAE and Fast Free Form Fabrication" are alternative titles for a project, briefly known as CARP[1], researching these subjects to reduce the time to prototype automotive powertrain components.

This paper discusses some of the issues and some pieces of the solution.

1   BACKGROUND

In order to prototype on the computer an increased integration of the activities involved in evaluating products is needed to facilitate communication between them.   For example, analysis of the behaviour of a component may suggest alterations in its shape and also the shape of other components with which it interacts.   With current technology, this is inhibited by data conversions between different representations and lack of appropriate associative data. More important, shape should be related to functional requirements - not usually represented in the computer.

The word, integration, is widely and loosely used.   The complex association of people, processes, materials and information that is the product introduction process needs decomposition to aid understanding so that possible improvements can be identified, controlled and are achievable.   This is the role of a reference model.   In considering integration in the product life cycle, the authors view the whole process as a distributed information system and find the ODP-RM helpful.   This model advises the consideration of such a system from five viewpoints:-

a) business or enterprise, that is, the things to be done (activities or processes) to achieve the objectives of the system;
b) information, that is, its form and the flows to achieve necessary communications and,
c) computing, system engineering and technical implementation often classed together as the software support environment.

The advantage of this model is that the integration of the enterprise with respect to its objectives and of the information requirements can be addressed independently of and preferably before the computer implementation issues.   This can help to avoid the trap of being driven by current computer based applications.   Gielingh [Gielingh 1994] breaks the integration of product data into the problems of enterprise integration, application (computer) integration and data integration.   Application integration is interfacing today's software using data exchange technology.   The solution to data integration, sharing, is seen in terms of networks, files and data bases software suport in ODP-RM terms rather than the appropriate integrated product representation.

---

1   CARP is a three year collaborative research project organised under the auspices of the DTI EUREKA programme. The primary objective is the integration of design, analysis and rapid manufacturing methods such that prototype components can be produced rapidly and with greater confidence of "right first time", the secondary to realise the potential of layered manufacturing technology. The consortium is lead by Ricardo Consulting Engineers Ltd, of the UK, and includes CADDETC, Delcam International plc, Webster Mouldings Ltd, and the University of Leeds, all from the UK, Volkswagen AG of Germany and Dott. Vittorio Gilardoni SpA of Italy.

STEP, the STandard for the Exchange of Product model data [Owen 1992] addresses computer applications in the product life cycle. It was initiated in 1984 to enable better interchange of data between computer aided activities (CAx) than that provided by the current data exchange "standards", e.g. IGES [Smith 1991], SET: 268-300 [1989]. Many hundreds of man years have been voluntarily contributed under the aegis of ISO TC184/SC4.

STEP is a methodology and data models. These latter are loosely based on the ANSI SPARC, Application, Logical and Physical layers [ANSI/X3/SPARC 1975] and the methodology suggests understanding the engineering requirements for a particular exchange by studying the activity and information flow first. This is encouraging and we use the tools adopted by STEP contributors for these operations.

Implementations of STEP will have a collection of Application Protocols, data models to which exchange software must conform. Exchange software converts between the data representation of the CAx system and a neutral representation in the form, currently, of a physical file. Application Protocols are constraining in order to enable data exchange, i.e. so that conforming processors cover the same range of entities and use them in the same way. The requirement to enable data exchange enforces an emphasis on what current CAx systems represent. The Application Protocol 203, Configuration Controlled 3D Design, has a wider view embracing product structure and configuration. Unfortunately supposed prototype implementations by vendors, do not impose conformance which is required by STEP if it is to revolutionise data exchange. Current attempts at STEP based data exchange can be categorised as "IGES STEP" with much of what was wrong with IGES and its contemporaries remaining. The current debate is the perceived need for data sharing which we define as applications viewing, or dipping into, parts of a single representation of the product model conforming to an integrated product data model. Intuitively STEP should enable this but current thinking is confused. The contribution of the Integrated Resources, particularly the Generic Product Description Resource [ISO 10303-41 1994], is explained in this paper.

Sections 2 and 3 use this background to progress towards the type of product data model we need to integrate the information and enable sharing. We, then, complete the picture with a discussion of a novel geometry representation.

2.   ACTIVITIES AND INFORMATION MODELLING

Models based on SADT(IDEFO) [Marca 1988] cemented understanding of the issues among the technically diverse group of collaborators in CARP. Although IDEFO has its critics as a communication medium, the thought and rigour required to understand the inputs, controls, outputs and mechanisms and agree viewpoints and purpose is a necessary prerequisite to understanding the required *information* and software support. At this stage the depth to which one expands such diagrams should be compatible with understanding the product development activities and the information requirements, not the details of computer processing and data requirements. As a simple example, we identify the geometric functionality that the activities themselves require, i.e. we do a stress analysis of a solid part, a flow past a surface or in a cavity, we machine a surface, not the data requirements imposed by current shape representations or the computational methods we normally use to simulate behaviour.

3.   INFORMATION

One is then in a position to understand information flows, forms and content. In choosing a formal notation, differences between the engineer and the information technologist appear. Our contention, not novel, is that the activities which the information system supports and the information flows should be clearly understood before the role of the computer is defined. An information model can be evolved as is, for example, the Application Reference Model of STEP.   The direct use of STEP's Data Definition Language (DDL) Express [ISO TC184/SC4/WG7 1991] may suit those who know the computer applications for which they are catering and hence the data requirements, but is not appropriate as a first step to an information model to support product development. It is, however, very difficult to separate the product information from those representations, data, which are needed to support the computer processing and the form in which results are presented, e.g. stresses on meshes rather than stresses available throughout the body.   Today information systems are usually studied with preconceived views of the computer processing to be applied because commercial systems, e.g. CAD, FE, are already in use.   Collaborators selling CAD or CFD systems are interested in improving current systems and thus start from current technology.

3.1   The integrated product data model

The information model in EXPRESS-G (Figure 1), the graphical form of EXPRESS [ISO TC184/SC4/WG7 1991], from the CARP project falls into this trap because implementation is driven by existing CAx systems. However, we do try to identify the conditions to be applied to geometry or shape description rather than to the discretisations required by particular computational techniques. The model also uses STEP models where these are available, e.g. the advanced BRep AP204 for solids.

More significant to the CARP model, and a feature of all STEP APs, is the influence of the GPDR, the generic product description resource. This imposes a root and a framework to the data model which models product and some accompanying structure. This structured form is essential if sharing is to be enabled. Our research proposes a framework for product data which supports all phases of the life-cycle, specification to disposal, and is detailed only as users and applications require. This is a top-down approach not the bottom-up approach which pragmatic data exchange requirements encourage. This top-down integrated product data model is essential for sharing.

The STEP GPDR distinguishes between the description of and the representation of the properties of products. For example, the STEP Integrated Resources, of which the GPDR is a part, has a geometric description which is shape definition information. We would suggest this should be shape information that comes from the functional requirements. The shape representation itself is what computer processing/applications require, generally the primitives (lines, arcs, polynomial patches, blocks, spheres, cylinders) or discretisations of which computer representations are constructed. From the perspective of the engineering process such representations are of secondary importance.

It is interesting and understandable that many of those in STEP concentrate on data exchange of representations - the area which CAx systems embrace and on which STEP could be best demonstrated - and see the geometric description and the product oriented framework as an unnecessary overhead.

The authors believe it sets the necessary framework for STEP. The authors are using the ideas behind the GPDR that is exploiting a framework to support specification, functional structures and assembly information [McKay 1993].

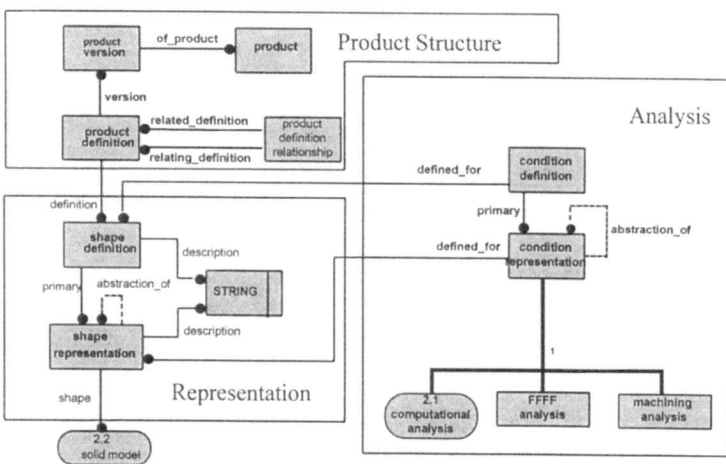

Figure 1 Root of Information Model (in EXPRESS-G) GPDR, Geometric Representation and Analysis.

4. COMPUTER PROCESSING

Having understood the activities and the information requirements, ideally one is in a position to think about the software support environment, firstly the processing that the computer will do and for which we must design software and, lastly, the computer system's architecture and implementation.

By this time one knows the observations, analyses and manipulations that are significant in the development of a given product or type of product, i.e. for what one needs a prototype. Whether this is a physical model or whether the observations etc. can be adequately performed on the computer will depend on the testing to be performed and the possibility of adequate computer simulations. Most parts can now be realistically visualised including movement. Assembly, to test fit and maintainability, can be performed for many functional mechanical parts where the solids involved can be represented. Analyses of stress (commonly using finite element (FE.) methods) can be confidently performed in many situations. In many companies FE based stress analyses are performed by designers, not analysts, using FE codes validated for strictly defined physical situations. Flow analyses are less reliable. Although codes for complex 3D situations are sold, confidence in their use requires experienced fluid dynamicists backed by experimental confirmation. Additionally the time taken to simulate (and convert) can inhibit realistic visualisation which many see as an essential of virtual prototyping. The display of complex models, in digital pre-assembly of aeroplanes or examination of oil rigs today, is built on static computer representations of admittedly complex geometries and relies on view changes to model the human moving through the scene. Visualisation of dynamic environments and those involving models of physics is not yet possible because of accumulations of times built up by various aspects of the computation.

The overhead in conversions is a secondary problem with this virtual prototyping scenario, e.g. from CAD representation to analysis discretisation and, if the prototyping leads to suggested improvement, back from the results of the analyses to improved shape.

As explained in previous sections these conversions can drive the information system rather than being relegated to a simple reliable routine process. However, the associativities required to respond appropriately to the results of analyses demand an integrated top-down product information model as an immediate priority.

5.   REPRESENTING SHAPE

One aspect of the work reported here which will help virtual prototyping and also shape synthesis is the definition of geometry where the functional constraints define the shape description, where surface and solid representations have a unified definition and where discretisation suitable for analyses is implicit in the shape definition process. The over-riding objective on the work reported here is not the downstream processing that the method facilitates, but minimising the number of parameters that define the shape and relating those to the design specification not solely the geometry. The aim is optimisation and shape synthesis where the shapes have complex sculptured surfaces. The speed and reliability with which discretisations for layer manufacture, for CFD and for FE based solutions are produced has been remarkable in contrast with current CAD systems.

A new method for the efficient definition of complex shapes has been developed at Leeds. It is based upon a boundary-value approach, in that a surface is defined in terms of boundary-conditions specified along the curves which form its boundaries. A consequence of this boundary-value approach to surface generation is that the CAD geometries produced by the Partial Differential Equation method are described in terms of surface patches that meet perfectly. This facilitates both the computer-aided analysis of an object's physical properties and its manufacture. Furthermore, a design may readily be changed and yet this property - of perfect connectivity - is still maintained.

The definition of the surface shape between defined boundaries by partial differential equations [Bloor 1989] was introduced as a blending problem. The technique has been extended to cover free-form surface design where the aim has been to generate functionally useful surfaces (e.g. propellers, ship-hulls) from patches of PDE surface [Lowe 1990].

The PDE method is not primarily a method for surface representation, but a method for surface generation. It is envisaged that objects whose surfaces are generated by the method are designed to serve some function, and hence it is crucial that the design of an object's geometry is integrated with functional considerations. What makes the PDE approach especially suitable for this is that it can parametrise complicated free-form surfaces in terms of a relatively limited parameter set. This is possible because of its boundary-value approach: surfaces are specified in terms of data distributed around curves, rather than across the surface itself. In practice, this means that the number of shape parameters specifying an object's surface is often small enough for the task of optimising its shape to be computationally feasible; yet at the same time the method is sufficiently flexible for a wide range of shapes to be accessible to it.

Work has been carried out on the integration of an object's geometric design with its functionality in a number of applications, ranging from heat transfer, stress minimisation, and the design of a surface for its hydrodynamic properties [Bloor 1990] [Wilson].

When creating complex surfaces, it is generally necessary to construct the surface from a number of parametric patches, and we need to be conscious of the fact that each patch must be bounded by specified curves with adjacent surfaces having the required degree of continuity across their common boundary. However, when constructing a free-form surface from a number of PDE patches, one still has considerable freedom with which to choose the boundary conditions, despite the continuity requirements in order to achieve the desired shape. The solution remains sensitive to the choice of boundary conditions and this fact is turned to the designer's advantage as it is the boundary conditions that provide him or her with a powerful tool for surface manipulation. A convenient way to demonstrate the relationship between parameters and surface changes it through the series of examples given in earlier papers to which the reader is referred rather than repeating here.

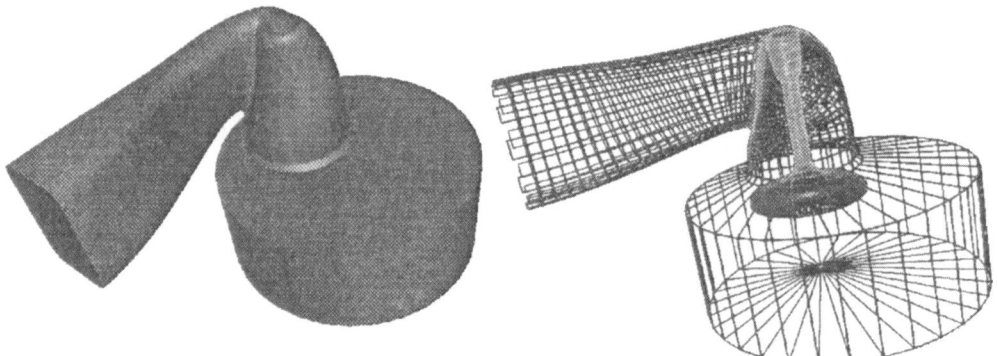

Figure 2 Inlet Geometry for a Diesel Engine.

As an example which combines most of the likely complication which may arise in multiple patch surface generation we consider the inlet geometry for a diesel engine, as shown in Figure 2. Notice that this entails creating the space within the inlet port through which the gas flows before entering the combustion chamber. For this reason the volume forms part of the domain over which a fluid dynamical analysis needs to be carried out. It follows that once this has been done, the definition of the mechanical part surrounding this space requires further, though generally less critical, surface generation to complete the definition of a solid object.

In obtaining the surface as a combination of PDE patches, each of which is a solution of a boundary value problem, there is no problem in ensuring that adjacent patches meet perfectly with the desired degree of continuity. In other words, there are no holes in the final surface, which if need be, can be represented in discrete form by a set of quadrilateral surface patches. This is an extremely important point, from the point of view of both analysis and manufacture through such techniques as selective laser sintering. For instance, in CFD analysis, conservation laws are discretised and fluxes of physical quantities across cells are calculated. Clearly if there are holes in the cells, inaccuracies can arise. More importantly, holes in the surface cause difficulties for inside/outside testing which may be needed for automatic mesh generation. Also, in layered manufacture, holes in the surface description give rise to flaws in the final object.

As it happens, the CFD code (VECTIS) which we use to demonstrate the physical analysis of our model requires a closed surface of triangular facets bounding the solution domain before the automatic mesh generation and CFD analysis can be carried out. Thus, this code provides us with an ideal test both for analysis and rapid fabrication. The geometry for the inlet port, which we have created, is supplemented by additional geometry of the combustion chamber and a valve (created using the PDE method) to make a physically realistic model for analysis. The surface data was put into STL[2] format and was immediately ready for the automatic mesh generation of VECTIS without the necessity for surface 'stitching', i.e. the closing of surface holes.

6.   A LAST WORD ON STEP

The virtual prototyping environment requires a shared information model. Although the STEP resources are enabling the data exchange for which STEP is designed, currently it concentrates on the detailed data representations rather than an overall framework for product information. This is emphasised because geometric representations, CAD data, are those most commonly exchanged. Attempts to discuss STEP in the context of sharing, demand solution to a currently unsolved problem [McKay 1994]. Shared data implies a model instance that agents dig into at will, but which inherently shares data and hence ensures consistency. We believe this needs an integrated product data model. How such viewing mechanisms are defined may be related to STEP Application Protocol (AP) interoperation as many believe. The authors fear AP interoperation is directly counter to the design aims of APs and could degenerate into the use of subsets[3] from a single large AP reminiscent of (old) IGES technology. Possibly, interpretation of STEP integrated resources [Ashworth 1993] is a way forward. An experiment which shows how sharing and data exchange can be achieved in the same environment is urgent.

7.   CONCLUSION

Building on a range of research this paper proposes an integrated product data model as a top-down a framework for information generated in the product introduction process. The model uses the STEP Generic Product Description Resource extended to cover specification, functional structures and assembly conditions. Additionally a novel representation of geometry designed to facilitate shape design and functional optimisation may prove a unifying link into shape representation and discretisation which will accelerate virtual prototyping for engineering - where models of physics are important.

---

2 STL - defacto standard for input to layered manufacture machines, a triangular facetted representation of a solid.
3   As currently proposed subsets are Units of Functionality or Application Integrated Constructs. These lack a context, an important part of the AP Development Method.

8. ACKNOWLEDGEMENTS

The authors wish to acknowledge the contributions of colleagues in the Departments of Mechanical Engineering and Applied Mathematics, and industrial collaborators especially in the CARP and the MOSES projects, and the provision of rapid prototyping technology by DTI and the University. The large and multi-disciplinary team at Leeds provides a challenging environment in which to consider tomorrow's information systems. Use of the Vectis CFD code from Ricardo Consulting Engineers Ltd has provided an ideal test of the PDE role in both analysis and rapid manufacture.

9. REFERENCES

ANSI/X3/SPARC (1975) Study Group on Database Management Systems Interim Report, FDT (ACM Bulletin) 7, No.2.

Ashworth, M. (1993) *A STEP Based Information Strategy and Tools for Engineering Organisations*. PhD Thesis, Department of Mechanical Engineering, University of Leeds.

Bloor, M. I. G. and Wilson, M. J. (1989) CAD, 21, No 3, pp. 165-171.

Bloor, M. I. G. and Wilson, M. J. (1990) CAD, 22, pp. 202-212

Gielingh, W. (1994) *Product Data Technology*, NAFEMS Newsletter, 1994

de Meer, J. Heymer, V. and Roth, R. (1992) editors, *Open Distributed Processing*. Elsevier Science Publishers, North Holland.

ISO 10303-11, (1991) *Industrial Automation Systems and Integration - Product Data Representation and Exchange - Part 11: EXPRESS Language Reference Manual.*

ISO 10303-41: Part 41, (1994). *Industrial Automation Systems and Integration - Product Data Representation and Exchange - Integrated Generic Resources* : Fundamentals of Product Description and Support.

Lowe, T. W., Bloor, M. I. G. and Wilson, M. J. (1990), in Advances in Design Automation, Vol 1, B. Ravani, ASME, pp. 43-50

Marca, A. D. and McGowan, C. L. M. (1988) *SHDT: Structured Analysis and Design Technique*, McGraw Hill

McKay, A. Bloor, M. S. and de Pennington, A. (1993) *A Framework for a Product Data Model*. Submitted to IEEE Transactions on Knowledge and Data Engineering.

McKay, A. Bloor, M. S. and Owen, J. (1994) *Application Protocols: A Position Paper*. International Journal of CADCAM and Computer Graphics, 9(3):377-389.

Owen, J. (1992), *STEP: An Introduction*. Information Geometers.

SET: z 68-300, (1989) *Industrial automation - external representation of product definition data*, Jrnl. L'association fran/c caise de normalisation (afnor), Tour Europe Cedex 7, 92080, Paris-la-Défeuse, France

Smith, B. and Wellington, J. (1991) *Initial Graphics Exchange Specification (IGES)*, Version 5.1, NIST

Wilson, D. R., Bloor, M. I. G. and Wilson, M. J. (1993) *An Automated Method for the Incorporation of Functionality, in the Geometric Design of a Shell*, 2nd Symp. of Solid Modelling and Applications, eds. J Rossignac, J Turner, and G A Allen, ACM Press, New York, pp. 253-259

7

Virtual Prototyping through Feature Processing

J. Ovtcharova, A. S. Vieira

Fraunhofer – Institut für Graphische Datenverarbeitung (IGD)
Wilhelminenstraße 7, 64283 Darmstadt, Germany
Telephone: +49 6151 155 207/246
Telefax: +49 6151 155 299
e – mail: jivka/vieira@igd.fhg.de

Abstract

For the integration of different product development tasks to be supported within a virtual prototyping process it is necessary to take account of semantic information, alongside the geometric shape information. The need of semantically endowed primitives leads to the use of features. In this paper we focus on one crucial important aspect in which virtual proto – typing systems should evolve: the *feature – based design of product prototypes*. This will offer possibilities to users to define *high – level semantic information* in the prototype model in a way which allows a high degree of correspondence between virtual prototype and physical product. Moreover, features should be used as *processing entities* for design, as well as for the information integration of design with different downstream phases of virtual proto – typing such as analysis, manufacturing simulation, cost and quality estimation.

Key Words

Virtual prototyping, semantic product modeling, features, feature processing, feature – based design module

"Virtual presence is the synthesis of a perceptual consistency that convinces users that either they are immersed in another space or that computer generated artifacts exist in the physical space (illusion)".

W. Chapin, T. Lacey, L. Leifer (Chapin, 1994)

1 INTRODUCTION

The virtual prototyping world gives us an unusual opportunity to compress time and to reduce production costs. The traditional limitations of the physical prototyping such as weight, measures and material treatment are avoided. The ability to use computer simulation implies that we have appropriate models of product prototypes. Appropriation means not only models which describe the geometry, but describe the complete product semantics in terms which designers understand and which reflect the design intent to downstream applications. As a consequence, the illusion of a prototype presence in a virtual space requires a comprehend and semantic correct computer−based model which fully complements a prototype existing in the physical space.

Some essential requirements to the computer−based representation of virtual prototypes are as follows:

A/   *Computer−Aided Engineering (CAE) system point of view:* The development of virtual prototypes requires an engineering environment based on an integrated product development process and an integrated product data model. One of the central components of a CAE system for supporting virtual prototyping is a sui−table design module which enable the development of prototypes corresponding to the customer needs and which can be developed easy, fast and cheap.

B/   *Design processing point of view:* First, supporting the intuitive and creative work of the designer in different design phases, such as conceptual, embodiment and detail design is one of the most important requirements for early and fast prototyping. The models which are generated in the design process have to be processed in downstream phases of the prototype development, such as analysis, manufacturing simulation, cost and quality estimation. Second, using data at different levels of abstraction is needed to build comprehensive semantic models (Ovtcharova, 1992). Third, a mechanism to check the consistency of the semantic model data based on feature constraints is essential (Vieira, 1994). Fourth, facilities for applying these constraint concepts for easy and fast redesign based on parametrics and variational design methods and tools must be provided. Fifth, easy and fast configuration by using pre−defined and user−defined modeling entities has to be supported.

C/   *Graphical presentation point of view:* New modeling and presentation techniques, such as virtual reality and graphical simulation are of great importance.

In this paper, we focus our attention only on the second topic, the design processing for virtual prototyping. Furthermore, to satisfy the requirements discussed above, we concentrate our research on the design processing using the feature paradigm.

Features have been identified as meaningful engineering aids for modeling products using high–level semantic data. Thus, a principal impetus for feature research was the observation that the successful computerization of product development needs augmented models that are capable of representing both geometric and non–geometric information. The main advantages of features include: 1) a vocabulary which is more natural for expressing the product prototype than are pure geometric models, 2) the possibility for using features as a basis for modeling product information in different phases of prototyping such as design, analysis, process planning and manufacturing, and 3) an increase in the designer's productivity and cost effectiveness. A vast amount of research activities has recently been focused particularly on the problems involved when integrating previously separate prototyping phases into one system that covers all major aspects of prototyping. Feature modeling is now regarded as a key technology in achieving high level of efficiency, automation and information integration in this broad area.

2   THE NOTION OF A FEATURE

The feature definition as adapted in our work deals with the assignment of features into feature types, as presented in (Ovtcharova, 1994). Here, we distinguish between *generic* and *application* features. Generic features, such as form, material, and precision features are used to describe properties of products in a general way. For example, *form features* are defined as *shape characteristics* of product parts which conform to some *shape configuration* of the product, like depressions (negative volume form features) and protrusions (positive volume form features), but have no *specific functional meaning*. Application features, such as design features and machining features, can be defined by using application–specific data over the definition of form features in the concrete application area. Such a feature object hierarchy is quite useful in the following sense: first, grouping features into classes with respect to given properties leads to development of *unified mechanisms* for modeling different products possessing these properties. Second, feature classification leads to a *common terminology* and development of product data exchange formats.

Design features are the set of application features used by the designer to model the a concrete product prototype. They encapsulate information about the semantic (or functional) meaning as well as the shape of a part which can facilitate communication of the design intent to downstream product prototyping phases.

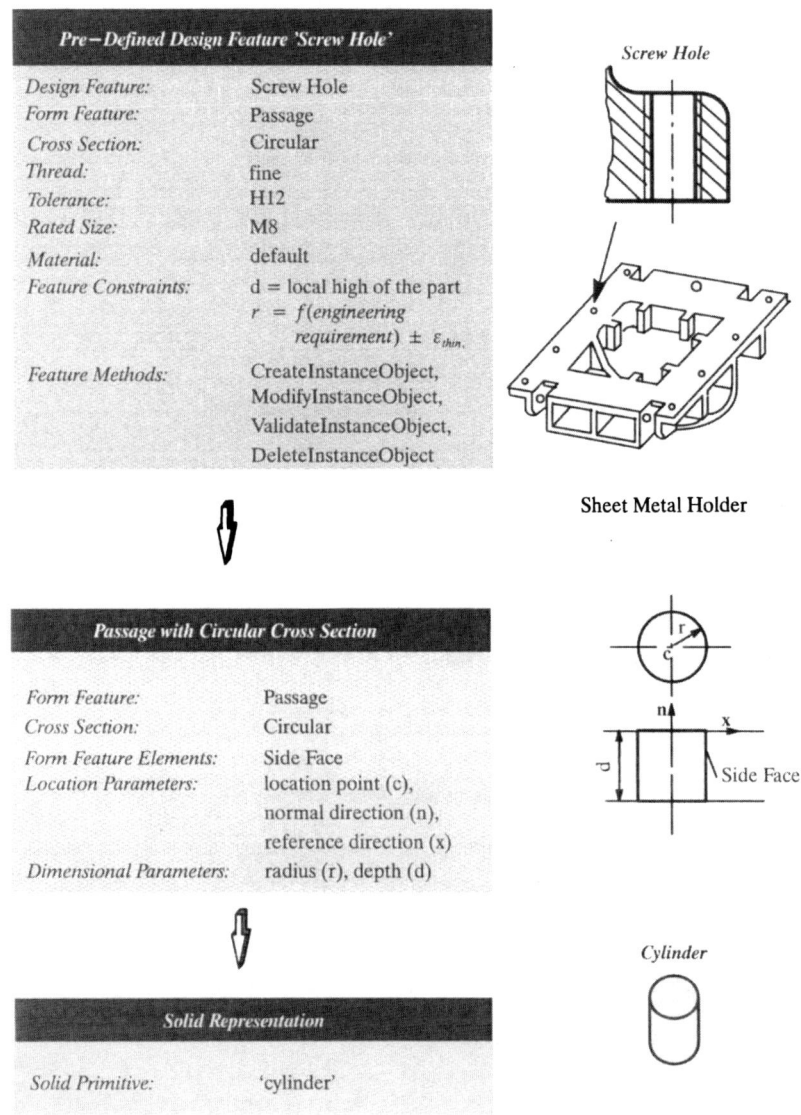

Pre−Defined Design Feature 'Screw Hole'	
Design Feature:	Screw Hole
Form Feature:	Passage
Cross Section:	Circular
Thread:	fine
Tolerance:	H12
Rated Size:	M8
Material:	default
Feature Constraints:	d = local high of the part r = f(engineering requirement) ± ε_{thin}.
Feature Methods:	CreateInstanceObject, ModifyInstanceObject, ValidateInstanceObject, DeleteInstanceObject

Screw Hole

Sheet Metal Holder

Passage with Circular Cross Section	
Form Feature:	Passage
Cross Section:	Circular
Form Feature Elements:	Side Face
Location Parameters:	location point (c), normal direction (n), reference direction (x)
Dimensional Parameters:	radius (r), depth (d)

Solid Representation	
Solid Primitive:	'cylinder'

Cylinder

Figure 1   An example of a screw hole definition

We define two main types of design features (Ovtcharova, 1993): (1) *pre−defined* design features which are always available and define the fixed set of most used design features, and (2) *user−defined* design features which complete the previous feature set and are dynamically created during the design session. The pre−defined design features are more often used design features, such as cylindrical holes, rectangular pockets, simple slots. Each pre−defined design feature is implicitly described by a name, the type of the corresponding form feature (depression, protrusion), a list of design parameters, constraints to express semantic information and a set of methods which are necessary for creating and manipulating the feature (Figure 1). If the set of pre−defined features is not sufficient, the designer can specify his own set of features, which will be included in a user−defined design feature library. This library is created dynamically (during the interactive design session) and complete the previous library with more complex design features.

In the design phase of the prototype development, the creation of semantically correct feature−based models not only increase the quality of the design but also influences the performance of the downstream phases. The certainty of the semantically correction of the model data, is also an important factor to achieve an integrated development process. Thus, resulting in faster and cheaper prototype development. Product semantics are expressed by different kinds of *features* and *constraints*. Therefore, we state that constraints are one of the major product information, besides features, used by designers to express their design intent. Constraints attached to features describe the consistent states that features can hold, and express the behaviour of the features in the prototype part model, as well as the behaviour of the model itself. For all these reasons constraints are used as crucial elements to preserve the semantic validity of feature−based models (Vieira, 1994).

In our approach, feature constraints are expressed on the basis of a full parametric definition of features. Parameters might be dimensions, locations, and functional or technological parameters without any direct geometric meaning. Parameters allow easier manipulation of feature entities, quick design changes, powerful engineering analysis of the prototype model, and improve the quality of the data transferred during the design process and to downstream applications. In association with parameters, constraints raise a potent design method to solve the design change problem, and to capture and maintain the design intent.

In our work, *three categories* of constraints exist. Two of them are based on the theory behind the parametric and variational design methodologies (Weissbarth, 1994), (Chung, 1990), (Roller, 1994) for solving geometry. The third category of constraints includes all the constraints which do not need to be solved, but only tested for validity.

In this paper, we deliberately omitted most of the definition and representation aspects of features and constraints which are described in details in our previous publications (Ov−tcharova, 1993), (Ovtcharova, 1992).

3 THE BENEFIT OF USING FEATURES FOR VIRTUAL PROTOTYPING

The intention of features is to provide the designer with a set of meaningful engineering aides, thus supporting him in his creative work at a higher level of abstraction. There is a common agreement that design by features has the potential to *improve* the quality of the design, to *speed up* the design process, to *reduce* the costs and to *shorten* the time–to–market. It also improve the link of the different phases of prototyping, such as design, analysis, manufacturing simulation, cost and quality estimation. The advantage of using features for virtual prototyping can be characterized by the definition of feature data at different levels of abstraction, the maintenance of constraints for preserving semantically correct feature–based models, and the development of an architecture for feature–based design adaptable to users and applications.

The purpose of this paper is to suggest essential directions for feature–based design for supporting virtual prototyping. Using a series of assumptions regarding conceptional, implementational, and applicational issues, we articulate the direction of design by features based on evidence through summaries of previous work and examples of design applications.

The following, presents essential challenges for research and development in feature–based design for virtual prototyping.

Challenge 1: Design of a methodology for developing comprehensive semantic models

>   For the integration of different virtual prototyping tasks it is necessary to take into account comprehensive semantically correct prototype models *through the entire prototyping cycle*. Features are used to define *comprehensive semantic mo–dels* of prototypes using data at different levels of abstraction in a way which allows a high degree of correspondence between the virtual prototype and the physical product.

Challenge 2: Support for configurable prototype models

>   Feature–based models offer facilities for the configuration of prototype models by using *pre–defined* and *user–defined* features. Tools and methods for graphical–interactive generation of features can be developed to support the designer in creating his own sets of features. User–defined design features will complete the set of standard (or pre–defined) features delivered with the system and can extend dynamically during the design session as required by applications and user needs.

Challenge 3: Support for easy and fast redesign

>   Design feature models offer facilities for easy and fast redesign based on parametrics and variational design methods which are used by feature constraint entities. Design features can be described by a *set of parameters and constraints* re-

quired to build the basis for consistency maintenance of features during the whole design process. The association of values to the parameters and the satisfaction of the constraints will offer useful and semantic correct feature representations.

Challenge 4: Development of an integrated feature processing environment

Features support the intuitive and creative working of the designer in different design phases, such as conceptual, embodiment and detail design. The models based on features can be processed in downstream phases of the prototype development, such as analysis, manufacturing simulation, cost and quality estimation. To take full advantage of features, an *integrated environment* supporting the design and the downstream phases has to be developed.

Challenge 5: Development of an open, modular and extensible system architecture

Feature – based design tools must have an open and modular architecture which allows them to be integrated in different CAE system environments. *CAE system architecture* can be developed by a complete separation of the functionality of the feature modelers from the underlying solid modelers. Thus, possibilities are offered to the users in modifying their feature modeler via integrating new feature modeling facilities configurating the feature modelers to new applications without changing the solid modelers.

4   FEATURE PROCESSING USING SINFONIA

For supporting virtual prototyping through feature processing we introduce SINFONIA, a module for *feature – based design* which is configurable to users and applications within diverse CAD environments, particularly in the area of mechanical engineering (Ovtcharova, 1994). The module has an open and modular architecture allowing the modification of existing functionalities and integration of new modeling facilities and application tasks.

The main modules of SINFONIA are the *Feature Modeler* and the *Design Feature Manager* (Figure 2). The Feature Modeler is responsible for the instantiation and modification of features and the creation of the feature – based model. The Design Feature Manager allows feature data and design processes to be managed in a uniform way. The CAD system environment in which SINFONIA is integrated consists of the following modules: the *User Interface System* and the *Application* modules (offering tools for interaction of the user with application specific part models and for communication with external systems and applications, such as NC modules, etc.), the *Solid Modeler* (responsible for creating the shape representation of the feature – based model), the *Consistency Manager* (providing services to handle all kinds of different constraints within the design environment) and the *Product Database* which includes all services for storing and retrieving various product data.

The central idea in SINFONIA is that the design features defined and instanced from the feature library are used to model a product part in a concrete application context. Their description derives specific meaning from the view of the function of the product part, includ-ing shape data as well as non−shape data. Moreover, SINFONIA supports the users to *define their own specific design features*, thereby allowing to represent new types of products. This is accomplished by a design that foresees the interactive specification of design features by the user. The starting point of the design process are *pre−defined design features* delivered with the module. During the design process it is possible to combine pre−defined design features or to define new features and store them as *user−defined design features*. By repea−ting this process more and more complex design features are immediately available during the design process.

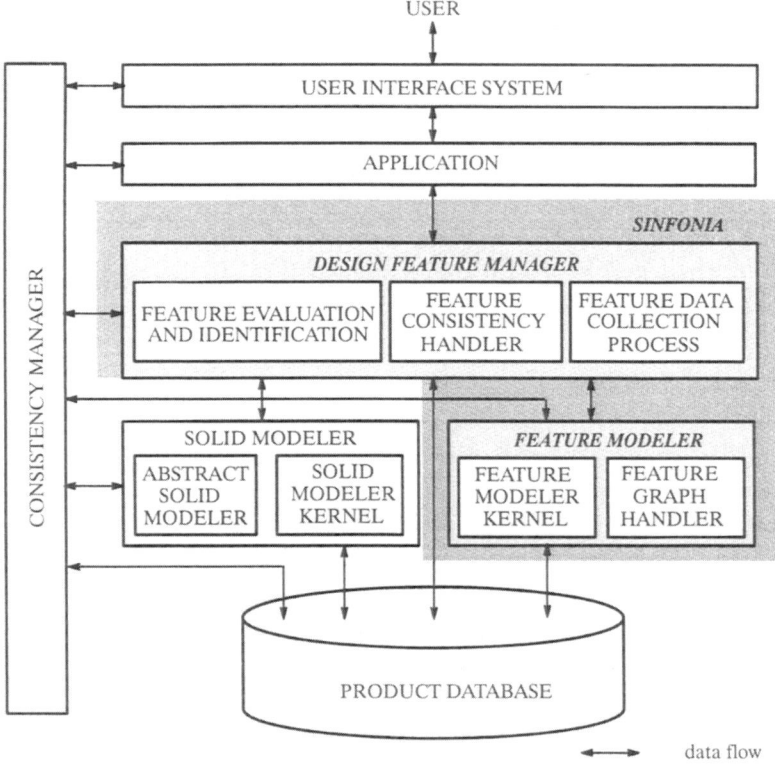

Figure 2 Feature−based design module within an open CAD environment

Figure 3 shows the feature processing pipeline for the creation and modification of pre−de-
fined design feature objects. The design process will be activated by the application or the
user interface level (1). This is done by selecting the correspond modelling operation such
as create, copy, or modify the feature object. To illustrate this process, the creation and mo−
dification of a cylindrical blind hole feature is presented.

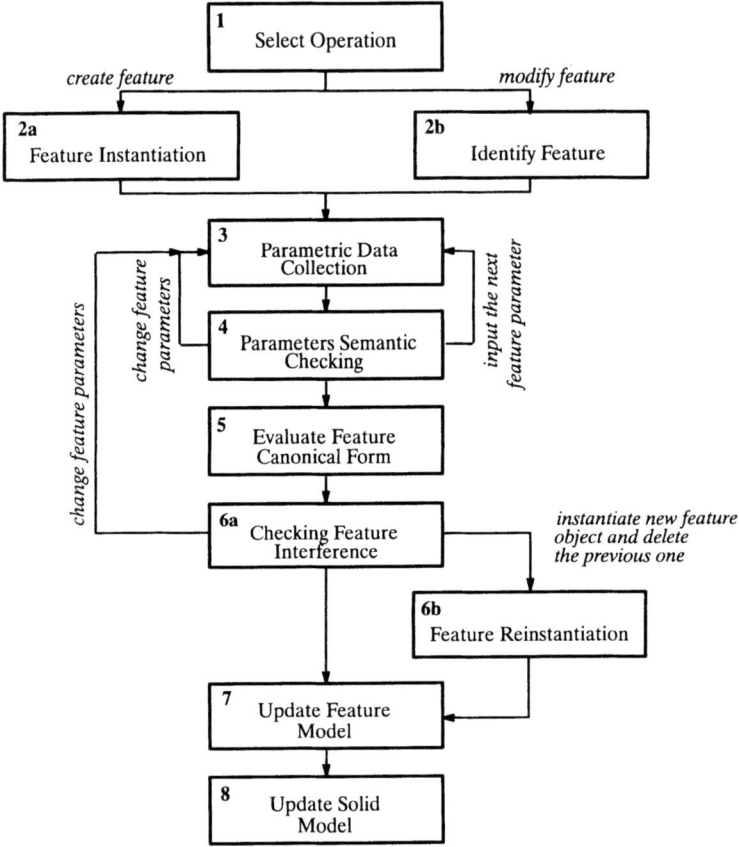

Figure 3   Feature processing pipeline for pre−defined design features

To insert a new design feature in the part model, an instantiation of the feature object will
be performed (2a). A set of default parameter values are parsed to this instance. For example, the instantiation of the cylindrical blind hole specifies automatically a form feature of
type depression and a cross section of type circular.

In the case of a feature modification, a shape entity is selected and the feature is identified from the feature graph representation (2b). This process of identification is handled by the feature evaluation and identification module.

The next step in this process consists of the complete collection and consistency checking of the semantic feature data (3,4), for example, the parameters defining the feature.

For our cylindrical blind hole, parametric data such as location (lcs) , radius (r), depth (d) and material (mat) must be supplied. These data will be checked (without interference on the semantic of other features of the part model), as shown in Figure 4a and Figure 4b, through geometric analysis techniques of the part model. The results of this geometric ana−lysis can be used as input for the constraint solving process. The data can be collected and checked either interactively while designing or, non−interactively after all parameters are collected. If a semantic error occurs, the process returns for a correction of the inputted data or is aborted.

After, the canonical form of the design feature is evaluated (5). This intermediate shape representation is used for the second phase of the consistency check: the detection of fea−ture interferences (6a). Figure 4b shows a simple situation where only consistency checking on the parametric data is not sufficient in order to detect this interference. Even if the part model is described parametrically, a check on the parametric model representation can often be too complex. Moreover, it is some times not solvable.

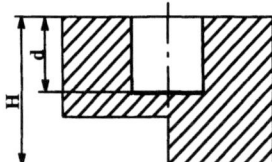

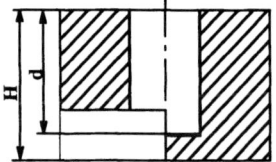

Figure 4a Creation of a semantically correct blind hole

Figure 4b Modification of the cylindrical blind hole

The parametric check (d<H) is not enough to detect the semantic inconsistency of the blind hole

At this stage of the design process, the following situation can occur:

- no interference is found and the process continues;
- interferences with other features where found but they do not change the semantic of the features in the part model;
- interferences where found and the user aborts the process; and finally,
- interferences allow a change of the object class and the process con-tinues.

This last situation is also illustrated on Figure 4b. The user had the creation of a blind hole in mind. However, the depth entered was to big and an interference was found. In this situation the system can analyze if the entered data fits semantically in some other kind of design feature available in the modeler. In our example the suggestion could be the reinstantiation of a cylindrical through hole (6b).

Finally, the design feature data are represented in the feature model. This representation consists of the creation of a node within the feature model and the insertion in the hierarchical graph structure (7). This process is performed by the feature modeler and the feature graph handler. The insertion process detects all dependencies of this feature object to all other features within the graph. The corresponding solid model representation will be updated (8).

5   CONCLUSION AND FUTURE WORK

In this paper, we focused our attention on the design processing aspect of the whole virtual prototyping process. Some essential requirements for the computer−based representation of virtual prototypes using the feature paradigm have been identified as follows: First, to support the intuitive and creative work of the designer in different design phases, such as conceptual, embodiment and detail design. The models generated in the design phases have to be processed in downstream phases of the prototype development, such as analysis, manufacturing simulation, cost and quality estimation. Second, to support different levels of abstraction to build comprehensive semantic models. Third, a mechanism is needed to check the consistency of the semantic model based on feature constraints. Fourth, facilities have to be provided for applying constraints for easy and fast redesign based on parametrics and variational design methods and tools. Fifth, easy and fast configuration by using pre− defined and user−defined modeling entities is needed.

To satisfy these requirements, an environment for integrated feature processing has to be developed. The system architecture of this environment has to be open, modular and extensible to allows the integration into different CAE systems.

As part of such an environment, we proposed SINFONIA, a feature−based design module. SINFONIA has an open and modular architecture that allows to modify and extend existing functionalities, and to integrate new modeling facilities and application tasks. Two main modules, the *Feature Modeler* and the *Design Feature Manager* have been presented in details. The Feature Modeler is responsible for the instantiation of features and the creation of the feature−based model. The Design Feature Manager allows feature data and design processes to be managed in a uniform way. Using SINFONIA, the users can work with standard pre−defined design features delivered with the module, or to define dynamically their own specific design features during the design session. Furthermore, SINFONIA allows the interactive definition of constraints concerning the product semantics. The CAD system environment in which SINFONIA is integrated has been outlined and the most important modules have been briefly described. At this stage of our investigation, we do not provide an overall implementation, but demonstrated that a sound conceptual foundation is necessary, achievable and realizable. Currently, an object−oriented version (C++ programming language) of the both modules, the Feature Modeler and the Design Feature Manager is in development.

6 REFERENCES

Chapin, W., Lacey, T. and Leifer, L. (1994) DesignSpace: A Manual Interaction Environment for Computer Aided Design, *Proceedings of the CHI'94 Conference Companion,* Boston, Massachusetts, USA, April 24−28, pp.33−34.

Chung, J.C. and Schussel, M.D. (1990) Technical Evaluation of Variational and Parametric Design, *Computers in Engineering,* Vol. 1, pp.289−298.

Ovtcharova, J. and Jasnoch, U. (1993) Towards a Consistency Management in a Feature− Based Design, *Proceedings of the 1993 ASME International Computers in Engineering Conference,* San Diego, California, USA, August 8−12, pp. 129−143.

Ovtcharova, J., Haßinger, S., Vieira, A.S., Jasnoch, U. and Rix, J. (1994) Sinfonia − An Open Feature−Based Design Module, *Proceedings of the 14th ASME International Computers in Engineering (CIE94) Conference,* Minneapolis, Minnesota, USA, September 11−14, Vol.1, pp.29−43.

Ovtcharova, J., Pahl, G. and Rix, J. (1992) A Proposal for Feature Classification in Feature−Based Design, *Computer & Graphics,* Vol.16, No.2, pp.187−195.

Roller, R. (1994) Foundation of Parametric Modeling, *Parametric and Variational Design,* B.G. Teubner, Stuttgart, pp.63−71.

Vieira, A.S., Ovtcharova, J. and Jasnoch, U. (1994) Consistency Management Aspects in Sinfonia, *Proceedings of the Workshop on Graphics and Modeling in Science & Technology,* Coimbra, Portugal, June 27−28.

Weissbarth, T. (1994) Associative Design. The Variational and Parametric Design Approach in the CAD/CAM/CAE Product EMS of Intergraph, *Parametric and Variational Design,* B.G. Teubner, Stuttgart, pp.73−80.

7 BIOGRAPHY

Dr. Jivka Ovtcharova graduated from the Department of Heat−Power Processes and Control of Nuclear Stations, Moscow Power Institute, Russia in 1982. From 1982 to 1987, she worked at the Bulgarian Academy of Sciences, Sofia. In 1987, she joined the Technical University of Darmstadt as a researcher. Jivka Ovtcharova received her doctor's degree in computer sciences from the Technical University of Sofia in 1992. She is currently a project ma−

nager at the Fraunhofer Institute for Computer Graphics, Darmstadt, Germany. Her current work is focused on the development of feature−based design modules adaptable to diverse CAD environments, in particular for supporting rapid prototyping and virtual prototyping technologies.

Ana Sofia Vieira graduated from the Department of Mathematics, University of Coimbra, Portugal in 1993. Since March 1992 she has been working as a guest researcher at the Fraunhofer Institute for Computer Graphics on the field of Cooperative & Hypermedia Systems. Since May 1993 she is working on the concept and development of a Consistency Mechanism for Feature−based Design Systems, based on the combination of three design methodologies: feature−based design, constraint−based design, and parametric design. Her research interests include feature−based modeling, algebraic and algorithmic constraint solving techniques and object−oriented analysis.

8

Dynamic and semantic modeling - a new approach to model products

K.-P. Greipel
Siemens Nixdorf Informationssysteme AG
Otto-Hahn-Ring 6
D-81739 München
Tel. ++49 89 636 49852
Fax ++49 89 636 47413

J. Colpaert
Siemens Nixdorf Informationssysteme AG
Siemenslaan 1
B-8020 Oostkamp
Tel. ++32 50 83 2739
Fax ++32 50 83 2488

Abstract

We present a new modeling approach that transfers the power of object orientation from the programmer to the end user. The approach is non-geometric and independent of the application. The keywords are *semantic* (supporting conceptual reasoning), *dynamic* (supporting creativity and flexibility), *relations* (supporting constraints) and *structure* (reducing complexity). The main innovation is its dynamic object technology: the end user can extend applications at runtime with her/his problem-specific notions, constraints and structures. Coupled with modern intelligent user interface techniques this technology realizes a new big step towards intuitive graphical interactive example-based programming. This opens the horizon for a new generation of easy-to-customize CAD systems that exceed the possibilities of current parametric and feature-based shape modeling and introduces more logic into modeling. New system can be realized that support early design and branch-specific product modeling.

Keywords

dynamic object technology, early design phase, semantic product modeling

1   INTRODUCTION

Current CAD tools are essentially geometry-oriented. Although concepts as driving dimensions (parametrics) and regions of interest (features) have been introduced in industrial CAD systems, the modeling focus is still the shape of a product. The core of a product description still is geometry: points, curves, surfaces and solids are the building blocks of current CAD systems. In advanced system some relations are imposed to geometry, for example: construction rules (parallel, orthogonal, aligned, etc.), variant dimensions, feature-to-part structures. Product performance aspects and product manufacturing are derived from these geometric descriptions that already detail the product to the final shape.

The support during early design phases remains completely unresolved. Continuous reengineering and the use of better and better tools have optimized manufacturing since the begin of industries by factors up to thousand whereas the product design process has essentially remained the same. Production industries are increasingly recognizing that the design process has to be optimized to reach time-to-market requests. Thus investigations of companies and scientific experts say:

- 2/3 of product development time are spent in the design phase
- 80% of the cost of a product is determined in the early design phase
- only 20% of the product knowledge is reached in the early design phase.

CAD as electronically drawing board with 3D extrusions (at the high end side) cannot help to solve these problems.

2   NEW MODELING PARADIGM

We propose a new modeling approach to solve these problems. This approach does not regard geometry as center of a CAD system any longer. It propagates a dynamic logical data scheme called dynamic object technology as new core. This innovative scheme is capable to describe the semantics, relations, structures and multiple representations of a product. The shape aspect is merely handled as one specific view that is nevertheless of greatest importance as natural interaction environment, for visual and physical verifications and for shape design reasons.

2.1   Example

In the following we consider a simple example. It illustrates how additional views like functional specification view, principle solution view, structure view and layout view can enhance the early design process. We consider the functional requirement to be 'combine two plates'. A possible solution is to 'connect both plates with bolt and nut'. This results in the following structure: two 'plates' with subfeature 'countersink hole' and a 'bolt' with 'nut'. The layout of all features and parts is shown in the picture below. In addition these components may have many interdependencies. So the bolt and its nut may be read from a discrete standard part table according to the thickness of the two plates. The diameters of the subfeature countersink hole are again dependent on the diameters of bolt, bolt head and nut. If the designer has modeled with these relations, then a simple editing of say the thickness of one of the plates will result in an automatic consistent update of all other components.

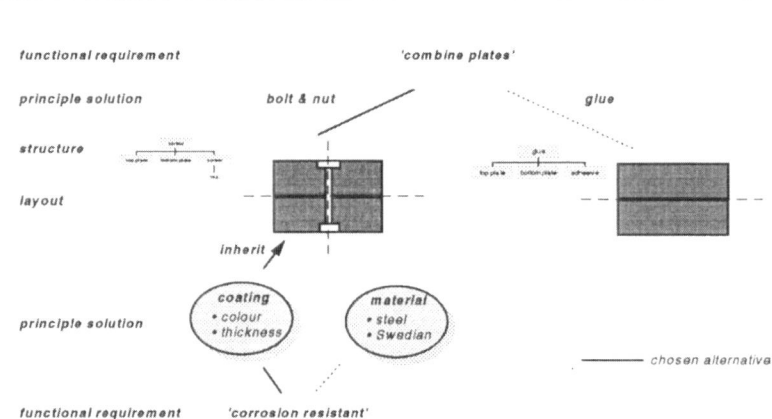

Figure 1 A simple design example capturing full design intent.

An alternative solution may be to 'glue' the two standard plates leading to a complete different structure and bill of material consisting of two plates and the 'adhesive'.

Lets assume the designer decides for the first solution for ease of maintenance. In a concurrent engineering environment a new additional requirement considering the corrosion resistance of the bottom plate may arise. Again the designer may consider two solutions: a special material or a special coating for the bottom plate. A completely new notion 'corrosion resistance' will be added to the design model with attributes that have never been considered before.

This very simple example already reveals

- how the additional semantic modeling domains function, principle solution, structure and layout help to capture design intent and help to approve design concept as early as possible
- how dynamic modeling helps to react to new requirements without loosing already designed elements.

2.2 Semantic modeling

Semantic modeling complements the geometric view of products with views needed to support conceptual reasoning and early product knowledge. In the early design stages a functional requirement is fulfilled with a coarse technical solution. These are described in a structural way (the core components of solution), with relations (what are the dependencies between the components) and layout (how do these components fit together). This conceptual product description   can already be used to derive most of the later manufacturing, maintenance and disposal aspects of the product. The layout also serves as a geometric base on which the detailed shape of the product can be modeled.

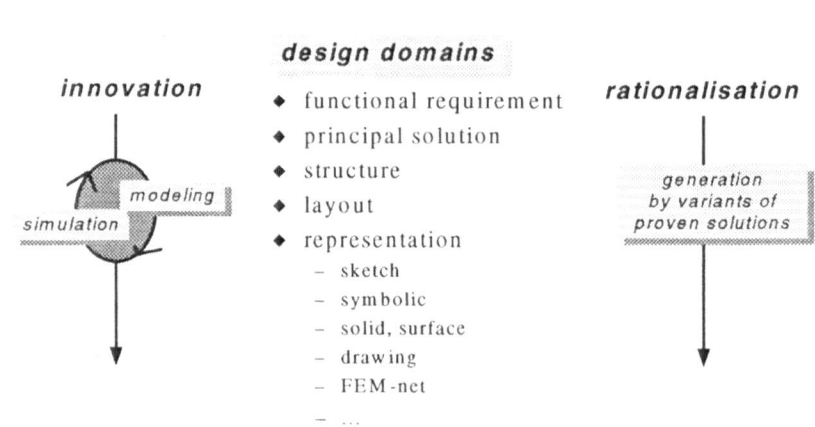

Figure 2   New design domains for semantic modeling.

This modeling approach supports innovative design (with early simulation cycles on not fully detailed models) and reuse of proven solutions (generation by variants of proven solutions).

2.3   Dynamic modeling

In current CAD systems the software engineer determines the objects and methods by which a product can be modeled. The end user can only design in the restricted views (in modern systems: sketch, 3D feature, assembly, drawing) and elementary functionality (in modern systems: parametric geometry operations) programmed in the CAD system. The designer has no possibility at all to add her/his problem-specific views, object types and methods to the system. Dynamic modeling frees the designer's creativity and provides the flexibility needed in concurrent engineering environments. The designer can customize the system to her/his specific needs at runtime. She/he can learn the system to communicate with the user using the semantic notions that are best suited to describe a problem and its solution. The base for this revolutionary modeling approach is the innovative dynamic object technology.

3   DYNAMIC OBJECT TECHNOLOGY

The current object-oriented database systems and programming languages operate with a very rigid relation between the class definition and the instance object of a given class. The class definition exactly describes how each instance object will look like. In CAD systems, this is a severe handicap for the creative mind of their users and the flexibility of concurrent development processes. To overwhelm this handicap we propose to open object oriented techniques to support end user creativity and flexibility and to bring object technology to end users. This will be provided by the following characteristic of the dynamic object technology:

- dynamic instances
  The content of an object instance can be extended without having to extend the class definition. An object type definition must be like a template. Instances may contain more information than what the type prescribes. When one starts designing an object, one is only aware of the key characteristics it has to satisfy. This can be seen as functional specification. During the design process this specification is enriched to a technical solution within an object instance.
- multiple inheritance
  Multiple inheritance is provided at the object instance level, not only at the object type level. This means that a specific technical solution can fulfill several orthogonal functional requirements (for example regard our example above where the bottom plate is both of type 'plate' and 'corrosion resistant'). In most programming languages one can only inherit from class A and B if an intermediate class that inherits from both A and B has been defined.
- introspection
  The system as well as the models designed in the system are fully transparent. The end user can query system state and data model. Typical questions of the user are: give me all operations I can perform on this object; how has this object been designed (browse the object history)? what other users of the system are using this object type?
- deferred generalization
  Design of technical solution often happens inductively: the designer realizes after constructing an object that she/he has defined a reusable solution. So the technology must be able to derive the class definition (specification) from an object (example solution) that is already there. This behavior has been called deferred generalization. In conventional object oriented systems, one has to define the class before one can create the first instance.
- persistent actions
  As we have seen with objects above, dynamic extensibility is the key to more creative and flexible design. This in addition also implies that at runtime also new commands acting on these objects can be easily defined. In most cases these commands combine as parameters objects of different classes. We therefore introduced the notion of an 'action' that is independent of an assignment to a specific class as methods do. These actions are independent entities that take objects as parameters.

4   OBJECT LAYERS

The innovative object technology is realized in three layers. The core is an object repository. A second layer called dynamic object driver provides the runtime extension method. The last layer called constraint manager is responsible for maintaining consistent relations. The technical concept is based on two advanced techniques exceeding current oodb techniques:

- a general identification scheme guaranteeing unique reproducibility of dependent objects according to whatever changes of their determining environment
- an elaborated basket scheme which collects objects and actions at runtime and builds new types out of them.

The identification scheme manages the associatively and consistency of the data model. The basket scheme delivers the kernel functionality of the dynamic object technology described above. It supplies dynamic instances, multiple inheritance, introspection, deferred generalization and persistent actions.

♦ object repository: oodb

♦ dynamic object driver: basket scheme

♦ constraint manager: identification scheme

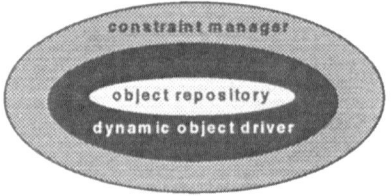

Figure 3   Object layers enabling dynamic and semantic modeling.

5   MODELING ARCHITECTURE

Introducing the described concept into the environment of a CAD system we identified three kinds of different knowledge for a modeling application. These different knowledge types again can be assigned to different software architectural layers.

5.1   IT knowledge

In the 1st layer the information technology knowledge that is common to all CAD is provided in a predefined runtime kernel. It especially handles data management. So the first layer realizes the identification scheme and the basket scheme. The basket scheme presents several predefined generic data structures to the user, e.g.: object and action, user object and user action, layer and group. These predefined structures are again connected to generic user interface elements so that new CAD applications can be built very quickly.

5.2   Domain knowledge

The 2nd layer realizes the basic notions and the basic operators of a CAD application. These are programmed by using the generic objects and actions of the first layer. Typically the geometric functionality of a CAD application is provided in the second layer, e.g.: points, curves, surfaces and solids. This functionality can be realized by integrating industrial standard libraries for geometric computing.

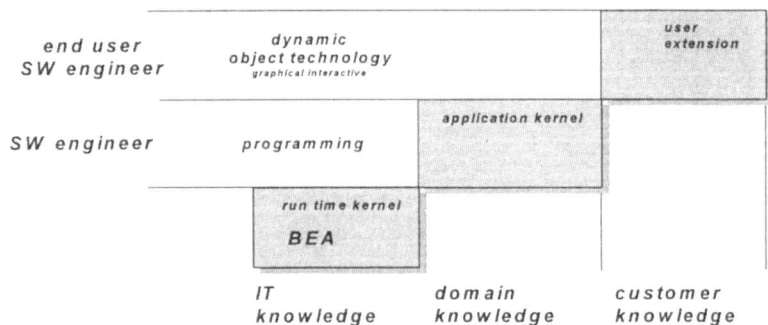

Figure 4   A modeling architecture mapping specialization of knowledge.

5.3   Customer knowledge

The 3rd layer realizes domain-specific and company-specific knowledge. The dynamic extension techniques of the 1st layer and the basic objects and action of the 2nd layer are used to build high level applications with advanced domain know how. Typical examples are sheet metal, driving gears, conveyor equipment. The main advantage of the dynamic object technology is that the domain specialist himself can teach the system the special semantics. There is no more translation to be done between the domain engineers' language and the software engineers' language needed to realize specific applications.

6   STATE AND PROSPECT

The core ideas of dynamic and semantic modeling have already been proven: first advanced prototypes show the promising potential of this new modeling approach. A new CAD direction overwhelming current geometric thinking and capturing full design intent can be developed.

BIOGRAPHY

K.-P. Greipel Doctorate in mathematics at University of Munich. Assistant professor. At Siemens Nixdorf since 84. Senior developer and project manager. Design of innovative CAD systems for electronics and mechanics. Central consultant for logistics and production. Manager of international large scale projects.

J. Colpaert   Electronic engineering degree from the KIH Ostende. Engineering degree in computer science from the University Leuven. Key member of start up team of the electronic imaging systems department at AGFA-Gevaert. Designer of laser printer architectures. Manager of the printer software department. Group leader for innovative CA-technologies at Siemens Nixdorf since 84.

9

Solid Modeling as a Framework in Virtual Environments

M. Figueiredo, J. Teixeira
Grupo de Métodos e Sistemas Gráficos
Departamento de Matemática - FCTUC
Apartado 3008, 3000 COIMBRA
email: mauro/teixeira@mat.uc.pt

Abstract

The new technology of virtual environments provide to a visitor of a virtual world a high degree of immersion. On the other hand, the interaction techniques available in these virtual worlds are still rudimentary. Frequently, the visitor is enabled to grab 3D objects by gesturing, but unable to touch a face, edge or a vertex, neither edit its shape interactively.

In this paper we present the reasons that let us believe that boundary representation data structures should be integrated in virtual environments. They can be used not only to provide a fast rendering, but mainly to take advantage of the explicit geometrical and topological data of 3D models. This data enabled us to improve the interaction component of a virtual environment prototype by letting visitors to locally edit in real time the shape of virtual objects and, in this way, augmenting the visitor participation in a virtual world.

Keywords

Virtual environments, interaction techniques, solid modeling, boundary representations.

1   INTRODUCTION

Virtual environments try to provide to the user the illusion of being immersed and participating in a virtual scenario. To achieve these goals, the visitor is enabled to navigate inside a virtual world (that looks like a real one) and to interact with virtual objects.

Many experiments have been reported about successful prototypes that generate the illusion of immersion. Remaining problems to be solved, as lack of resolution, are mainly technological. In general, a high degree of immersion can be achieved by providing stereoscopic views of a given environment.

The sensation of participation is currently generated by providing to the visitor a set of gestures to perform actions, such as grab, push and release. In this way, the participant is enabled to manipulate 3D objects in a direct format by grabbing, pushing or releasing them.

In some virtual environment systems, these 3D interaction techniques are only supported by bounding volumes. In this case, visitors manipulate 3D virtual objects by interaction with bounding boxes parallel to the coordinate axes. Thus, the visitor is unable to touch the surface of the virtual object and his actions upon the environment (grabbing, pushing, others) are inaccurate.

However, in many real tasks the user need to touch 3D objects or select its geometrical entities (vertices, edges and faces). For example, for tele-presence applications it is extremely important to let users accurately grab, push and position objects. In CAD, it is also important to select vertices, edges or faces of a 3D virtual object (for example to simulate the interactive distortion of object's surfaces done by a virtual hand).

This paper addresses these questions. We explain the approach that we have taken to enable the visitor to directly interact with the surface of virtual objects or to select its vertices, edges or faces, in a virtual environment.

2   REQUIREMENTS FOR GEOMETRICAL MODELS IN VIRTUAL ENVIRONMENTS

The representation of 3D objects' geometry is an important step in the graphical simulation of reality, since they have to describe real objects and provide the required data for the realistic simulation of the original entities and their behaviour.

The geometric modelling and its related representation schemes are already well studied (Baer, 79), (Requicha, 80), (Weiler, 86), (Mäntylä, 88): wire frame, surface, constructive solid geometry (CSG), manifold and non-manifold boundary representations (Breps). Each of these representations have pros and cons that characterise and enable them to be used in a specific application context: product modelling, virtual environments, etc. In order to distinguish them, Requicha (Requicha, 80) proposed a set of properties to characterise solid representation schemes: domain, validity, completeness, unambiguity, uniqueness, conciseness, efficiency, ease of creation.

These properties are extremely important in the geometric modelling area and should also be considered when studying and determining the representation scheme to integrate in a virtual environment. However, we have to emphasise that some of these properties are not equally important for a virtual environment application. For example, in a virtual space for arts it is not essential to ensure the geometrical validity of a representation, since the artist might be interested in the creation of surrealistic models.

In this case, we also have to study formal and practical properties of each representation that can enhance the three virtual environment components (Foley, 87): visualisation, behaviour and interaction.

Thus, a suitable representation scheme for a virtual space should include those properties studied by Requicha and the ones that will enable such representation to fulfil the virtual environment paradigm requirements:
- the creation of convincing and realistic environments (visualisation component);
- the intuitive interaction with visitors (interaction component);

- and the realistic simulation of objects' behaviour (behaviour component).

3   MODELS FOR THE REPRESENTATION OF VIRTUAL SCENES

In this section we discuss strengths and weakness of the most important representation scheme based on the set of properties proposed by Requicha (Requicha, 80). We will analyse wire frame, surface, constructive solid geometry, manifold and non-manifold boundary representation schemes to describe virtual scenes. Based on this stuff, we will show in the next section why manifold and non-manifold boundary representations are suitable and the most adequate choice for virtual environment applications.

3.1 Wire Frame Models

A wire frame model is a collection of lines, defining the edges of an object (figure 1). This representation can be used to generate object's drawings (2D wire frames), but a 3D wire frame model has no information about the surface boundary and this information cannot be generated from the geometry of vertices and edges, restricting the use of these models to certain applications. For example, 3D wire frame model have many drawbacks for product modelling (Goldman, 87), since he they can be *impossible* (figure 1-a), *ambiguous* (figure 1-b) or *incomplete*. The interpenetration of faces in a way that it makes impossible to represent a physical shape of a solid object and the multi-interpretation are two common problems of 3D wire frames.

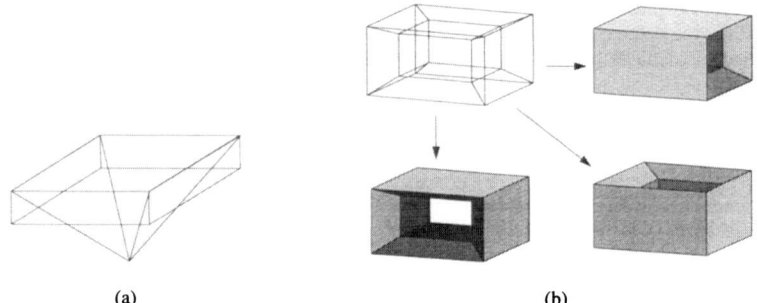

(a)   (b)

Figure 1   (a) *Impossible* model (Mortenson, 85); (b) *ambiguous* and *incomplete* geometry (Goldman, 87).

3.2 Surface Models

In a surface modelling scheme, objects are represented as a collection of surface elements that describe the boundary of the object (figure 2). Several types of surface elements can be used: polygons (flat surfaces bounded by straight lines), analytical surfaces (for instance natural quadrics surfaces) and more general free-form surfaces (for example parametric surfaces, such as defined by the Bézier and B-spline method) (Foley, 90).

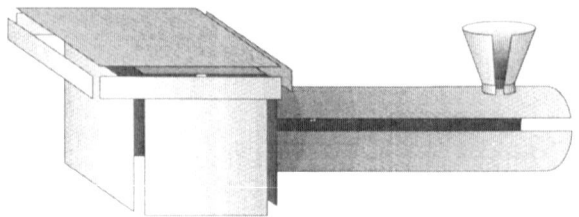

Figure 2 Surface model.

Surfaces are explicitly described in the computer model, providing some advantages over wire frame models for geometric modelling. With these representations interpenetration of faces can be identified, and the designer can be supported on preventing the creation of *impossible* objects. The introduction of faces removes the *ambiguity* from the wire frame models. Holes can be located easily and unambiguously, but not all objects which can be modelled with surfaces are realisable solid objects. A well known example is the Klein bottle. However, surface models could be *incomplete*, since there is no guarantee that the surfaces bound completely the model.

3.3 Solid Models

Wire frame and surface models do not guarantee the representation of *valid* physical solid objects, therefore these geometric modelers must rely on human assistance to supply missing data and clear inconsistencies, such as, ambiguity in wire frame models. Since complete and consistent 3D models are very important in many scientific and engineering applications, such as mechanical and civil fields (Requicha, 77), it has been widely identified that instead of representing drawings in a computer and incorporating their inherent limitations, computers should explicitly represent 3D objects as solids (Baer, 79], (Requicha, 80), (Mäntylä, 88).

The major driving force for the development of solid modelling techniques has been the need to provide *complete* descriptions of the shape of physical solid objects, that means, representations which can answer arbitrary geometric questions automatically. These descriptions enable the computation of *geometrical* and *inertial* solid object's properties, *reliably* and *automatically* (Requicha, 80).

Next, we will concentrate on the three most important solid representation schemes: constructive solid geometry, manifold and non-manifold boundary representations.

3.3.1 Constructive Solid Geometry

Constructive solid geometry represents complex solids by applying Boolean operations and transformations on parametrized instances of solid primitives, such as block, sphere, cylinder and cone for example (Requicha, 77). Each solid primitive could be represented as a combination of half-spaces defining a bounded point set of $\mathbb{R}^3$. Thus, the constructive solid geometry representation can be said to be supported by a two-level scheme (Encarnação, 90): On the lower level, bounded volume primitives are defined on the basis of half-spaces; On the upper-level these primitives are combined by Boolean set operators.

The CSG defined object is internally represented as a binary tree, called CSG-tree (figure 3): The primitive solids are positioned at the leaves or terminal nodes; the internal or non-terminal nodes contain the Boolean operators of union, difference and intersection; transformation data for rigid-body motions, such as translation and/or rotation, can be stored both at the leave nodes as well as at the internal nodes. The internal nodes represent a solid defined by performing the transformations and the set operations of that node to its two subsolids indicated below it. The root node represents the resulting composite objects.

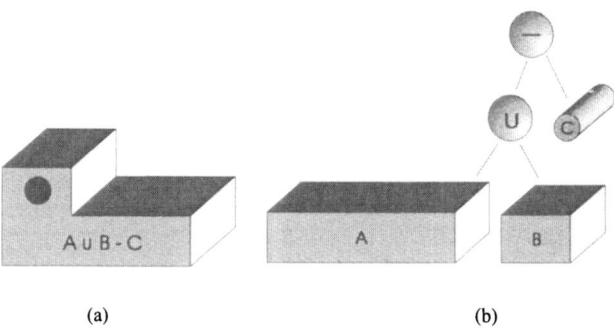

(a)   (b)

Figure 3   a) Solid model; b) and the corresponding CSG tree.

Constructive solid geometry is a very powerful representational scheme and several researchers (Requicha, 80), (Mortenson, 85), (Mäntylä, 88) had already outlined its advantageous and problems: every CSG tree *unambiguously* models a physical solid, but they are not *unique* since in general it is possible to construct the same solid in other ways. However, if the primitive elements provided by a solid modelling system are valid bounded solids and the set operators are regularised, then the resulting solid models are guaranteed to be *valid* and *bounded*. This is an important property because it ensures that the validity of a CSG representation, based on bounded primitives, can be guaranteed by evaluating primitive leaf validity.

Nevertheless, CSG schemes produce *unevaluated* models, that is, they contain data that must be further processed in order to perform basic operations. For instance, for displaying and interaction with a solid model, details of the edges and faces of the object are required. Since these details are not explicitly present in it, the CSG representation must first be converted into a boundary representation, which can then be displayed with standard hidden-surface algorithms or used for the identification of interactions. This conversion known as *boundary evaluation* and it may be time consuming.

3.3.2 Manifold Boundary Representations

Boundary models represent a solid by describing *geometrically* its surface which is constructed as a closed boundary of surface elements, "faces" or "patches" (Baumgart, 74). In turn, planar faces can be represented by their bounding edges and vertices for example.

Additionally, it is provided a *topological* description of the connectivity and orientation of vertices, edges and faces. Topology specifies how bounding surfaces of a solid model are joined together.

Clearly, the boundary representation must satisfy certain rules in order to be able to represent physical solids and to reject boundaries that do not enclose volumes, as for example, the surface of a Klein bottle. The basic ideas of boundary representations are that the boundary should be *closed* (i.e. the boundary cannot have a boundary), *orientable* (the object must have a consistent inside and outside) manifold embedded in 3-space, and should *not self-intersect* (Baer, 79), (Hoffmann, 89). A two-manifold surface has the property that, around every one of its points, there exists a neighbourhood that is homeomorphic to the plane (Weiler, 86), that means, the surface exists in three-dimensional space but it is topological "flat" when the surface is examined closely enough in a small area around any given point.

Figure 4 illustrates a typical BRep data structure architecture.

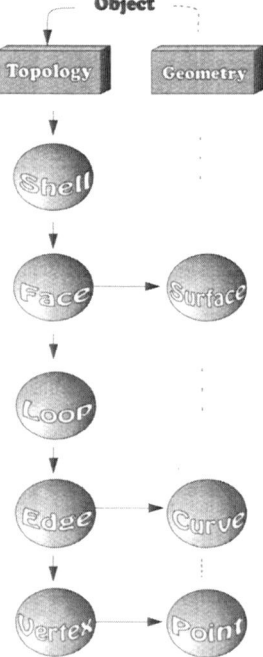

Figure 4 BRep data structure.

The solid object representation is defined as an hierarchical data structure of topological entities with shell, face, loop, edge and vertex nodes. Entities are linked by means of pointers, for example, from a face to each of its bounding edges, from an edge to its two ending vertices, and so on. This description that specifies vertices, edges and faces abstractly, and indicates their incidence and adjacencies relations is the **topological information**. Geometric information is attached to each of the three object types: *vertices* are defined by coordinate triples; *edges* in general are represented by a parametric equation; the portion of the curve that forms the edge is defined by its two ending vertices; each *face* in the solid model lies on a single planar, quadratic, toroidal or parametric surface that supports it and its bounding edges. All these data, about the geometry of the entities of a solid model, we call the **geometry** of a boundary model.

A boundary representation scheme is *valid* if it defines a closed, orientable and not self-intersect boundary in order to guarantee that the model is representing a solid object. Boundary representations are *complete* and *unique* if the Brep data structure includes enough data about: (i) adjacency relationships between topological entities (vertices, edges, loops, faces, shell); (ii) orientation and closed properties (Gomes, 92). These properties are extremely important to those concerned in the design and implementation of a solid modeler. In addition, boundary representations provide *explicit representations* for the geometry of faces and edges and for the relations between them, which are quite useful for visual and interactive operations: the computer can automatically, by using the surface geometry, generate realistic pictures of the objects represented from any desired point of view in real time.

3.3.3 Non-Manifold Boundary Representations

Boundary based solid modelling techniques have found wide applications. However, conventionally they are restricted to representing only two-manifold domains. This disables to represent such conditions as two surfaces touching at a single point, two distinct enclosed volumes sharing a face as a common boundary, and a wire edge emanating from a point on a surface (figure 5). These conditions are known in the geometric modelling field as non-manifold conditions (Weiler, 86). Furthermore, common modelling operations, such as the Boolean operations can produce non-manifold results, and therefore not representable under manifold representations, even with strictly two-manifold input.

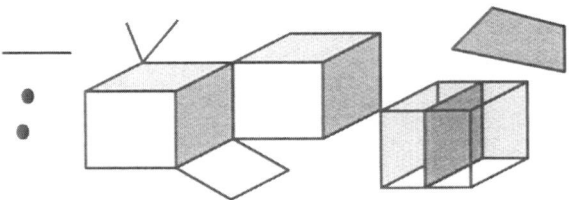

Figure 5 Non-manifold topology (Gursoz, 90).

Research on non-manifold surface topology led to the development of a new representation that expresses the non-manifold topology in order to expand the representational domain to cover such cases. Whereas manifold based solid modeler either give up some adjacency data whenever a non-manifold data occurs, treat it as a special case, or simply fail to perform the operation, in a non-manifold topology, we can represent all possible adjacencies among the basic topological elements (Hoffmann, 89).

Non-manifold geometric modelling domain encompasses both manifold and non-manifold conditions and is therefore quite general. An edge might bound no faces (wire edge), one face (lamina face), two faces (manifold edge) or more then two faces (non-manifold edge). A major advantage introduced by non-manifold boundary representations is the fact that they provide a single unified representation for any combination of wire frame, surface and solid modelling forms (Weiler, 86). In this way, these geometric modelling approaches can exist under the same representation scheme in pure or hybrid form. This uniformity offers significant advantages to the staging and delivery of geometric modelling systems as well as providing enhanced functionality and simplicity.

4 A SUITABLE REPRESENTATION SCHEME FOR VIRTUAL ENVIRONMENTS

Polygonal representation schemes (commonly used in virtual environment's prototypes) are perfectly adequate to generate the illusion of immersion on a virtual scenario. However, they do not provide complete geometrical and topological data required to enable the visitor to perform actions as those described above: touching object's surface, selecting its geometrical entities, identifying the neighbouring geometrical entities. In fact, there are many data structures that can generate data for real time realistic visualisation, but are unable to provide complete data for the interaction and behaviour component of a virtual environment.

Our goal was to provide more advanced interaction capabilities in a virtual environment to enhance the visitor's participation involvement, but the polygonal representational scheme that we were using was too restrictive. Therefore, we should review the representations schemes presented in section 3 and discuss their properties in the specific context of the virtual environment paradigm to find out a suitable representation that provide all the required data for this paradigm.

We can start with the wire frame models. The visual component of a virtual environment undertakes an important part in providing to the visitor the illusion of immersion in a virtual world by the creation of photo-realistic scenes. However, the lack of facet information in the wire frame model implies that there is no capability for the automatic generation of realistic shaded images that make visualisation so much easier.

Furthermore, wire frame models do not provide the required data for the interaction and behaviour component. It is the object's boundary that interact with any other objects in an environment. But, in a wire frame model we cannot locate its boundary and therefore, in a wire frame virtual world we cannot identify any interactions between virtual objects (figure 6). On the other hand, we cannot simulate physical object's behaviour. These models do not provide the required data to compute volume properties, such as weight and mass, and in this case, wire frame virtual models do not follow gravity laws, for instance.

Thus, we can say that wire frame models have drawbacks that make them unuseful in a virtual environment.

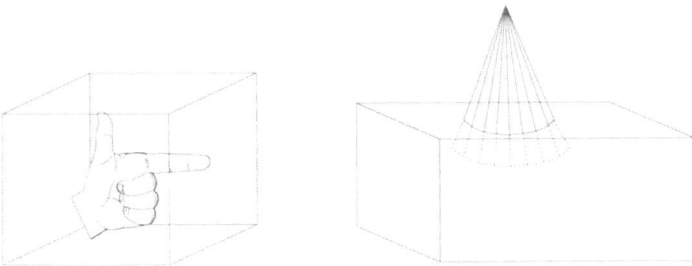

Figure 6   No object interactions identified in a wire frame virtual world.

For the surface models, we have to find out, as we have done above, if these representations provide the means to generate to the visitor the illusion of immersion in a three-dimensional space. In fact, surface models describe the solid object's geometry (faces, edges, vertices) required for the visible-surface determination, illumination and shading algorithms and, in this way, they can be used to generate realistic 3D scenes. In particular, polygon modelling schemes, initially developed for rendering (Mortenson, 85), describe the surface geometry by a cross-referenced list of vertices, edges and planar faces, enabling the creation of photo-realistic images and in real time. Thus, we can understand the massive use of polygon representations in virtual environment toolkits, since they provide the means to present to the visitor the illusion of a world that does not exist out of his perception.

The availability of surface's geometry in object's models introduces remarkable advantages for the interactive component of a virtual world. In generic terms, the surface can be used to find out the intersection of two objects. This will suffice for the identification of interactions between virtual objects (figure 7). Certainly, it is a powerful feature, auspicious for the recognition of grabbing or pushing operations upon virtual objects enabling the visitor to interact with 3D models through a hand cursor (figure 7-a). In the same way, the identification of interactions can be used to prevent two objects of sharing the same spatial region (figure 7-b).

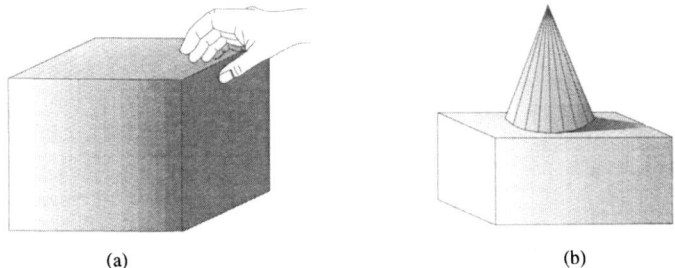

(a)   (b)

Figure 7   (a) In a surfaces' virtual world the visitor can grab virtual objects; (b) and the system can simulate objects' impenetrability.

On the other hand, we have to remember that it is users' responsibility to warrant the creation of a closed surface model, with a finite volume. Therefore, it is clear that incomplete models of solid objects can coexist in some cases and cannot be used to determine algorithmically its mass properties. This insufficiency leads to a virtual environment where its virtual models cannot algorithmically simulate the behaviour of their real counterpart objects.

In the end, we can succinctly say that surface models provide proper data for the generation of convincing interactive 3D environments. Nevertheless, they cannot be considered the best choice for virtual environment applications. In fact, some faults stand out immediately when we want to use them for behavioural simulation and in more advanced interactive features.

Constructive Solid Geometry schemes provide *valid* and *complete* solid models which can be used in virtual environments, especially to determine object's volume and mass properties, providing a preliminary evaluation of its performance and for the behavioural simulation of 3D virtual objects.

Nevertheless, CSG schemes produce *unevaluated* models, that is, they contain data that must be further processed in order to perform basic operations. For instance, in displaying and for interaction with a solid model, details of the edges and faces of the object are required. Since these details are not explicitly present in it, the CSG representation must first be converted into a boundary representation, which can then be displayed with standard hidden-surface algorithms or used for the identification of interactions. This conversion known as *boundary evaluation* and it may be time consuming. Thus, CSG representation cannot maintain in real time basic operations for a virtual environment and therefore with no practical interest for the moment for this paradigm.

Boundary representations provide *explicit representations* for the geometry of faces and edges and for the relations between them, which are quite useful for the visual and interactive components of a virtual environment. The computer can automatically generate realistic pictures of the objects represented from any desired point of view and in real time using the surface geometry. In this way, boundary models provide to the visitor the illusion of immersion in a real world.

In the same way, the boundary data available in a Brep model can be used for the identification of interactions, in particular for: (i) letting the visitor to grab, push or release virtual objects; and (ii) preventing two virtual objects of sharing the same region.

On the other hand, the availability of the geometrical data of faces, edges and vertices, makes it possible to find out precisely when two virtual objects are colliding (Figueiredo, 93). In fact, the collision detection can be implemented using bounding boxes (to filter out pairs of faces that cannot intersect, speeding up this process) and calculating the cross sections of the two objects. In this way, we can let the visitor to touch the surface of a virtual object and grab it with his finger tips (figure 8).

In addition, boundary models have unique advantages quite important for the geometric modelling field, since they explicitly represent topological adjacency of a solid model, which are also send back into the virtual environment paradigm.

Topological data is relevant in the resolution of certain problems automatically, such as, interactive manipulation of the shape of a solid object. Consider, for example, a can described by a collection of surface patches (figure 9). A designer then decides to alter the shape of one of these patches. If a boundary is affected, then adjacent patches must also be changed; otherwise these panels will not join evenly and they will separate and tear. The problem for the computer is to fix these adjacent patches automatically or at very least to cue the designer to the location of the problem patches. If topology is not available directly into the geometric model, the computer will be unable to determine connectedness and juxtaposition and therefore it will be unable to solve this question.

Figure 8 The visitor is enabled to touch the virtual cone's surface.

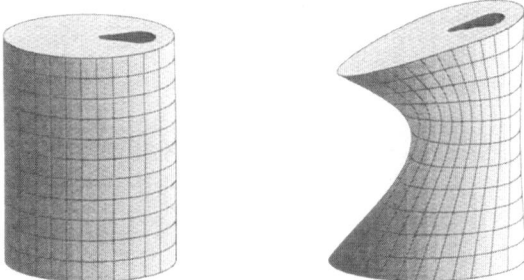

Figure 9 Patches manipulation.

Thus, a quite convenient advantage of a BRep is that it provides explicit topological data that can be used in general to enhance the interaction capabilities in a virtual environment.

We can use adjacency data in the implementation of intuitive 3D interaction techniques for selecting geometrical identities (e.g., faces, edges, vertices) and for local operations. For example, as presented in figure 10 the topological data enables the visitor to select all the edges of a face by selecting the face directly with a finger. Then, we can edit interactively the surface of the virtual object (figure 11-a-b), using the topological data to guarantee that all patches maintain jointly.

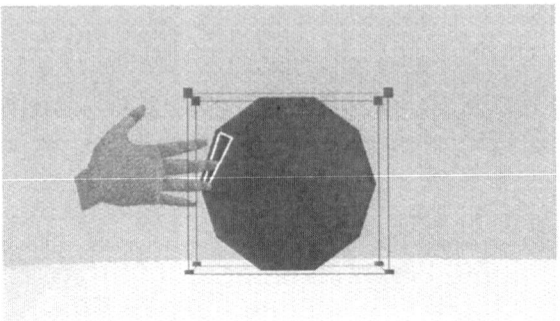

Figure 10   Face and its boundary edges are selected.

In fact, in a virtual environment where virtual objects are represented by Brep models, the visitor is enabled to interactively edit the model's shape with a 3D cursor (virtual hand). In this way, it allows the modification of a localised region of the data structure in an efficient manner and with greater naturality. For example, the geometry associated with a single face can be redefined and the result evaluated quickly. In this way, a Brep data structure can contribute to augment visitor's capabilities under a virtual environment enabling the implementation of intuitive 3D interaction techniques to let the visitor touch the surface of 3D objects, or select geometrical entities or performing local operations.

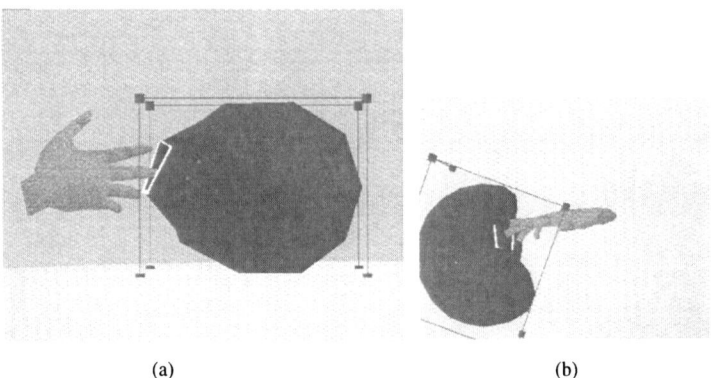

(a)   (b)

Figure 11   The visitor interactively edit the shape of a 3D object via a hand cursor.

5   CONCLUSIONS

We have shown that boundary representation schemes not only have important formal properties, but they also have practical characteristics that are suitable for the goals of the virtual environment paradigm; they guarantee *valid* and *complete* representations of solid object's boundary and provide *geometrical* and *topological* data *explicitly*. Thus, the visual, interaction and behaviour components of a virtual environment can be improved, improving the overall experience of a visitor.

To conclude, it is presented a table that compares the representation schemes described above and clarifies the advantages introduced by Brep models. We should emphasise, that *validity*, *completeness* and *explicit* data of the boundary surface of an object, are extremely important properties that make boundary models suitable for virtual environments.

Table 1 Representation schemes' classification procedure for virtual environments.

Models	Valid	Unique	Complete	Explicit data	
				Geometry	Topology
Wire frame					
Surface				✓	
CSG	✓		✓		
BRep	✓	✓	✓	✓	✓

ACKNOWLEDGEMENTS

We would like to thank Klaus Böhm, Volker Kühn for allowing the use of GIVEN (Böhm, 92) in our experimental work. We also would like to thank to Rosario for the support in the implementation of part of this work.

REFERENCES

Baer, A., Eastman, C. and Henrion, M. (1979) Geometric Modeling: a Survey. *Computer Aided Design*, **11**, **5**, 253-272.

Baumgart, B.G. (1974) *Geometric Modeling for Computer Vision*. PhD Thesis, Stanford University, Palo Alto.

Böhm, K., Hübner, W. and Väänänen, K. (1992) GIVEN: Gesture Driven Interactions in Virtual Environments, a Toolkit Approach to 3D Interactions. *Proc. of the Interfaces to Real and Virtual Worlds Conference*, Montepelier, 243-254.

Encarnação, J.L., Lindner, R., Schlechtendahl, E.G. (1990) *Computer Aided Design — Fundamentals and Systems Architectures*. 2.ª ed., Springer-Verlag, Berlin.

Figueiredo, M., Böhm, K. and Teixeira, J. (1993) Advanced Interaction Techniques in Virtual Environments. *Computer & Graphics*, **17**, **6**, 651-661.

Foley, J.D. (1987) Interfaces for Advanced Computing. *Scientific American*, **257**, **4**, 126-135.

Foley, J., Dam, A., Feiner, S.K., and Hughes, J.F. (1990) *Computer Graphics — Principles and Practice*. 2.ª ed., Addison-Wesley Publishing Company, Massachusetts.

Goldman, R. (1987) The Role of Surfaces in Solid Modeling, in *Geometric Modeling: Algorithms and New Trends* (ed. Gerald Farin), SIAM, 69-90.

Gomes, A. (1992) *Modelos Algébricos de Sólidos e Morfologia*. Master Thesis, Coimbra.

Gursoz, E.L. and Prinz, F.B. (1990) A Point Set Approach in Geometric Modeling, in *Advanced Geometric Modeling for Enginnering Applications* (ed. F.L. Krause, H. Jansen), Elsevier Science Publishers B. V., 73-88.

Hoffmann, C.M. (1989) *Geometric and Solid Modeling — An Introduction*. Morgan Kaufmann Publishers, San Mateo, California.

Mäntylä, M. (1988) *An Introduction to Solid Modeling*. Computer Science Press, Rockville.

Mortenson, M. (1985) *Geometric Modeling*. John Wiley & Sons, New York.

Requicha, A. (1977) *Mathematical Models of Rigid Solid Objects*. Production Automation Project, Technical Memorandum 28, University of Rochester, New York.

Requicha, A. (1980) Representations for Rigid Solids: Theory, Methods, and Systems. *ACM Computing Surveys*, **12**, **4**, 437-464.

Weiler, K.J. (1986) *Topological Structures for Geometric Modeling*, PhD Thesis Rensselaer Polytechnic Institute, Troy, New York.

BIOGRAPHY

Mauro Figueiredo is an Assistant at University of Coimbra. His research interests are in virtual environments, 3D interaction techniques, cooperative work, solid modeling and user interfaces.

Figueiredo received his degree in Computer Science from University of Coimbra in 1990 and his MS in Industrial Automation from University of Coimbra in 1994.

José Carlos Teixeira is an Auxiliary Professor at the University of Coimbra in the areas of Computer Graphics and Geometric Modelling, head of its Computer Graphics Research Group and President of the CCG/ZGDV Executive Board. His main research interests are in Geometric Modelling, Virtual Environments, CSCW and new Interaction Techniques.

Responsible and involved in different European and Portuguese Projects, is member of the Editorial Board of the journal "Computer & Graphics" (Pergamon Press), founding member of the WG 5.10 on Computer Graphics of IFIP TC 5, President of the EUROGRAPHICS Portuguese Chapter and head of the Portuguese Technical Committee for Standardisation on Computer Graphics - CT 109. He is a member of EUROGRAPHICS, ACM, ACM-Sigraph and IEEE.

10

Virtual prototypes and product models in mechanical engineering

M. J. Pratt
National Institute of Standards and Technology,
Manufacturing Systems Integration Division,
Building 220, Room A127,
Gaithersburg, MD 20899-0001, U.S.A.
Tel: +1 (301) 975-3951, Fax: +1 (301) 258-9749
E-mail: pratt@cme.nist.gov

Abstract

This paper gives an overview of some of the modelling and virtual prototyping techniques used in product realization, with emphasis on the mechanical engineering field. It is pointed out that virtual prototypes, in the commonly accepted sense of computer models permitting realistic graphical simulation, represent only one class amongst the many types of computer models used in design and planning for manufacture. Each such model is usually created for some comparatively narrow purpose, and one of the major problems faced by developers of integrated computer-aided product realization systems concerns the transmutation of one type of model into another. A related problem is that of interpretation by any model of information generated by interrogations of another model. These difficulties are compounded by the increasing presence in such models of semantic information concerning different aspects of the intended functionality or manufacturing requirements of the modelled artefact.

Keywords

Models, virtual prototypes, product realization, mechanical engineering, CAD/CAM.

1   INTRODUCTION

A model is an abstraction or representation of some real thing. It may take many different forms. For example, a mathematical model of the economy of a country may consist of a set of differential equations, while a model of the exterior shape of a new car may be sculpted in clay.

Engineers construct models throughout the product realization process to obtain answers to questions. Sometimes qualitative answers are required; in the car body case the clay model is used to assess the general appearance and attractiveness of the body shape. Other applications, a structural analysis of the car body for example, may require precise numerical results and demand the use of some other type of model.

Given the wide diversity of different types of query arising, for example, in designing and planning for the manufacture of a new airliner, it is inconceivable that any one model can serve for all purposes. Specialized queries demand specialized models; only the real thing – the airliner which has not yet been made – holds the answers to all possible queries.

This paper is concerned with computer models, which reside in a computer and provide support for the mechanical product realization process. In order to set the scene for the discussion of various types of computer models a brief summary will initially be given of the major activities making up that overall process.

2   THE PRODUCT REALIZATION CYCLE

The product realization process can be divided into three stages: design, manufacturing engineering, and production. The output of the design stage is a detailed specification of the product to be manufactured. This becomes the input to the manufacturing engineering stage, whose output gives detailed specifications of the intended manufacturing processes. These in turn are the input to the actual production process.

The three stages are separately described below, although in practice some of their activities may overlap. This is particularly so when modern *concurrent engineering* practices are used, in which case design and manufacturing engineering proceed to some extent in parallel, with frequent exchange of information (Nevins & Whitney 1989).

Much effort is currently being devoted to the use of computers in automating individual product realization activities, and in combining such automated processes into integrated product realization systems. Integration requires the smooth flow of appropriate information between activities, and progress in this area is hampered by the use of different models, each having its own informational requirements, for individual product realization activities. The focus of the present paper is the modeling aspect of automation, the intention being to highlight one of the major problems underlying the achievement of integrated systems for design and manufacture.

2.1   Design

Effective design is crucial to the success of any manufacturing organization, since a major fraction (up to 70%) of the total life cycle cost of a product is committed by decisions made in the early stages of design (Ullman 1992). The objectives of the design process are the attainment of a short development time with high product quality and low production cost. The use of computer models may help significantly in achieving these aims.

The product design function can be broken down into four phases (Pahl & Beitz 1984):

- Product planning

- Functional design

- Configuration design

- Detail design.

The activities actually undertaken in the design process vary considerably according to the nature of the product and the commitment of a company to the use of computer aids. Where families of essentially similar and comparatively simple products are concerned it is sometimes possible to encapsulate the basic design principles used in a few equations or design rules. These may then be used to drive the detail design process in such a way that the designer only has to enter values of a few key dimensions or other parameters to enable the design system to generate a complete specification of the product. The achievement of this situation requires considerable preliminary work in developing new software systems or configuring existing ones for the intended specialized applications.

On the other hand, the design of a more complex product such as a new passenger aircraft can require the individual design *ab initio* of many thousands of completely new components. The design activity can then extend over a period of several years, even with extensive use of computer aids. The overall process involves the extensive use of analysis and simulation in arriving at an optimal design solution meeting all the constraints imposed by conflicting requirements on payload, range, fuel economy, safety, noise generation, price and operating costs.

This very wide spectrum of possible approaches to design implies that any breakdown of the process into component tasks will almost certainly differ from the practice in any particular company. What follows is an 'averaged' breakdown, typical of the practice in companies manufacturing a diverse range of non-modular products.

Product planning: This first phase is essentially clarification of the design task to be addressed. Its initiation may be stimulated by the desire to improve upon an existing product, or by the identification of a new market niche. The latter may be stimulated in turn by new developments in technology. The questions arising at this stage are of a very broad nature – What is the purpose of the new product? What market sector is it aimed at, and what therefore should it cost? What will be the size of its market, and how many should be produced? The output of this phase is a set of constraints on the work of the next phase; in particular, the intended functionality of the product is defined and limits imposed on its development and production costs.

Functional design: This phase is concerned with the achievement of the desired functionality in the new product, subject to the constraints imposed at the product planning stage. There may be several solutions to this problem, possibly making use of different physical principles. An example of a design choice at this level is the decision whether a new aircraft will be powered by jet engines, turboprops, piston engines or some new and exotic form of propulsion. Initially, design choices are made at a high level, but each choice leads to a new set of design problems at a lower level which must be solved in turn. The process is therefore one of successive refinement; at each level, design possibilities are either rejected or followed down to lower levels of problem decomposition. Each new level poses a set of functional problems to which technical solutions must be found by the designers. What results eventually is a set of viable possibilities for achieving the desired functionality whilst satisfying the design constraints. The functional design phase is completed when the possibilities have been evaluated against each other and the one chosen which is optimal from the point of view of estimated cost, estimated performance, or some combination of these and other criteria.

Configuration design: Whereas the previous phase is concerned with a functional decomposition of the intended new product, the configuration phase deals with the mapping of the functional elements of the design onto mechanical systems and subsystems providing the required functionality. This phase therefore covers the specification and layout of assemblies and subassemblies. Once again the process is one of decomposition from higher to lower levels, and some iteration between levels may be necessary to obtain acceptable results. It is appropriate during configuration design to minimize the number of parts in assemblies, and to make preliminary decisions on part materials and manufacturing methods (Boothroyd 1994). As in the previous phase, the result is a multiple set of possibilities from which an optimal choice must be made. At this stage it is possible to make more accurate estimates of cost and performance.

Detail design: In the detail design phase the finally chosen configuration design is fully documented. Detailed drawings or product models are created for all components to be manufactured for the new product, and any standard components to be bought in from outside are specified. Once the detailed part designs are available, it is possible to generate detailed assembly models and to perform various computer-based analyses to determine whether the desired product functionality will be achieved. If not, a design iteration will be necessary.

2.2   Manufacturing engineering

The primary input for this activity is some representation of a product to be manufactured, and the output is a set of instructions for manufacturing it. Certain supporting resources are needed for the automation of manufacturing engineering. One is a database of available manufacturing resources, and another is a set of *process models*, i.e. computer models of the manufacturing processes which may be used in the production process. Most research to date has concentrated on the automatic generation of instructions for the production of machined metal parts (Alting & Zhang 1989, Eversheim & Schneewind 1993). However, there are many production methods other than machining. Some of the most important are stamping and other forming methods for sheet metal parts, die casting and injection moulding. Some attention has been given to process planning for these processes, but the technology is less advanced than for machining.

Another important production process, occurring after the individual parts of a product have been manufactured, is assembly. This activity also requires planning, and the development of automated assembly planning methods is a major current topic of research (Baldwin et al 1991, Sanderson et al 1990).

Various types of product models play an important role in the planning activities mentioned above.

2.3 Production

By the time the production stage is reached the product models have already played their major part. However, they still have some remaining roles, for example as specifications of 'nominal' parts against which measured data from inspection and testing processes can be compared.

3 COMPUTER MODELS USED IN PRODUCT REALIZATION

Traditionally, the output of the design process is a specification of the product to be manufactured in the form of manually generated 3-view drawings together with supporting documentation. The use of CAD systems allows such drawings to be generated by the computer, but other more sophisticated types of geometric product descriptions are now routinely created by such systems, as described later in this section. These are models or representations of the design, whose key purpose is to act as substitutes for the real thing, in particular to provide answers to queries about the real product. Different types of models are generated by various classes of CAD systems, including the 2D drawing, the 3D wireframe model, the solid model and its enhancements containing parametric, constraint and form feature information with their associated engineering semantics.

The complexity of the product realization cycle for mechanical products often makes it appropriate to generate different models of the product, for use in different activities contributing towards the overall process. These models may be crude in the early design stages, but sufficient to provide rough-and-ready answers to the broad questions arising at the time. Clearly the output of detail design should include a fully detailed geometric description of the product; it may also contain a great deal of non-geometric information of various types discussed in the following sections.

3.1 CAD systems and their models

Historically, the first interactive graphical CAD systems were 2D drafting systems. These provided a means for the generation of drawings of the traditional kind, their primary advantage being that this could be done more quickly. The major time-saving resulted from the use of automated techniques for generating drafting symbols and for copying other recurring combinations of geometric elements. Many smaller industrial companies are still using systems of this kind, often running on PCs.

The next major development came in the early 1970s, with the introduction of the 3D wireframe model. This is a representation of the shape of a designed object as a set of edges in three dimensions; its primary significance is that it provides a unified model of the object rather than several partial models, as in the case of the traditional three orthogonal views of the engineering drawing. One immediate advantage of the wireframe representation of an object is that the computer can automatically generate drawings of it from any point of view and in any projection chosen by the user. Wireframe systems have been extensively used by industry for several years, but are now being rapidly superseded by more modern systems.

Most wireframe CAD systems also allow the attachment of surfaces to the edge-based model, and this enables the use of realistic shaded surface renderings. The geometry available generally includes complex doubly curved surfaces such as NURBS (non-uniform rational B-splines), whose use was pioneered in non-graphical systems developed in the 1960s, mainly in the aircraft industry.

The next development was the solid modeler, which brings together the advantages of the wireframe and the surface modelers in an optimal way. Like the enhanced wireframe model, the solid model contains information concerning all the faces of the object, including the surfaces they lie on and the edge curves which bound them. It also stores *topological* information indicating how all these elements are connected together in the model. One significant advance is that most of this information is now generated automatically and verified internally by the system, which can also automatically compute the volume, mass, and moments of inertia of the object. Most major CAD systems now possess a solid modeling capability, though this technology has only recently become widely used in industry.

During the 1970s it was thought that the existence of a complete computer model of the geometry of an object would enable the automation of many activities downstream of design, such as process planning. Unfortunately, during the 1980s this proved not to be so, and further developments in CAD systems have been made and are still being made since that time. There are several different but related thrusts, which are beginning to converge in the CAD systems available today. The aim is to generate not merely a *solid model* (i.e. geometry alone) but a *product model*, containing additional engineering semantics.

Some of the major areas of new development in CAD modeling are briefly summarized below:

Parametric modeling: Here the intention is to allow the design of a product in which certain dimensions are not fixed, but can be varied for purposes of design modification or to generate different members of the same family of products. This capability has existed in a limited form for several years.

Constraint-based modeling: This is related to parametric modelling but is more powerful. It allows the specification of constraints on elements of the design, such as 'these two plane surfaces are parallel', or 'Circle A is concentric with Circle B'. Such constraints are usually driven by the intended functionality of the product, and once defined they are required to hold when any design modifications are made. The provision of this capability is giving rise to many technical problems, but most major CAD systems now offer at least limited 2D constraint modelling.

Feature-based modeling: In the mechanical engineering context a *feature* (or more fully a *form feature*) is a local geometric configuration on the surface of a manufactured part which has some engineering significance. Design features are related to the intended functionality of the product; examples include cooling fins, gear teeth and holes for bearing housings. Other product realization activities may have different feature-based views of the same part. For instance, features for machining processes are simply volumes of material which must be removed, such as holes, pockets or slots. Research has shown that form feature information provides the 'natural' input required for manufacturing engineering applications. It has proved difficult to generate this information automatically from the shape representations used by the purely geometric type of solid modeler mentioned earlier. For this reason, many CAD systems are now providing facilities for 'design-by-features', though few of them currently have any means of automatically generating manufacturing feature models from design feature models.

The most significant aspect of the historical progression of CAD system development is the increasing potential for interpretation of the model by the computer. The manually produced drawing was intended exclusively for human interpretation, whereas the design systems of the future will generate information that will directly drive automated processes downstream of design. In particular, these systems will be capable of creating models that not only provide geometric product descriptions, but also richly augment them with engineering semantics. One current research problem concerns the capture of 'design intent' or 'design rationale', i.e. the retention with the product model of the reasons why particular design decisions are made.

In addition to the essentially geometry-based graphical systems of the kind discussed above, which are what generally come to mind when CAD is mentioned, there is a variety of other types of systems providing additional support for the design process. Some of these are briefly discussed below.

3.2 Modeling for engineering analysis

Analysis and simulation tools provide support for the design process. They aid designers by providing information about functional behavior, cost and other concerns pertinent to the design process. Many computational tools are currently available for structural, thermal and fluid flow analysis and associated simulations. Another widely available form of engineering analysis system provides a means for modeling kinematic assemblies and allowing dynamic simulations of their motion. Such a system often provides an additional capability for vibration analysis of mechanical systems.

Analysis and simulation tools are most frequently used in the detail design phase, after the part is fully described. However, as emphasis shifts toward concurrent engineering (Nevins & Whitney 1989), where decisions must be made earlier in the design cycle, these tools will need to be developed to support the design in its earlier phases as well, for example by providing approximate results on the basis of incomplete design information

(Dabke 1994).

One of the most common types of analysis model is the finite element (FE) model, a specialized approximate representation of a part in terms of a mesh of simple geometric elements, used as the basis of structural and other types of analysis (Armstrong 1994). The elements are usually either triangles or quadrilaterals in 2D (e.g. cross-sectional) analysis, and tetrahedra or hexahedra in 3D analysis. In the structural analysis case, loads are specified at the nodes of the mesh (usually at the corners of elements where they connect to each other), and the resulting displacements of the mesh are calculated, again in terms of the nodes. Although FE models appear to be purely geometric in nature, there is also a partial differential equation or variational principle underlying the analysis which makes use of them, and this must also be regarded as an implicit component of any such model.

A major current problem with FE analysis is that, although the process is automatic once the mesh is set up and the loading conditions imposed, a 'good' finite element model cannot in general be created automatically from a detailed geometric product model. There are several reasons why this is difficult, especially in 3D. Some of them are concerned with problems of generating the preferred hexahedral meshes whilst satisfying certain criteria on mesh topology or connectivity. Others are concerned with the avoidance of long, thin element shapes, whose presence leads to inaccurate computed results. The fact is that the setting up of good FE models is an activity generally requiring the knowledge and experience of a highly trained human operative, and it has been found difficult so far to encapsulate the necessary knowledge in a rule-based system. Consequently, the interface between CAD and finite element analysis is at present far from fully automated, and the setting up of analysis models is a lengthy and painstaking task that sometimes creates bottlenecks in the design cycle.

A further aspect of the mesh generation problem is the desirability of idealizing regions of 3D models as thin shells, plates or beams. This allows simplification of the FE model through the use of 2D or 1D elements. The resulting reduction in size of the system of equations to be solved may lead to greatly reduced solution times and possibly also to improvements in accuracy. Advantage can additionally be taken of symmetry of geometry (provided it is associated with corresponding symmetry of loading conditions), since this often permits the results of a full analysis to be inferred from the analysis of only part of the model. This again reduces the size of the computational problem. Full automation of mesh generation therefore requires the automatic identification from a CAD model of symmetries and regions where idealizations can be used. These capabilities currently exist only in certain university research projects (Dabke et al 1994).

Another major problem at present relates to the reverse interface between FE and CAD. The results of FE analysis are in the main human-interpretable, the provision of automatic feedback into the design process being in the very early stages of development. The optimization of designs with respect to functionality and cost is essentially an iterative process, and this paucity of feedback puts the human very firmly in the loop. Optimization can therefore be quite a labor-intensive activity.

This particular type of model has been dealt with at some length because it provides good illustrations of some of the difficulties facing researchers trying to develop integrated product realization systems.

3.3   Virtual prototypes

Virtual or computational prototyping is generally understood to be the construction of computer models of products for the purpose of realistic graphical simulation, often in a 'virtual reality' (VR) environment. This provides the ability to test part behavior in a simulated functional context without the need to manufacture the part first. It is one of many strategies aimed at reducing design cycle time. However, a 'virtual prototype' in this sense is only one amongst many different types of model having value in the design process – the name given to it reflects the fact that this type of model originated in the computer graphics community whilst most of the others discussed above were developed by the engineering community. There is no clear-cut distinction; they are all models, and in the sense that they can be used to provide answers to engineering queries they are all virtual prototypes.

Virtual prototyping also lends itself to realistic process modeling. The availability of a graphical model of a part or product in course of manufacture allows simulation of the effects of manufacturing processes. For example, it is possible to generate animated simulations of material removal during machining processes.

The advantages of using virtual prototypes in an 'immersive design' virtual reality environment are currently being studied by a few large manufacturing companies. Boeing uses it for 'fly-throughs' of complex structures in visual checks for interference of parts, and Caterpillar as an aid for the design of cabs for earth-moving equipment. Other reported users of VR in vehicle design are the Daimler-Benz group (Haban 1996) and PACCAR (Jayaram 1996).

Such simulations rely on the ability to generate realistic graphical representations at real-time speeds, and to this end the true 3D shape of artefacts is usually approximated for rendering purposes in terms of a large number of planar tiles or facets. Interestingly, this type of model is also routinely generated for quite another purpose – it forms the input to a range of processes variously referred to as *solid free-form fabrication (SFF), layered manufacturing, rapid prototyping* or (more recently) *holoforming*. Stereolithography is an example of such a process, whose intention is the rapid generation, directly from CAD data, of a non-functional physical prototype of a part or assembly. This can be used to judge appearance or to test assemblability of a designed part into an assembly, for example. Many CAD systems generate a faceted representation of a part in an industry standard format known as a .STL file, to provide input to SFF systems. Workers in VR have also found that .STL files provide suitable models for generating animated visualizations.

3.4   Knowledge-based analysis

Knowledge-based systems use expert knowledge bases and inference engines. Their automated use in design requires the provision of interfaces to design systems that convert certain design data to 'facts' comprehensible to the inference engine. The inference engine then uses these facts or assertions in the knowledge base to deduce other facts, a process which may ultimately lead to important deductions about the characteristics, quality, and functionality of the design. In a system of this kind the design model is reduced to a set of assertions in the knowledge base, and depending on the particular application concerned these may be either quantitative or qualitative. The automated use of systems of this kind before the detail design stage is problematical, since design information may still be largely on paper or in the designer's head. However, the importance of advisory design systems is highlighted by the significant advantage to be gained from their use in early design with manual entry of product data (Boothroyd 1994).

A few cases exist where feedback from knowledge-based systems into geometry-based systems occurs automatically, but there is currently no standard allowing the automation of such interfaces in a general way.

3.5   Other examples of non-geometric models

Other kinds of non-geometric models also have a role to play in the product realization process. A model used for estimating production cost, for example, is likely to have the form of an algorithm or set of formulae, taking into account the time needed for manufacturing operations, the operational and depreciation costs of the equipment used, costs related to tool wear and so on.

4   PRODUCT MODELS IN MANUFACTURING ENGINEERING

The type of model required for manufacturing engineering depends upon the nature of the manufacturing process to be employed. There is an immediate difficulty here, in that the process may not be known at the time the product is designed. A subsequent decision on process may necessitate changes to the design to make it more suitable for manufacture by the chosen means. This is just one of many examples of feedback between the various stages of the overall product realization cycle.

For purposes of illustration it will here be assumed that a designed part is to be machined from solid material. Experience has shown that the most suitable type of model for planning this process is one based on form features. For this application the features will be material removal features such as pockets, slots and blind or through holes. The machining strategies available for generating each such feature type are relatively few in number, and they differ primarily in the accuracy and surface finish they are capable of achieving. The choice of strategy for any particular feature may then be made on the basis of the feature type and the required engineering tolerances and surface finish associated with it in the part model. Normally, the cheapest operation meeting the desired criteria will be chosen. If this procedure is repeated for all the machining features exhibited by the part, the resulting set of machining strategies forms the basis of a process plan for its manufacture. They must be sequenced in some logical manner to give the final plan; this requires complex reasoning, but much of the required information is of the same kind as is needed for the earlier stage of the process.

Other manufacturing processes, including assembly, may also be decomposed into feature-based sub-processes, but it is important to realize that different processes will require different feature models of the same part. For example, in machining, the features are all subtractive, but if the part is to be built up by (for example) welding together several originally separate components then the features of the final part are additive. It is possible to arrive at the same final geometry by either method in some cases.

An equally important point is that, if the part is designed in a feature-based design system the designer's feature model will almost certainly not be the most appropriate model for manufacturing planning. The design features are created to provide functionality in the part; they may be either additive or subtractive features, as in the case of a locating pin and the hole into which it fits. However, as we have seen, some manufacturing processes require features which are either all additive or all subtractive. There are also more subtle differences between the feature models appropriate for different applications (Pratt 1991).

A further possibility is that the part is designed in a pure geometry-based system, so that the design model contains no feature information at all. Since a model based on manufacturing features is the prerequisite for the automated generation of a manufacturing plan, the essential problem in both this and the previous case is, how is the manufacturing feature model generated? Some partial answers are provided in Section ?? below.

5 TRANSMUTATION OF MODELS

The creation of feature models for processes downstream of design is one of the major problems impeding the building of integrated product realization systems for industrial use. The automated generation of a manufacturing feature model is discussed in some detail below, since this is currently a major emerging area of research. However, this is just one of many feature model transmutation problems, and some other cases are also given some attention at the end of the section.

5.1 Feature recognition

The initial motivation for working with features came from a growing realization that part models of purely geometric types do not readily provide the kind of information most immediately useful to a process planning system. At one time it was thought that the solid model would be able to do this, but experience proved otherwise. There are two main approaches to solid modeling: a boundary representation (B-rep) system represents a part as a connected collection of faces with specified geometry, while a set-theoretic or constructive solid geometry (CSG) system represents it as a set of points in 3D space, expressed in terms of combinations of simple volumetric primitives such as blocks and cylinders. It was found that B-rep and CSG modelers provided information respectively at too low and too high a level for easy interpretation by a process planning system. The appropriate median level proved to be that of the form feature, expressed as a (usually connected) set of faces in a B-rep model, or as interactions between two or more primitive volumes in a CSG model. It should be mentioned in passing that despite the popularity of the CSG approach some years ago all existing commercial CAD modeling systems are now based primarily on the B-rep methodology.

Much attention has been given to the problem of automatically recognizing form features for manufacturing processes (machining in particular) from a model of a part, usually in the form of a solid model of one of the types discussed above (Shah 1991). In a B-rep context this involves identifying a set of part faces which match some predefined sets of rules characteristic of each recognizable feature type. For example, a rectangular pocket consists of five faces: a rectangular floor, perpendicular to four walls connected at right angles to each other at the corners (and therefore forming two mutually perpendicular parallel pairs). This has proved to be an easy configuration to recognize in isolation, but a much more difficult one where features overlap and their characteristic face patterns are modified as a result. The first commercial generative process planning systems for machined parts based on the automatic recognition of manufacturing features from a solid model are now available. However, they are only successful for a limited part domain, and their capability needs to be extended to cover other types of manufacturing processes.

5.2   Feature model transmutation

Many modern CAD systems allow the designer to design in terms of form features. These systems provide a range of frequently occurring functional features, and also offer the facility for extending this range with user-defined features to meet the specialized requirements of any particular product range. The design process with such a system results in a product model containing design feature information; the problem for process planning is that design features and manufacturing feature are in general not the same. It is only necessary to think of a rib of material created by the designer as a strengthening element. If the rib exists on a machined part then it defines two machining features, one to remove material on either side of it. Whereas feature recognition takes as its input a pure geometric model, the corresponding process when the input is a design feature model is known as *feature model transmutation* (also *feature mapping, feature conversion, feature transformation* – there is no agreement yet on the terminology). Here the problem is to input a design feature model and output the corresponding feature-based model for some other activity such as process planning or inspection.

Although not much has yet been demonstrated in this area (Bronsvoort & Jansen 1994, Falcidieno & Giannini 1990, Shah et al 1994, Wozny et al 1994), feature model transmutation should ultimately prove to be easier than feature recognition, since the input model contains more information. An essential preliminary will be to check each design feature present to see whether it is also a manufacturing feature; if it is, the scale of the remaining problem is reduced. No commercial systems yet provide a capability of this kind. Those having the capacity for automatic feature recognition simply ignore any feature information present in the input model, and use methods based on geometry and topology alone, as described in the previous section.

5.3   Other examples of model transmutation

Other examples have in fact been given earlier in the paper. In all cases quoted, the CAD model has provided the primary or canonical representation, and the other model has been generated from it, generally on the basis of geometric and topological information alone. The generation of an FE model from a CAD model is one example, and in this case human intervention is still generally necessary to achieve the process. The generation of a faceted model for SFF or VR purposes is another example, though it has proved relatively easy to automate this process using an original CAD solid model with exact geometry. Despite this, 'bad' faceted representations with missing or unconnected facets are often encountered by organizations using SFF (Barequet & Sharir 1995). Knowledge-based models can sometimes be generated automatically, but other types of non-geometric models generally require human input.

6 FEEDBACK OF INTERROGATION RESULTS BETWEEN MODELS

As stated earlier, models are created for purposes of interrogation. The interrogation results are usually readily interpretable in the context of the model used to obtain them, but for most purposes it would be much more useful to have them interpreted in the context of the original, primary or canonical model, i.e. the CAD model. This was mentioned previously in connection with FE analysis. If this detects an unacceptably high level of stress at a certain node in the FE model, what is the implication on the CAD model? It may be that simply moving that particular node, and some of its neighbours, will lower the stress; the corresponding interpretation in the CAD model might be a thickening of material in a certain region. But in most cases the automatic generation of solutions in the CAD model to problems detected in the FE analysis is far from reality.

Similar problems exist in other cases. VR models, like FE models, are based on rather crude geometric approximations. Thus the accuracy of processes such as collision detection in simulated assembly may not be very high. This makes it desirable to check that a collision detected in the VR environment really exists in the more accurate CAD model environment. However, the links between the elements of the VR model and those of the CAD model are usually non-existent (or at best indirect), which makes automatic feedback of VR results into the CAD environment far from straightforward.

As a final example, a CAD/process planning dialog will be considered. Suppose the CAD model of a part to exist, and suppose also that no decision has yet been made on how it will be manufactured. Possibly there are several alternatives, such as sheet metal stamping, injection molding and die-casting. The original design is probably not ideal for any one of these processes. Ideally, a flexible planning system should be able to evaluate the cost of making the part as designed, using any one of the processes, but also to recommend design changes which will not change the functionality of the part but will make it cheaper to manufacture. In some cases we are currently fairly good at estimating manufacturing costs, but feedback of recommended design changes from the planning environment into the CAD environment is still some way in the future. One of the major barriers appears to be the requirement for the planning system to have some understanding of the design concept of functionality, which does not exist in the current conception of a planning model.

7 CONCLUSIONS

The paper has attempted to make and to illustrate three main points:

1. Multiple different types of product model are generated and used for different purposes in the course of the product realization process. Most of them are generated from a primary CAD model, which usually has a higher level of detail and geometric accuracy than the other types of model, some of which are in any case not geometric in nature.

2. The process of generating the secondary models is in most cases not completely automated, and in many cases is not even well understood. Nevertheless, strenuous efforts are being made to automate the interrogations and processes making use of those secondary models.

3. The information generated by interrogating the secondary models is readily interpretable in the context of those models, but it is often desirable to interpret it in the context of the primary model. We are currently in the very early stages of tackling this problem of information feedback between models.

Taken together, these points lead to an important conclusion regarding the development of integrated product realization systems. Significant advances have been achieved (ISO 1994) in developing standard means for importing, exporting and sharing the data required and generated by individual modules of such a system. However, the problem remains that each module functions in terms of its own internal model. Thus the data exported by one module is often not immediately comprehensible to another, since it is generated in a different context and has different semantics. Full communication between any pair of modules requires not only the *representation* and *transmission* of product data (the problems addressed by current standards), but also its *interpretation* by the receiving module, based on knowledge of both the old and the new context and semantics. The requirement is analogous to that of computer translation between different natural languages such as English and Japanese, a notoriously difficult problem. Much work remains to be done in this area.

The models discussed in the paper may actually be implemented in various ways. At one extreme is the case where all models are completely separate from each other, and communication is through the medium of file transfer or via calls to application program interfaces (APIs). At the other extreme, all the models are in some sense constructed on top of the original CAD model, with built-in associative links between related entities in the various models. The second option appears to make life easier in some ways; for example, it is possible to arrange for a change in one model to lead automatically to consistent changes in all the other models. This is certainly not easy if the first option is adopted. On the other hand, the second option effectively requires the overall system to be integrated through the use of a shared database, with all software modules provided by the same supplier and consequently 'speaking the same language'. This makes it difficult to link other systems which may be needed for specialized applications not supported by that supplier. In practice, most major manufacturing organizations who set out to build integrated systems start with a set of modules performing different functions, chosen for the effectiveness of their performance of those functions, and usually from *different* suppliers. Each module will then generate its own internal models, and the problems described earlier will have to be overcome. There is clearly at present no ideal solution to the integration problem.

As a closing note, the author would like to reiterate the conclusion (generally agreed by the participants of the Providence Workshop) that almost any form of computer model will serve for some purpose as a virtual prototype. The use of this terminology should therefore not be restricted to the domain of virtual reality; the VR community is undeniably doing exciting things, but there are many parallel fields of endeavour in product realization which make use of essentially the same principles; modeling and interrogation are common to all of them.

8 REFERENCES

Alting L. and Zhang H. (1989) Computer Aided Process Planning: The State-of-the-Art Survey. *Int. J. Production Research* **27**, 4, 553 – 585.

Armstrong C. G. (1994) Modelling Requirements for Finite-element Analysis. *Computer Aided Design* **26**, 7, 573 – 578.

Baldwin D. F., Abell T. E., Lui M.-C., De Fazio T. L. and Whitney D. E. (1991) An Integrated Computer Aid for Generating and Evaluating Assembly Sequences for Mechanical Products. *IEEE Trans. Robotics & Automation* **7**, 1, 78 – 94.

Barequet G. and Sharir M. (1995) Filling Gaps in the Boundary of a Polyhedron. *Computer Aided Geometric Design* **12**, 2, 207 – 229.

Boothroyd G. (1994) Product Design for Manufacture and Assembly. *Computer Aided Design* **26**, 7, 505 – 520.

Bronsvoort W. and Jansen F. (1994) Multi-view Feature Modelling for Design and Assembly, in *Advances in Feature Based Manufacturing* (eds. J. J. Shah, M. Mäntylä and D. S. Nau), pp. 315 – 330. Elsevier.

Dabke P. (1994) Developing a Finite Element Analysis Agent. Report, Center for Design Research, Stanford University, Stanford, CA 94305, February 1994.

Dabke P., Prabhakar V. and Sheppard S. (1994) Using Features to Support Finite Element Idealizations. Report, Center for Design Research, Stanford University, Stanford, CA 94305, February 1994.

Eversheim W. and Schneewind J. (1993) Computer-Aided Process Planning – State of the Art and Future Development. *Robotics & Computer-Integrated Manufacturing* **10**, 1&2, 65 – 70.

Falcidieno B. and Giannini F. (1990) A System for Extracting and Representing Feature Information driven by the Application Context. In *Proc. IEEE International Conf. on Robotics and Automation, May 1990, Cincinnati, OH.* IEEE Computer Society Press.

Haban D. (1996) Cooperative Working on Virtual Prototypes. These proceedings.

ISO (1994) *Product Data Representation and Exchange,* International Standard ISO 10303. International Organisation for Standardisation.

Jayaram S. (1996) Feasibility of Virtual Prototyping for Automobile Interiors. These proceedings.

Nevins J. and Whitney D. E. (1989) *Concurrent Design of Products and Processes.*

McGraw-Hill.

Pahl G. and Beitz W. (1984) *Engineering Design.* The Design Council (London, England) and Springer-Verlag.

Pratt M. J. (1991) Aspects of Form Feature Modelling, in *Geometric Modelling: Methods and Applications* (eds. H. Hagen and D. Roller), pp. 227 – 250. Springer-Verlag.

Sanderson A. C., Homem de Mello L. and Zhang H. (1990) Assembly Sequencing Planning. *AI Magazine* **11**, 1, 62 – 81.

Shah J. J. (1991) Assessment of Features Technology. *Computer Aided Design* **23**, 5, 331 – 343.

Shah J. J., Hsiao D. and Leonard J. (1993) A Systematic Approach for Design-Manufacturing Feature Mapping, in *Geometric Modeling for Product Realization* (eds. P. R. Wilson, M. J. Wozny and M. J. Pratt). North-Holland.

Ullman D. G. (1992) *The Mechanical Design Process.* McGraw-Hill.

Wozny M. J., Pratt M. J. and Poli C. (1994) Topics in Feature-based Design and Manufacturing, in *Advances in Feature Based Manufacturing*, (eds. J. J. Shah, M. Mäntylä and D. S. Nau), pp. 481 – 510. Elsevier.

9 BIOGRAPHY

Dr. Mike Pratt is a Senior Research Associate in the Center for Advanced Technology at Rensselaer Polytechnic Institute, Troy, NY, but is currently visiting the National Institute of Standards and Technology, Gaithersburg, MD. Until 1991 he was Professor of Computer Aided Engineering at Cranfield Institute of Technology in England. He holds degrees from Oxford University (physics) and Cranfield (aeronautical engineering, mechanical engineering). He has worked in CAD/CAM for more than 20 years, mainly on applications of geometry in product realization, with particular emphasis on the use of form features in automating the interfaces between its various processes.

11

A feature-based framework for transforming and representing multiple format CAD for virtual prototyping

R. Ganesan
Research Associate, Automation & Robotics Research Institute
7300 Jack Newell Blvd., Fort Worth, TX 76118
Tel: (817)794-5900 Fax: (817)794-5952
E-mail: ganesan@eepost.uta.edu

V. Devarajan
Associate Professor, Department of Electrical Engineering
The University of Texas at Arlington, Arlington, TX 76019
Tel: (817)273-3485 Fax: (817)273-2253
E-mail: venkat@ee.uta.edu

Abstract
Although feature based modeling is increasingly becoming popular, there is still a huge backlog of designs created in conventional CAD. Therefore, there is a need to convert these old designs to the more intelligent feature format. In this paper we describe an integrated architecture of a feature-based framework capable of transforming between different design representations like wireframe, solid models, and feature-based models; making it an ideal virtual prototyping environment.

Keywords
Virtual prototyping, feature-based CAD, Computer Integrated Manufacturing, feature extraction

1   INTRODUCTION

Features have now been accepted as a potential common data format to both design and manufacturing because of the ease with which manufacturing systems can automatically access and interpret product design definitions. Conventional CAD (2-D or 3-D models) only examine raw geometry and are not capable of representing the underlying product information. Features are therefore regarded as the key enabling technology for many applications like automated process planning and machining, rapid prototyping, and virtual prototyping. Virtual prototyping environments demand a shared information model for product design that can be understood by all the design and manufacturing processes, and features have emerged as a potential standard suitable for virtual prototyping.

2   NEED FOR THE FRAMEWORK

The need for the proposed framework arises from several practical problems that have existed for several years in research related to features. Most of these problems have a direct bearing on virtual prototyping.

A need for this framework comes from the fact that there are a large number of designs today that are still in 2-D CAD. Most feature extraction systems fail to recognize the importance of legacy CAD and instead directly extract features from 3-D solid models. Furthermore, many of the "new designs " today are merely modifications and upgrades of older designs in 2-D CAD. This is particularly true of the aerospace industry where missiles and aircraft are frequently modified.

New design environments like virtual prototyping are leaning towards storing product designs in multiple design formats. By doing so, designs can be pulled up at any stage of the manufacturing process by multiple users for multiple applications and performing all the analysis in parallel on the same design. Users can then return and exchange suggestions on modifications required to a part leading to collaborative work. This in essence is Computer Integrated Manufacturing - or bringing together of design and manufacturing processes.

Storing designs in multiple formats requires transformation between the various design representations. For example, designs created by a feature-based modeling system may have to be converted to manufacturing features. The proposed framework can transform designs using a common feature extraction scheme. The strength of the whole architecture lies in the feature extraction algorithm that operates on 2-D designs as input. Therefore, all designs can be transformed through an interim 2-D representation. In (Ganesan and Devarajan, 1994), we discuss several other applications for the system.

Features have been classified into several types depending on the specified application. Design features are used to convey design intent and are merely a solution to a functional problem. Machining features or manufacturing features on the other hand, are used for automated process planning and NC machining. Some of the other feature types are assembly features, tolerance features, and surface features.

The architecture has specific advantages over other systems researched in machining feature recognition. Most of these systems suffer because they work in 3-D space, attempting to recognize features directly in 3-D. Feature extraction in 2-D is much simpler and the proposed framework obviates the complex option of extraction in 3-D. Several review papers discuss the history of feature extraction in 3-D and the problems associated with it (Pratt, 1993; Case and Gao, 1993; Salomans, et. al., 1993). The hybrid nature of our framework combines methodologies from both feature extraction schemes and feature-based modeling schemes.

3   METHODOLOGY OF OPERATION

The overall methodology of the architecture is to first bring all input design representations to 2-D orthographic views in the Data Exchange Format (DXF). The feature extraction algorithm is then applied to the input to generate the necessary features. The extracted features can then be assembled to recreate the feature model of the part.

Fig. 1 shows the overall approach of the proposed architecture. A *feature library* is built into the system containing a list of the features that have to be recognized. In this instance they are parametric machining features. The input designs are brought into 2-D orthographic views in the DXF format. This is performed by the *converter and preprocessor*. 3-D solid models can be back projected to obtain the orthographic views. The *feature model determination algorithm* calculates the volume size needed to generate the feature model. The *feature extraction system* works as a two step process. Simple isolated features are first extracted and then intersecting features are differently handled. The resulting set of features is matched with the feature library to recognize and identify the extracted features. This is accomplished by the *feature identification system*. The *feature model reconstruction system* utilizes the inter-feature relationships to assemble the feature model using the feature model volume as the base. The *feature validation system* performs a check on the feature extraction system to make sure all features have been extracted. Graphical simulation and real-time visualization are incorporated to provide good user interface, although the system needs little human interaction. In depth details of the architecture can be found in (Ganesan, 1994).

4   COMPONENTS OF THE SYSTEM

4.1   Possible inputs to the system

The various forms of design representation inputs that the system can accept are :
• 2-D CAD (wireframe in orthographic views)
• 3-D CAD (wireframe models, CSG models, B-Rep models)
• Feature-based CAD with design features

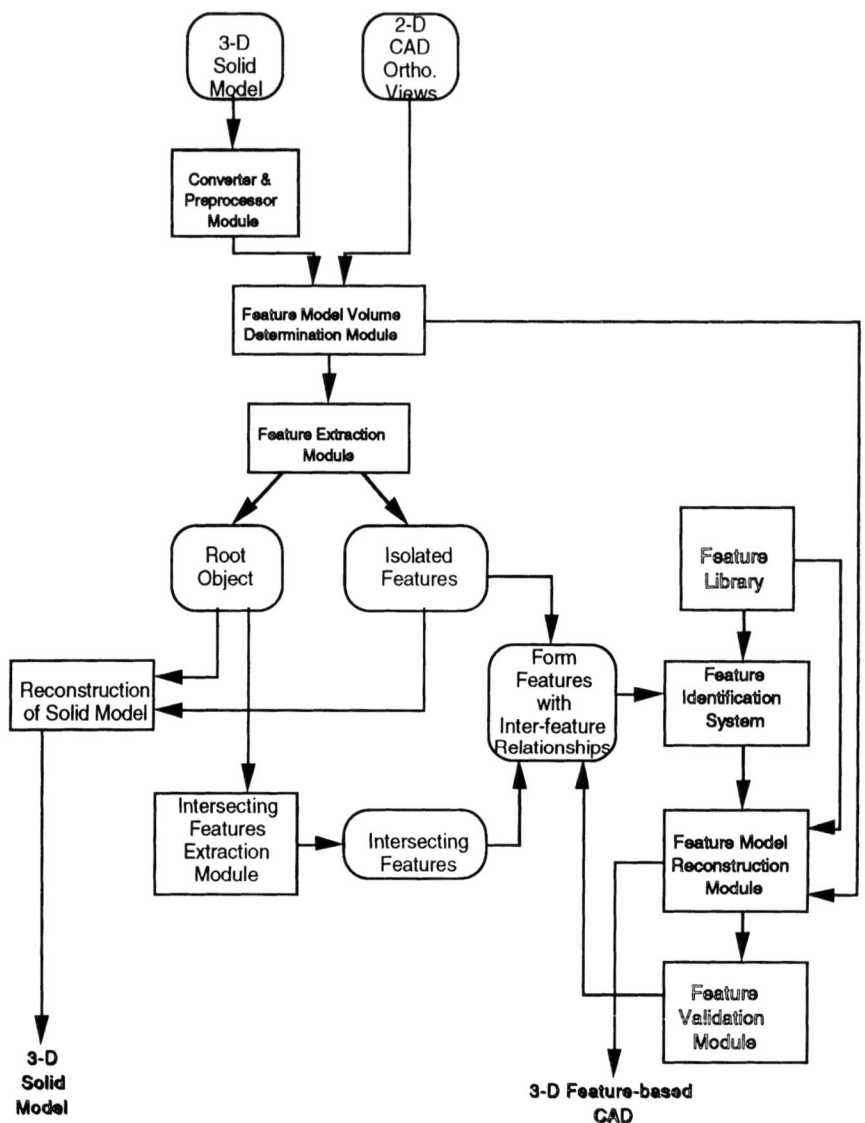

Figure 1 Overall methodology of the architecture

4.2 Neutral file formats

The data formats of the input designs differ considerably depending on the representation and the design software used to create the original design. There could also be paper drawings and microfiche that must first be brought into a 2-D CAD format. DXF and IGES are currently industry standards to represent conventional CAD and there are preprocessors to convert from one neutral format to the other. However, feature information cannot be represented by any of these standards. The proposed system is based on the new object-oriented ISO standard 10303, also called STEP (Standard for the Exchange and Representation of Product Model Data). STEP, although still emerging, is the only standard that offers an information model that covers all aspects of the product life cycle.

4.3 The feature library

The feature library is built into the system using an object oriented paradigm and in conformance with STEP standards. The feature library is flexible and can be changed to suit specific applications, but only machining features are considered in this paper. The feature library created can also be used as a modeler when the final feature model has to be generated. This is explained under the feature model reconstruction module where identified features are pulled up from the feature library to reconstruct the feature model. Therefore, the system combines some of the strategies adopted by feature-based modeling systems.

Each feature is a combination of both Constructive Solid Geometry and Boundary Representation, defined parametrically in terms of edge and volume relationships. The parameterization of the feature database allows easy feature identification and also provides edit capabilities to the original design once the design has been converted to features. However, associativity checks cannot be performed on the parametric features.

Each feature is stored with both geometric and engineering information, supported well by the STEP standards. Feature properties cannot be autonomously extracted from any conventional CAD and would therefore have to be manually entered. Some of the attributes that can be tagged to each feature are -
• Geometry - origin, length, width, thickness, diameter, depth, etc.
• Surface finish requirements.
• Dimensioning and tolerancing information.
• Machining and production requirements.
• Miscellaneous information like user name, revision number, department name, etc.

Some of the machining features stored in the library are : blind hole, closed blind pocket, through hole, open pocket, closed through pocket, T slot, square slot, step, wedge, fillet, etc. Figure 2 shows a tentative list of the machining features to be included in the library.

As part of creating the feature library, will also be the creation of a Standard Data Access Interface (SDAI) to access property information for each feature from the library.

The SDAI can be used as a bus to transfer information between the feature library and the feature identification module and the feature model reconstruction module. Since SDAI is an standard defined by STEP it can also be used as an interface between the output of the system and other commercially available CAD software

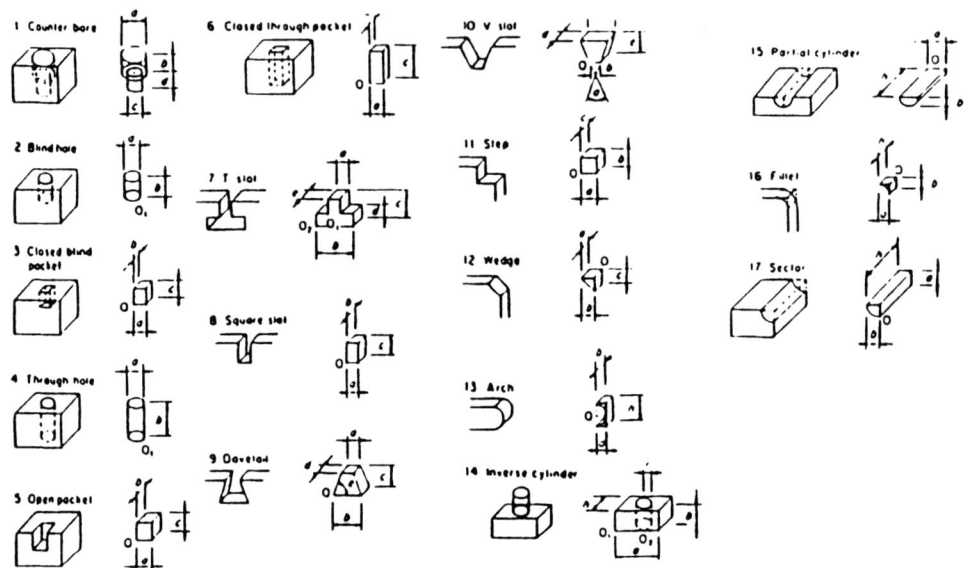

Figure 2 List of typical manufacturing features in the feature library

4.4   Converter and preprocessor module

The feature extraction system accepts only 2-D orthographic views in the DXF format. This implies that all other design inputs must first be brought to the input specifications. This is accomplished by the converter and preprocessor module in the system. Preprocessing is quite simple in most currently available CAD systems that can convert between DXF and IGES formats. The system will call the CAD software externally to perform this task.

If the input designs are 3-D solid models or feature-based models then they have to be back projected in three perpendicular planes to obtain the orthographic views. This is also accomplished external to the system by the host CAD software and then transported back to the system in DXF format.

4.5   Feature model volume generation module

The feature model volume is defined as the volume generated as a result of putting together the bounding boxes that cover each orthographic view. The algorithm determines the minimum and optimum volume needed to recreate the feature model of the original design. The algorithm works as follows -
• The bounding box of the three orthographic views are obtained. The bounding box is a box drawn around each view so as to just cover the view on all sides. The boxes must be rectangular.
• The corner coordinates of the three boxes are determined and the dimensions of the block with the above corners are calculated.

The feature model volume calculated above is only for purposes of generating the feature model and has no significance in actual machining, although it does attempt to optimize material requirement and reduce waste.

4.6   Feature extraction module

The feature extraction system forms the core of the proposed architecture. The unique feature extraction in 2-D is briefly described in this section and is detailed in a typical example considered in the next section.

A divide and conquer approach has been adopted at both the macro level and micro level of feature extraction. At the macro level, huge drawings of complex assemblies can be broken into smaller subassembly parts and features extracted from them. At the micro level, each subassembly drawing is split into subparts formed by arcs and isolated closed loops. Fig. 3 illustrates the methodology of feature extraction.

Input to the feature extraction system is a DXF file of the orthographic views of a subassembly. The feature extraction algorithm works as follows :
• Arcs (including circles) are searched in each of the three views. The other views corresponding to the arc are then located. From these the sweep height needed to reconstruct the subpart containing the arc is calculated.
• The arcs are then deleted from the views and replaced with a virtual line connecting the ends. The arcs are stored as extracted entities. This process is called *arcuated subview extraction.*
• The algorithm then looks for isolated closed loops in each of the three views. Isolated loops are loops that do not touch any other edge in that view. From the other views corresponding to this loop, the extrusion height for the loop is calculated.

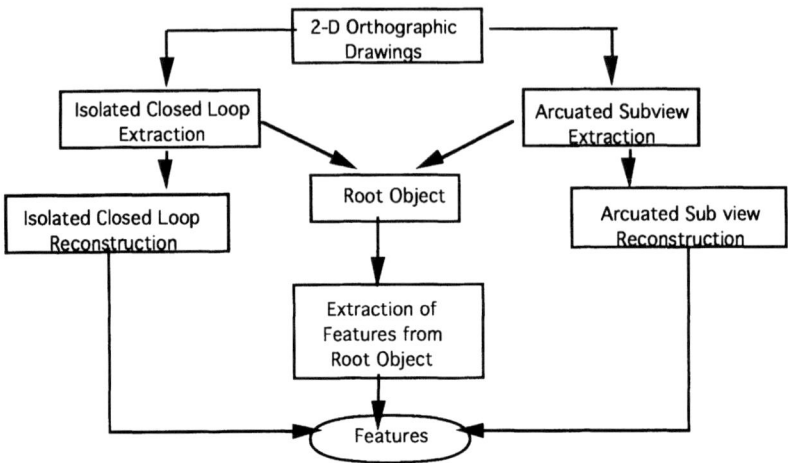

Figure 3 Methodology of feature extraction

• The isolated closed loops are then deleted from the views and stored as extracted entities. This process is called *isolated closed loop extraction*.
• Using the sweep height, the arcs are reconstructed separately and stored as a 3-D feature entity. This process is called *arcuated subview reconstruction*.
• Similarly, the isolated closed loops are reconstructed using the extrusion height of the loop. They are stored separately as 3-D feature entities. This process is referred to as *isolated closed loop reconstruction*.
• The part left behind after the arcuated subparts and isolated subparts have been isolated is called the *root part*. The root part is considered as a combination of intersecting features that could not be recognized by the above algorithm. The intersecting feature extraction system is used to extract the individual features and their inter-feature relationships.

The confidence that intersecting features can be solved easily in 2-D is strengthened by (Meeran and Pratt, 1993), who propose an algorithm capable of tackling simple intersections in 2-D. Inter-feature relationships are more easily determined in 2-D because all feature entities extracted and their corresponding 2-D shapes are referred to a common origin. Feature relationships are critical in the extraction of intersecting features.

4.7   Feature identification module

The feature entities extracted by the feature extraction module are in terms of reconstructed arcs and closed loops, and are not actual machining features. The feature identification module uses rule-based techniques to match each extracted entity with the feature library to identify the machining feature. Orientation and scale are of no importance here because the algorithm deals with a matching parametric definitions and do not consider the size or location of the extracted features.

To understand the need for a feature identification module consider a simple 'closed through pocket' (A) in a block (B) shown in Figure 4.

When A is reconstructed and extracted as an entity A, it cannot be identified automatically as a closed through pocket. But by analyzing the feature entity and checking for various volume enclosure relationships it can be recognized as a depression feature of type 'closed through pocket' enclosed in B. Details of algorithms to test for depression and protrusion features are discussed in (Balachander, 1994).

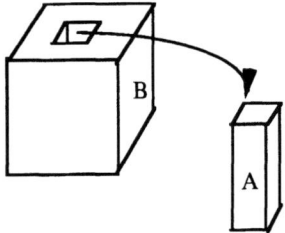

Figure 3 Example for feature identification

One of the major advantages of the architecture is that other feature types like assembly features can also be extracted using the same feature extraction module, by simply modifying the rules to identify the features.

4.8 Feature model reconstruction module

Reconstructing the design in terms of features after they have been extracted is an open research area. The reconstructed feature model, in terms of machining features, is necessary for various downstream manufacturing activities like process planning or machining and also for updating an older design.

In the architecture discussed in this paper, the feature model is reconstructed using methods very similar to those used in feature-based modeling. When features are extracted and identified, they are stored as specific instances of the corresponding parent feature present in the library. The geometric dimensions and inter-feature relationships are carried over to the feature instance, but other feature attributes like tolerancing and surface finish cannot be detected. The user has to attach these attributes to the extracted features to completely define the machining feature.

Once all the features have been defined, the feature model reconstruction module takes over from the user to recreate the design in terms of machining features. The reconstruction starts with the volume generated by the feature model volume generation module as the base stock. Since the exact positions of all the extracted features in the design are known in terms of the inter-feature relationships, the feature instances are automatically pulled up from the feature library and placed on the feature model volume. When the process is complete, the base stock now completely represents the machining feature-based design.

4.9   Feature validation module

Although the architecture calls for little by way of highly skilled human interaction, it is always a good idea to perform some level of semi-automated validation check. The feature validation system has been incorporated to check if all the features have been correctly extracted and identified. The validation scheme works as follows :

   The reconstructed feature model is back-projected to obtain the three orthographic views. Vector edge matching algorithms are used to match these views with the original orthographic views input to the system. When an error is detected in terms of missing or spurious edges in the new orthographic views, the system signals for human interaction. The user can then manually identify the missing feature and add it to the feature library as another feature instance.

   The feature model is again reconstructed with the new set of features and the validation algorithm is run to detect any further errors.

   If the input design is available in more than three orthographic views, the additional views (for e.g.. auxiliary views and sectional views) are of great help in the validation of subpart reconstruction from the orthographic views.

4.10   Graphics and user interface

Unlike most research systems, the proposed architecture will have good interface for user interaction. Real-time visualization of all processes will be incorporated in one corner of the screen and Windows-based pull down menus will allow users to enter feature attributes for the features recognized. Since the architecture is not a self-standing CAD system, the host CAD software will still be available for many of the other graphics utilities.

5   A DETAILED EXAMPLE

In this section, two examples are considered to illustrate the operation of the system described. The first example depicts how the system can be used to convert 2-D CAD drawings to a solid model (Sundaramurthy and Devarajan, 1993), and the second example shows how the same feature extraction methodology can be used to convert 2-D CAD drawings to a feature-based CAD format in terms of manufacturing features

(Ganesan, 1994). Both examples are not real life examples, but have been considered to prove the capabilities of the system.

5.1   Example 1

Figure 5 shows the first example part in terms of the three orthographic views. Although the 2-D drawings are shown with dimensions, the input to the system is a neutral file that represents the design in terms of lines, arcs and circles with dimensions.

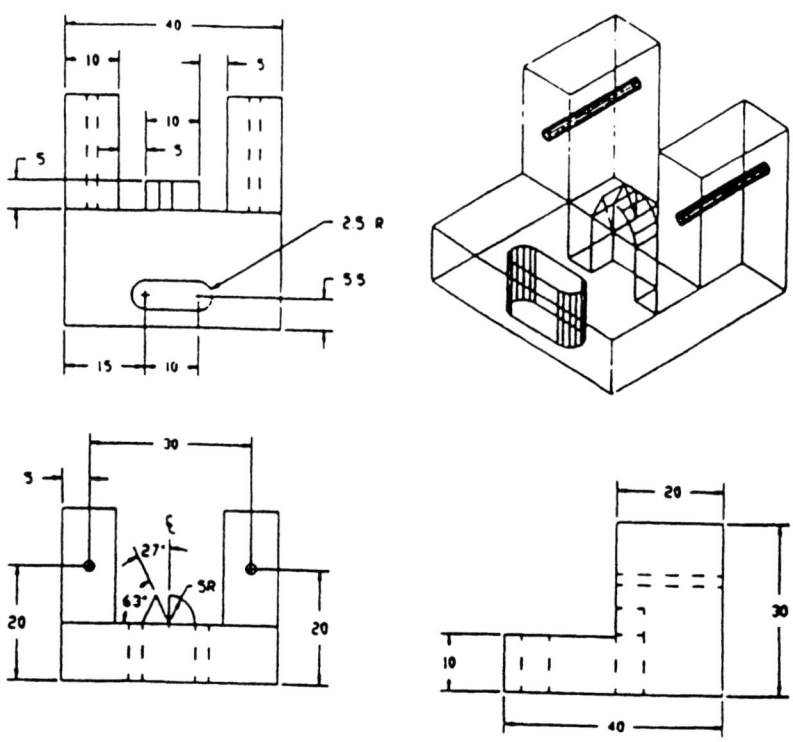

Figure 5 Example Part 1 with orthographic views

When the feature extraction algorithm is applied to the neutral file of the design, the arcs in every view are located. In this example, three arcs are found - two associated with the ends of the through slot, and one isolated arc in the front view. The other two views corresponding to these arcs are located and the edges are removed from the file. The arcs are replaced by a straight line between the same two points. Also, two circles are located in the front view, associated with two through holes. The circles and their corresponding edges in the other two views are removed from the file. The extracted 2-D entities are stored separately for later reconstruction.

The algorithm now looks for isolated closed loops in the orthographic views minus the arcs and circles. One isolated closed loop is found in the top view corresponding to the through slot. The other two views associated with the isolated loop are located and are removed from the file. These 2-D entities are now stored in a separate file for reconstruction. The orthographic views that result after all the above 'features' have been extracted is shown in Figure 6. They correspond to the root object - a polygonal outline of the part.

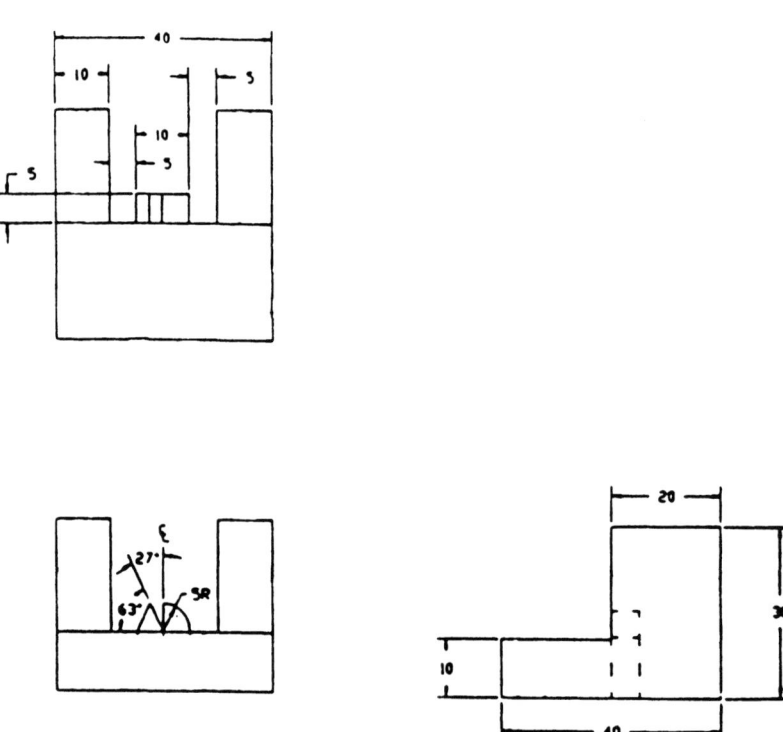

Figure 6 Orthographic views of Example Part 1 after features are extracted

In the reconstruction phase, the extracted entities are reconstructed individually. Figure 7 shows the reconstructed subparts.

Figure 7 Reconstructed Subparts (or feature entities)

The root object is reconstructed by a volume intersection technique. Each view is extruded in the perpendicular direction and then all the three extruded views are intersected to obtain the reconstructed root part. This is illustrated in Figure 8.

The reconstructed subparts are now assembled with the reconstructed root object to generate the solid model corresponding to the three orthographic views given as input.

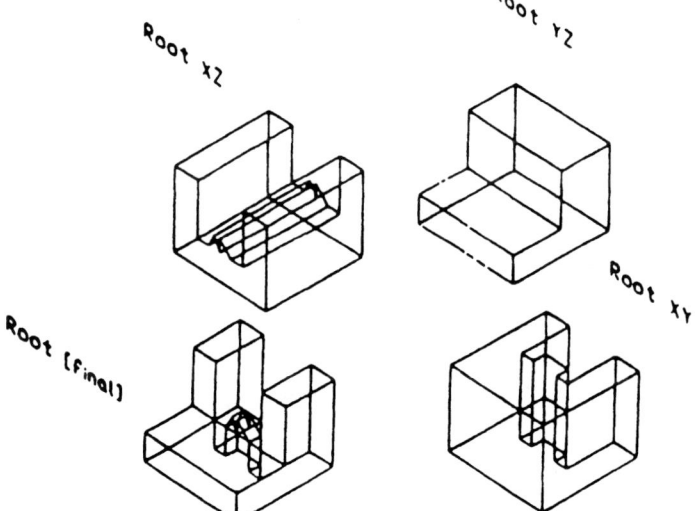

Figure 8 Reconstruction of the root part

5.2   Example 2

Figure 9 shows the orthographic views and the solid model of the example part 2. If the solid model was input to the system then the orthographic views would have to be first obtained by back projection.

The first step in the operation is the feature model volume determination. When a box is drawn around each of the three views, a volume of (100X100X100) is obtained. This will be used later for reconstructing the feature model.

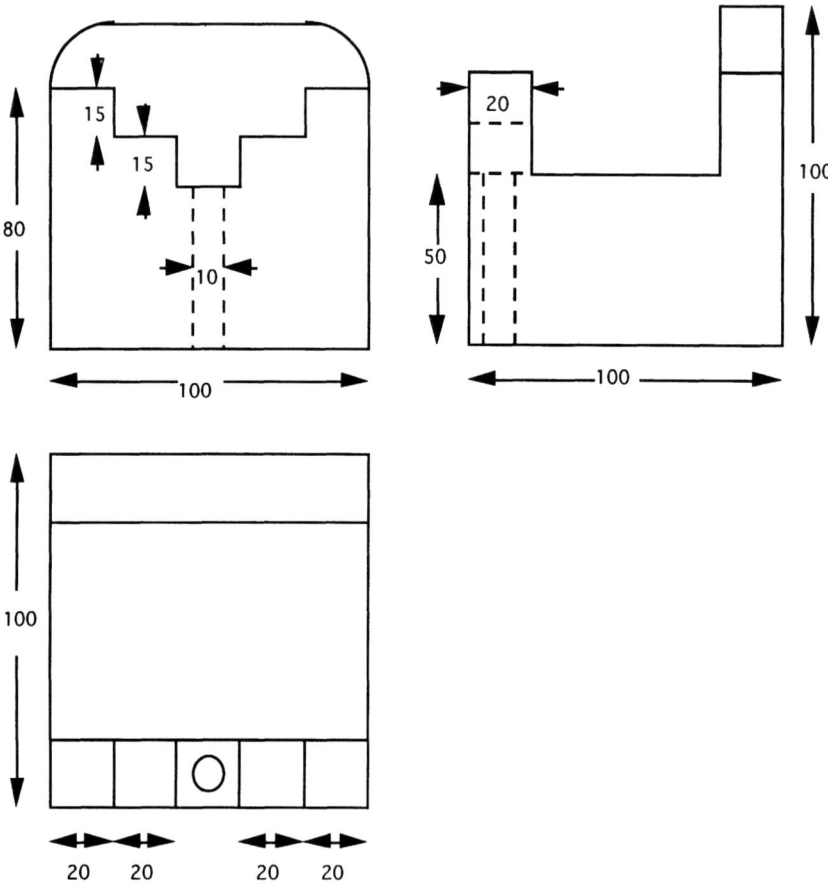

Figure 9 Orthographic views and solid model of Example part 2

In the feature extraction module, two arcs are located in the front view and a circle is located in the top view. The arcs and the corresponding edges in the other two views are isolated from the drawing and stored separately for reconstruction. The reconstructed entities are named A1, A2 and A3. The orthographic views obtained after the isolated features are extracted is shown in Figure 10. This is the outline of the root object.

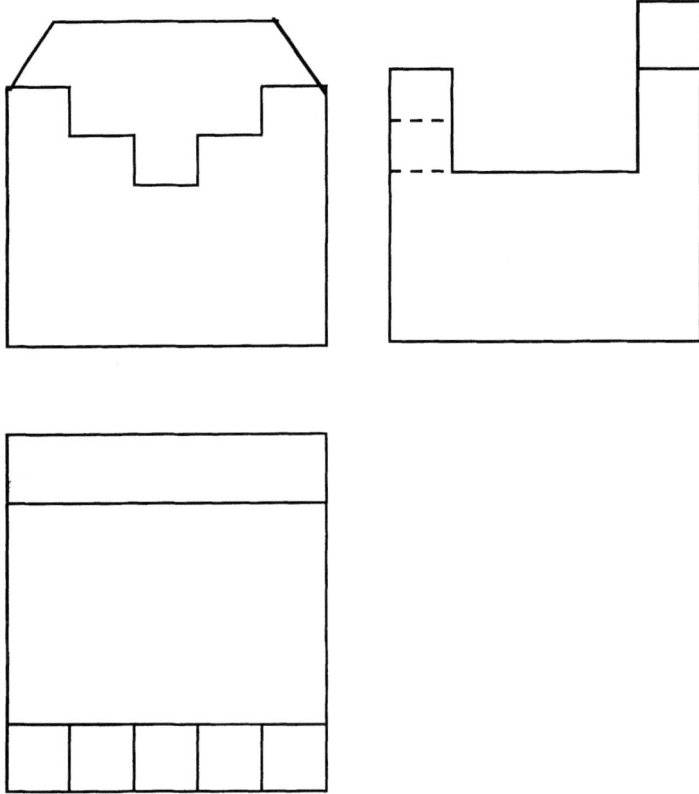

Figure 10 Orthographic views of the root object

The above orthographic views of the root object are given as input to the intersecting feature extraction module to extract the other features. Four feature entities can be extracted from the root object and they are named B1, B2, B3 and B4 and are stored separately along with their geometry.

In the feature identification module, the isolated feature entities A1, A2 and A3, and the intersecting feature entities B1, B2, B3 and B4 are identified and recognized as the following manufacturing features

A1 : Through hole
A2 : Fillet
A3 : Fillet
B1 : Square slot
B2 : Step
B3 : Step
B4 : Open pocket

In the feature model reconstruction module, the features are assembled with the feature model volume to recreate the design completely in terms of manufacturing features. The feature model of example part 2 is shown in Figure 11.

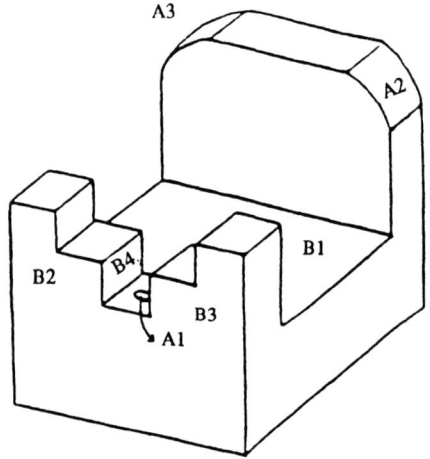

Figure 11 Feature model of Example part 2

6   CONCLUSIONS

In this paper, we have briefly outlined the architecture of a feature-based architecture ideal for a virtual prototyping environment. The architecture combines methodologies from both feature recognition and feature-based modeling to convert 2-D legacy CAD to either solid models or to a more intelligent feature-based CAD.

The system is currently under implementation at the Virtual Environments Lab of UTA and Automation and Robotics Research Institute (ARRI). The conversion of 2-D orthographic views to solid models is complete and is an integral part of ARRI demos. The features part of the system is a large group research with several students working on different aspects of the problem.

7 REFERENCES

Ganesan, R. (1994) Architecture of a flexible feature-oriented CAD framework for interactive product modeling (FlexiCAD). Masters Thesis, The University of Texas at Arlington.
Ganesan, R. and Devarajan, V. (1994) FlexiCAD: An architecture for integrated product modeling and manufacture by features. *Fourth International Conference on Computer Integrated Manufacturing and Automation Technology*, New York.
Pratt, M. J. (1993) Applications of feature recognition in the product life cycle. *International Journal for Computer Integrated Manufacturing*, **6**, 13-19.
Meeran, S. and Pratt, M. J. Automated feature recognition from 2-D drawings. *Computer Aided Design*, **25**, 7-17.
Balachander, R. (1994) Form features extraction from 2-D orthographic views. Masters Thesis, The University of Texas at Arlington.
Case, K. and Gao, J. (1993) Feature technology: an overview. *International Journal for Computer Integrated Manufacturing*, **6**, 2-12.
Salomans, O. N. et. al. (1993) A review of research in feature-based design. *Journal of Manufacturing Systems*, **12**, 113-132.
Sundaramurthy, V. and Devarajan, V. (1993) Reconstruction of 3-D CAD models from 2-D orthographic CAD views. *International Conference on CAD, CAM, Robotics and Autonomous Factories*, New Delhi.

8 BIOGRAPHY:

Rajan Ganesan is Research Associate at the Automation and Robotics Research Institute, a premier automation research, education and technology transfer center. He is currently pursuing a Ph.D. in Electrical Engineering with research interests in feature-based CAD for virtual prototyping. He earned his M.S.E.E. from the University of Texas at Arlington in 1994.
Dr. Venkat Devarajan performed his doctoral dissertation on image compression. Thereafter he worked for LTV Missiles and Electronics Co. where he led the development of US Navy's mission rehearsal system called TOPSCENE. This system uses real world input imagery and terrain elevation data to give pilots real time fly-through capabilities over country-sized data bases. Since 1990, Dr. Devarajan has been at the Electrical Engineering department of the University of Texas at Arlington where he conducts research into Virtual Reality applications including Virtual Prototyping.

12

A Reference Model
for Product Information Sharing in
Concurrent Engineering Environment

T. De Martino*°, B. Falcidieno*, F. Giannini*

* Istituto per la Matematica Applicata, C.N.R.
Via De Marini 6, 16149 Genova, Italy
tel. +39 10 64751 Fax. +39 10 6475660
Email falcidieno(giannini)@ima.ge.cnr.it

° Fraunhofer Institute für Graphische Datenverarbeitung
Wilhelminenstraße 7, 64283 Darmstadt, Germany
tel. +49 6151 155208 Fax. +49 6151 155299
Email teresa@igd.fhg.de

Abstract

In this paper a reference model for sharing product information in a concurrent engineering environment and suitable for virtual prototyping is presented. The model, called *Feature Kernel Model*, is feature oriented. It is able to support different functional viewpoints, i.e. different application dependent sets of features, and propagation of modifications from one view to another one, thus, allowing different experts to work simultaneously on the product definition and analysis.

Key Words

CAD, Concurrent Engineering, Feature-based Modelling

1. INTRODUCTION

Increasing global competition leads industries to bring competitively-priced, well-designed and well-manufactured products to market in timely fashion. Since decisions made during the design stage can have significant effects on product cost, quality and lead time, increasing research attention is being given to speed up part development taking into account all the considerations related to the product directly at the early design stage (Gupta et al. 1993). The design of a product part is the result of an interactive decision-making activity organised to conceive the idea for, to prepare the description of, and to produce the plans by which resources are converted into artefacts or devices to meet the human needs (Peña et al. 1992). Traditionally, it is the result of the work of a team of experts. Firstly, the designer produces a part description according to what may be wanted and which needs should be satisfied. Then, this description is evaluated by the experts of the different activities involved in the production and use of the part. This evaluation activity makes designer go back to revise his design according to the suggestions of the different experts, thus requiring long time for the product definition.

In order to avoid or at least to optimise this definition loop, all production requirements should be considered during the initial design stage. A help for the solution of this problem comes from the great advances gained in networking technology, groupware graphical user interfaces, multimedia, and the decrease of computing cost, that make possible real-time communication among team members working in a heterogeneous computer equipment and geographically distributed environment. These tools give the possibility to create a virtual environment in which all the experts involved in part definition and production can work together at the product development by performing all the necessary analysis in parallel and thus immediately returning suggestions of part modifications that can be simultaneously evaluated by the different members of the team and then accepted, refused or negotiated.

For the development of a system based on such new design perspective, i.e. working in a concurrent engineering environment, it is possible to identify two main key points:

- the problem of data sharing between the users of the system: all the knowledge related to the product development has to be stored and should be visible, but each user should have the access rights only to what is meaningful for his specific application. Moreover, data modifications should be handled in a consistent way by the different viewpoints;
- a concurrent processes management: the processes that contribute to the product specification and have access to the product data base have to be synchronised, taking into account that the modifications should be proposed to each of the experts who can be interested in, and then, if accepted, they should be propagated to all the contexts.

The first point is a big problem, since it involves the definition and the modelling of all the knowledge related to the product life cycle, from information strictly related to the object, like its shape and its final behaviour, to information more related to the enterprise in charge of its construction as, for instance, manufacturing processes, tools, materials and other resources that are available. All these data should be modelled and encoded in a way usable by the involved experts. This means that each expert should have the possibility to look at the object model in terms of elements that are meaningful for his specific context.

The second point is related to the problem of synchronising the access to the part description for performing analysis and inserting modifications, since the different actions may influence each other. A good co-ordination is necessary to give rise to a high-quality product specification in the shortest turnaround. A particular attention should be paid to the part model access: it is important to give a priority order of access to the experts and to define tools for negotiating the modifications to be performed. This means that the evolving design must be visible globally and its ramifications to any interested team member must be highlighted (Reddy et al. 1993).

As a consequence, what seems crucial in a concurrent engineering design perspective is a complete but also high level definition of the part model and a mechanism supporting inter process communication.

This paper focuses on the requirements that should be satisfied by the product representation in a concurrent engineering environment. In particular, a feature oriented reference model is presented, which is suitable for different application dependent views and for supporting a mechanism that propagates modifications to the different contexts involved in the design activity.

2 FEATURE BASED MODELLING FOR CONCURRENT DESIGN

The realisation that many important design tasks could actually be performed with the help of CAD systems has led to the development of several geometric modelling techniques. However, traditional solid models can only capture the information about the geometry of the parts. Furthermore, these models are built by non-intuitive means, such as Boolean combinations of geometric primitives or Euler operators, completely unrelated to the real engineering significance of the parts. As a consequence, the structure of these geometric models and their information content not always reflect the design intent and meet the application needs.

The request for more powerful models has led to the emergence of *feature* technology. Since features of a part are shape elements with some functional meaning, they have been identified in the engineering community as the key entities that provide a convenient language for modelling product parts using both geometric and functional information and that allow each application to have its own view of the product. Depending on which analysis has to be performed, the engineering meaning of a feature may regard the function that the feature serves, how it can be produced, which actions its presence must initiate, etc. (van Houten 1991).

This means that the concept of feature strictly depends on the application context, thus, different sets of features are necessary for describing a product from the different applications viewpoints. Moreover, considering multiple viewpoints, the same part of the object can be interested by more than one feature, thus, in general, these feature sets are not completely independent, in the sense that features identified in a specific domain may be partially or fully mappable to features in other domains (Shah 88).

As an example, the object depicted in Figure 1.a is described in terms of features partially overlapping, like slot1 or slot2, meaningful in the machining context (see Figure 1.c), which have some common boundary entities with the T-rib feature, meaningful for the assembly context (see Figure 1.b).

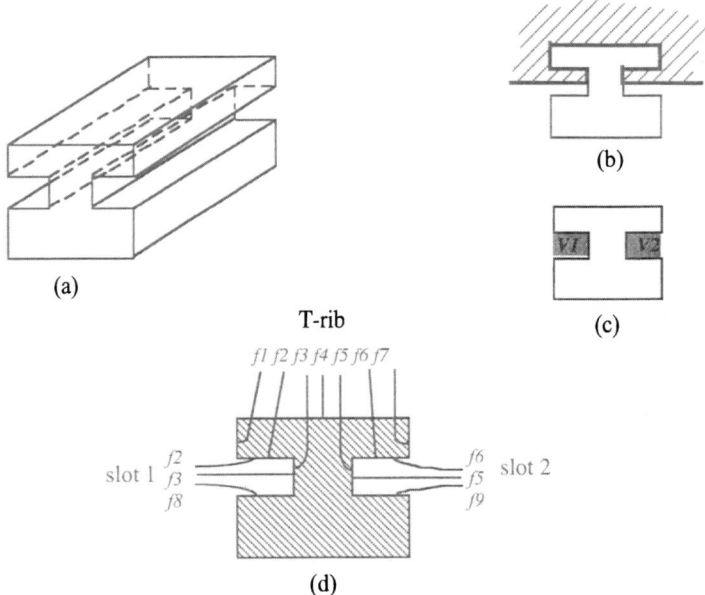

Figure 1 An object (a), its features meaningful for assembling (b) and for machining operations (c) and the classification of feature boundary entities (d).

The problem with the concurrent engineering design perspective is that all the features of interest should be represented at the same time in order to allow each expert to contribute at the right moment to the product design.

Crucial for the solution of this problem is a mechanism for mapping feature based models from one context to another, allowing the propagation of modifications between the different models and the communication between processes.

In our opinion, the best solution is to define a common shape feature oriented model, where all the geometric and topological information about shape features are collected and organised in a way that allows dynamic management and easy updating. This representation can be considered the reference model for the product data base, since the context dependent technological information is externally stored and refers to the associated form features represented in the model.

Concerning with the management of this structure, its application interface includes a mechanism that allows each expert to filter only the information of interest for his specific action and a mechanism of communication with all the other team members for the propagation of modifications.

2.1 The reference geometric model

A geometric model able to support concurrent design and virtual prototyping, must satisfy the following properties:

- to be able to support multiple levels of geometric and functional abstraction with different amount of details in order to capture part information at different stages of the design process;
- to be multiview, in order to be mapped into different application contexts;
- to be general, in order to represent any kind of aggregation of geometric entities;
- to be expressive enough to completely represent any information about feature structure (simple, compound, interacting,..);
- to be independent from a specific solid modeler;
- to be extensible in order to meet the requirements of future applications;
- to be able to support different methods of feature-based modelling: the top-down approach, when feature information is directly available and can be used to design the part, but also the bottom-up approach necessary when features are the result of some modelling operations and have to be recognised and explicitly represented in the model.

Based on these requirements, we have defined a new representation scheme for modelling with form features called *Feature Kernel Model* (FKM).

The structure of this model is an adjacency graph, where each node corresponds to a set of boundary entities, and each arc connecting two nodes represents an adjacency relationship between the corresponding object parts. In this structure, set of components can be grouped together in order to represent the features of the object. The components which belong to more than one feature, are labelled and express the intersection between overlapping features. The FKM is able to represent features of non-homogeneous dimensions, in particular volume and surface features, but also open face sets. In other words, it can handle *non-manifold geometry*, that allows to better satisfy the requirements of the different applications. For example, it is meaningful to represent features with volumes for machining analysis, since in this way they can be associated directly to the extent of material to be removed and to the corresponding machining processes. On the other side, surface features are more suitable for boundary-oriented applications, that is those applications that must associate information to a face, an edge or to a set of them. For example, in the handling context pairs of parallel faces can be relevant for robot arms, while in the assembling context, the mating features are the contacts among assembly components and can be represented by faces or edges of the object. This multiple choice for feature representation permits also to describe compound as well as intersecting features, keeping hierarchical relationships between components, where meaningful.

The main advantage of the model is its dynamic structure that can be updated through the use of operators that insert or delete a feature, taking care of the model consistency and isolating the regions of interest for more than one viewpoint, where there can be conflict between different functional interpretations.

2.3 Facilities for Collaborative Design

The Kernel Feature Model is a good support for the complete description of the physical components of the product in the collaborative design activity. It is the means which relates the object geometric description to the domain-specific features, that incorporate process-related information that has not to be lost.

In order to produce a neutral reference model, the technological information is stored in external structures and refers to the features represented in the model. In this way, the semantic, i.e. the high level interpretation of the part, is separated from the low level description, i.e. the geometric and topological information.

In Figure 2, it is shown how the information related to the object depicted in Figure 1, is organised. The geometric and topological information is represented within the object B-rep and in the FKM graph, while the technological and functional information is separately stored and refers to the groups of components in the FKM.

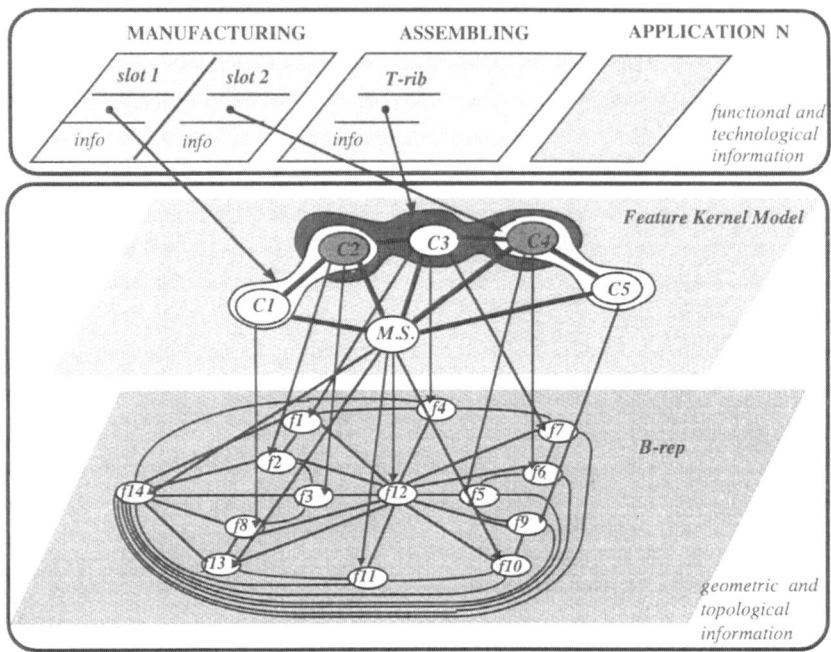

Figure 2 The different levels of information for the object in Figure 1.

Each application domain maintains visibility on the global model but it has access rights only to the features that derive from that domain.

The set of geometric entities that are referred by more than one context, are fundamental for the communication between the different viewpoints. The intersection between feature are explicitly represented in the FKM and are associated to special nodes, which express an overlapping relationship between features. In Figure 2, for example, the nodes C2 and C4 express respectively the boundary intersection between the slot 1 and the T-rib and between the slot 2 and the T-rib.

When there is the need to modify a feature, a check on its components set is performed in order to see if the feature is fully or partially overlapping with other features. Through the components expressing the features intersection, it is possible to communicate the modification to the other contexts interested in the same part. During this phase, all the processes that are not involved in the negotiation can continue their work, if it does not influence the part proposed for modification.

In Figure 3, some examples of modifications, which can be performed on the object depicted in Figure 1, are shown. From the context, where it is meaningful to describe the object in terms of two slot features, as for machining processes, the expert can require to modify the dimensions of the width *d1* of the slots (Figure 3.b), or the depth *d2* (Figure 3.c). In both cases, the proposed modifications affect the T-rib in a different point of view, the proposed changes should be validated in the other interested contexts, before their acceptance. As a matter of fact, an analysis for assembly will accept only the situation in Figure 3.b, since the situation in Figure 3.c does not fully satisfy assembly functional requirements.

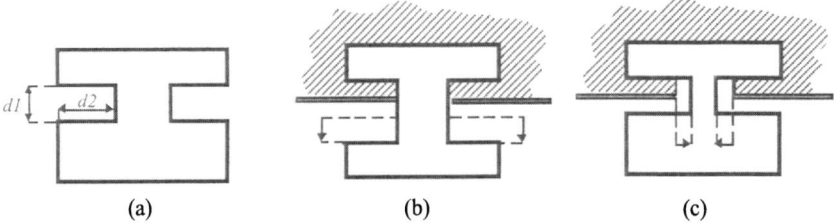

(a)   (b)   (c)

Figure 3 Examples of feature modifications.

3. CONCLUSIONS AND ONGOING RESEARCH

In this paper a model for representing shape features of different dimension has been described. This model provides a good means for sharing product information and for supporting the propagation of modifications in concurrent design and virtual prototyping . It has been implemented in C++ language on the top of the non manifold boundary model provided by ACIS™ geometric modeler tool kit (ACIS 1994). The implementation is currently in progress.

The model interface supports operations like creation, deletion and modification of the feature entities, it also provides tools to calculate feature intersection and dependencies. Through this model it is possible to define a mechanism that permits each application to have visibility on the global model and to have access rights only on what is meaningful in the specific context. The access on the information can be performed at different levels of specification, from the functional interpretation to the low level geometric description.

A modifications proposed by one viewpoint is immediately highlighted to the other viewpoints, through a propagation mechanism that works on the structure, and then performed only after the acceptance of all the other experts.

Since this model has been developed with the aim to be extensible by applications, an attribute handling mechanism is available, that gives the possibility to each application to define its own set of attributes and to associate them to the shape features in the model (De Martino et al. 1994a).

Our future work will mostly focus on the integration of the FKM in a system that combines different feature modelling techniques (i.e. design by features and automatic feature recognition) sharing the same product database (De Martino et al. 1994b), and on the extension of the model for better representing parametric information.

4 ACKNOWLEDGEMENTS

This work is partially supported by the Italian National Council of Research (CNR) within the National Project "Applicazioni della matematica per la tecnologia e la società" and by the EC within the Human Capital & Mobility Project N. ERBCHBGCT930380.
The authors are grateful to Stefan Haßinger, Dr. Jivka Ovtcharova and Dr. Joachim Rix for the helpful discussions on the topic.

5 REFERENCES

ACIS (1994) *ACIS Geometric Modeler Interface Guide,* Spatial Technology Inc., 2425 55th Street, Building A, Boulder, Colorado 80301

De Martino T.,Falcidieno B., Gamba F. and Giannini F. (1994a) Feature Kernel Model: a reference model for integrated feature based modeling systems, *proc. AICA'94,* Palermo, Italy, October 1994

De Martino T.,Falcidieno B., Giannini F., Haßinger S. and Ovtcharova J. (1994b) Feature based modeling by integrating Design and Recognition approaches, CAD August 1994

Gupta Satyandra K. and Nau Dana S.(1993) Generation of Alternative Feature-Based models and Precedence orderings for Machining Applications, *2nd ACM Solid Modeling,* May 1993, Montreal, Canada, pp. 465-466

van Houten F.J.A.M. (1991) *PART: A Computer Aided Process Planning System,* PhD Thesys 1991

Peña F. and Logcher R. (1992) Capturing design assumptions for rational coordination, integration and negotiation in the design process, *Computing Systems in Engineering,* Vol.3, No. 6, pp.651-660,

Reddy Y.V. Ramana, Srinivas Kanakanahalli, Jagannathan V. and Karinthi Raghu (1993) Computer Support for Concurrent Engineering, *Computer,* January pp.12-15

Shah Jami J. (1988) Feature transformations between application-specific feature spaces, Computer-Aided Engineering Journal, December, pp. 247-255

6 BIOGRAPHY

Teresa De Martino is currently persuing her PhD in Engineering at the Technische Hochschule in Darmstadt, Germany. She received her Master Degree in Mathematics from the University of Genova in 1990. Since then, she is doing doing research, at first at the Istituto per la Matematica Applicata del C.N.R., Genova, and since 1993 at the Fraunhofer Institut für Graphische Datenverarbeitung in Darmstadt, Germany, within a cooperation agreement between the two Institutes. Her research interests include Geometric and Feature-based product Modeling, Concurrent Engineering.

Bianca Falcidieno is currently research director at the Istituto per la Matematica Applicata of C.N.R.. She now acts as head of the Computer Graphics Group and as leader of national and international projects on Computer Graphics and its applications. She has written more than 70 refereed technical publications. She is member of the Editorial Board for Computers and Graphics and Computer Graphics Forum, and coeditor for the Technical Report series in EUROGRAPHICS. As a member of EUROGRAPHICS, ACM, AICA, IEEE, IFIP, she served on numerous program committees for several international conferences. She is coordinator of the national activity in Computer Graphics of the Italian Association for Computer Science (AICA). Her research interests include Computer Graphics, Geometric Modelling and Computational Geometry.

Franca Giannini graduated from the Department of Mathematics, University of Genova in 1986. She joined the Research and Development Group at Italcad Technologie e SISTEMI. In that period she participated to some national projects and to ESPRIT 865 . From1989 she is doing research at the Istituto per la Matematica Applicata of C.N.R., Genova. She has been involved in some national and international projects and in industrial argeements on user interface and geometric modeling.
Her research interests regard solid modeling and product modeling using Boundary Representations in Concurrent Engineering environment.

PART THREE

Architectures for Distributed Systems
(Architectures for Virtual Prototyping Environments)

13

Supporting Multidisciplinary Teams in Concurrent Engineering

J.H.Maxfield, L.T.P.Fernando and P.M.Dew
The Keyworth Institute,
Virtual Working Environments Laboratory,
School of Computer Studies,
University of Leeds,
Leeds LS2 9JT, England.
Tel: +44 532 336819
email: {max, ltpf, or dew}@uk.ac.leeds.scs

Abstract

This paper describes an architecture for supporting real-time collaboration in concurrent engineering within a distributed virtual environment. This architecture integrates a multiperspective distributed virtual environment with a standard product model and supports interaction among users through shared objects. The initial implementation of this architecture supports accurate assembly modelling and kinematic simulation for virtual prototypes and runs on a network of SGI Indy workstations over an ATM network. This environment enables designers to test the assembly and disassembly task within a distributed virtual environment. The realistic manipulation of the assembly models within the virtual environment is supported by constraint-based 3D manipulation techniques. The shared objects, that support collaboration, encapsulate product information in a standard format based on the developing draft ISO standard STEP (a standard for the exchange of product data).

Keywords

Distributed virtual environments, concurrent engineering, STEP, constraint-based assembly modelling, shared objects.

INTRODUCTION

The pressures for modern design and manufacturing companies to remain competitive in today's world markets has led many to investigate the adoption of concurrent engineering to reduce the lead time for new products and improve their quality. Concurrent engineering is a systematic approach to the integrated concurrent design of products and related processes, including manufacture and support. When using concurrent engineering specialist knowledge and expertise from downstream tasks of a sequential design process, such as manufacturing and maintenance, is introduced during the early design phases. The largest percentage of design and manufacturing costs are allocated during the first stages of a project. As a result decisions made during these early stages are the most difficult and expensive to correct at a later time. The basic philosophy behind concurrent engineering is to encourage the consideration of as many product development issues as possible, during these crucial early phases. Such considerations should result in fewer unexpected problems during subsequent development and consequently fewer design iterations, reduced development time and costs together with a better quality design.

Given sufficient resources, experts from several product development stages can be introduced by simply making the individuals available for consultation during the design process. An effective way of managing such consultation is through the creation of multidisciplinary teams. When employing such teams, the members will usually need to convene in a single location for regular meetings to review progress and make important decisions. Such meetings will normally be arranged in advance to give, the individuals involved, time to prepare necessary documentation and travel to the location of the meeting. During the meetings the experts from different areas of product development will be able to offer advice and suggestions from their own perspectives and can ensure that important issues are not overlooked. However, the physical co-location of a team is no longer a trivial issue because many companies are now exploiting the opportunity to trade in the global world market and consequently are becoming more decentralised in their activities. The travel, time and expense lost as a direct consequence, can inhibit the regularity and spontaneity of team interactions and act as a direct barrier to the successful implementation of concurrent engineering techniques.

High speed networks, distributed virtual environments and multimedia communications are now commercially available and have a potential to support interaction between geographically dispersed users in a much more dynamic and synchronous nature than traditional file exchange and electronic mail. Such advances have made it possible to develop collaborative working systems supported by networked computers in which geographical dispersion is transparent. A team that conducts its work in such a way is *virtually co-located* and thus called a *virtual team*, since interactions only require the participants to be available at the same time, but not necessarily at the same place.

This paper discusses a system architecture that supports interaction between members of a geographically dispersed multidisciplinary virtual team who are engaged in product development activities. We call such a system a *Distributed Virtual Engineering Environment*. In particular the environment allows the team to interact, reach a common understanding of the problems quickly and then make decisions regarding those problems from multiple perspectives using a distributed virtual environment. Each user accesses the distributed virtual environment through a user configurable interface which has been designed to allow the easy integration of further engineering applications to support different engineering perspectives. The applications interoperate using shared objects that encapsulate product information in a standard format based on the developing draft ISO standard STEP (a standard for the exchange of product data).

BACKGROUND

Concurrent engineering has received a great deal of attention in the engineering and management research communities since its introduction over a decade ago and much work has been published on its advantages over more traditional sequential design processes. Major research projects such as DARPA DICE (Cleetus 1993), SHARED (Wong *et al* 1993) and PACT (Cutkosky *et al* 1993) are all addressing issues involved in computer support for managing and co-ordinating multifunctional, cross-disciplinary teams in this context. However these projects have mainly concentrated on the asynchronous activities of such teams. Although this is important, teams must also be able to communicate effectively during meetings. This is especially difficult if the team is geographically dispersed. Certain issues that still need to be addressed for synchronous working in concurrent engineering have been indicated in several papers (Prasad *et al* 1993 and Clausing 1993). In summary these issues are :

- The establishment of virtually co-located multidisciplinary teams with integration of and mapping between individual view points.
- The ability to share and exchange standard product data and tools.
- The ability to perform collaborative decisions in a single trade-off space with a common understanding of the problems.

Some concurrent engineering projects are beginning to address some of these issues. The SHARE (Toye *et al* 1993) project is addressing negotiation and trade-off in real-time through videoconferencing. Support for synchronous collaboration in virtual teams has also been explored within the DICE project through the use of MONET, a teleconferencing system, and COMIX, a system for transparently sharing X-Windows applications (Cleetus 1993). However, these projects are not investigating the use of distributed virtual environments for supporting collaboration over a virtual prototype of a product. Neither do any of the collaborative tools that are used tackle the problems of real-time multiperspective meetings as required by multidisciplinary teams.

There now exists many commercial and non-commercial tool kits for the creation of distributed virtual environments, for example dVS (Division), World Tool Kit (Sense8), MR-Tool Kit (University of Alberta), DIVE (Swedish Institute of Computer Science). Many research projects are using such tool kits to develop distributed or single user virtual environments for specific engineering applications. For example, Bayliss *et al* (1994) are investigating a virtual manufacturing environment consisting of a machine shop in which engineering components can be made. The VirtuOsi project is investigating the organisational issues involved in the formation of a virtual factory (Benford 1994). NASA have conducted a successful experiment in which they used a virtual prototyping system called Preview to assist the correction of the Hubble Space Telescope (Hancock 1993).

These projects are just a few examples of many that are investigating support of specific engineering applications in virtual environments. However, the effective integration of these engineering applications in a multiperspective distributed virtual environment has not been addressed. The integration of international product data standards such as STEP in distributed or even single user virtual environments has not yet progressed beyond the ability to access IGES or DXF (AutoCAD) geometric definition files. Another limitation in current virtual environments is the lack of support for accurate positioning of objects in 3D space. At present, most of these systems employ crude collision detection techniques based on bounding boxes. Such techniques fail to provide powerful and accurate 3D manipulation methods necessary for exploiting the virtual environment technology for engineering applications such as solid and assembly modelling. Techniques such as accurate constraint-based 3D manipulation (Fa *et al* 1993) are essential for supporting the realistic manipulation of solid models within virtual environments, especially when dealing with the integration of engineering application.

THE DISTRIBUTED VIRTUAL ENGINEERING ENVIRONMENT

The distributed virtual engineering environment has been developed to satisfy a number of requirements which will be discussed in section 3.1. The environment is based on a number of concepts which are outlined in section 3.2. A description of the detailed architecture and its current implementation is presented in section 3.3.

Requirements

A fundamental requirement for synchronous collaboration between participants of any distributed meeting, is the ability to share information in *real time*. This shared information will then form the basis for discussion within the meeting. A second fundamental requirement that is specific to collaboration in multidisciplinary teams is the ability to make collaborative decisions in a single trade-off space with a common understanding of the problems and with integration of and mapping between individual *perspectives*. A perspective defines a context within which the shared information can be manipulated in a meaningful way by an individual.

In addition, several requirements have been considered during the implementation of the distributed virtual engineering environment. These are that the architecture should :

- Be open and extensible so that different perspectives and engineering applications can be integrated easily.
- Have the ability to support user collaboration through sharing product data in real time.
- Be built on emerging and established standards where possible, including the product modelling standard STEP.
- Have the potential to be scalable for large meetings with a dynamic number of participants.
- Provide support for quality of service over high speed networks such as ATM, for real time interaction and communication.

A Conceptual View Of The Architecture

This section introduces a number of fundamental concepts that are used in the DVE environment to satisfy some of the requirements outlined in the previous section. For each concept, a high level description of the mechanisms employed by the DVE environment is also presented. A conceptual view of the DVE architecture is illustrated in figure 1.

The users *perspective* defines what information is meaningful to them and how they will manipulate that information. The DVE environment uses the concept of an *information mask* to define what information is meaningful to a user with a given perspective. The mask acts conceptually as a filter that defines what subset of the information in the shared space a user can actually access. The mask can also be used to filter an entire product model to define what information the user may add to the shared information space.

The perspective also defines how the user will manipulate the information they access. To support this, each user of the DVE environment has a user interface that they may *configure* with the operations they wish to perform and the visualisation style they wish to use. These operations are actually performed on the shared information by a set of distributed engineering applications that are invoked and controlled in the background automatically by the DVE environment. The users may modify their perspectives at any time during a meeting by changing the configuration of their interface. This will imply a change in the users information requirements and consequently a change to the users information mask. Such changes may also affect in the set of background engineering applications that support their perspective. The concept of the perspective and how it is supported within the DVE environment is illustrated in figure 2(a).

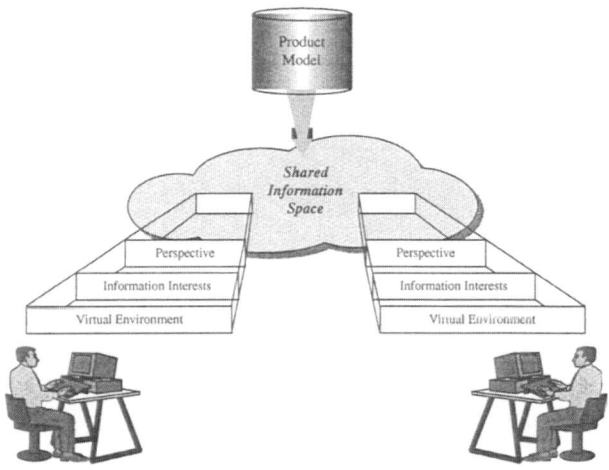

Figure 1   A Conceptual View of the DVE Environment

As a meeting progresses the discussion may change focus and more information may be added to the shared space from the product model. The shared information space can potentially contain a huge amount of information which may be meaningful to a user, and therefore satisfy their information mask, but may not all be of interest. The DVE environment allows a user to choose which subsets of the available information they are interested in and *register* this interest with the environment. Users may change their interest as the meeting progresses by registering an interest in further sets of information or discarding an interest in a particular set of information. Collaboration can occur when users register an interest in the same information, i.e. their particular interests intersect. Users may still collaborate over information that they do not share an interest in, if the information they view is related. In this case, changes can be propagated through the relationships using a constraint manager that will maintain a relationship (or dependency) graph for the information within the shared space. The users will therefore still be able to see changes that affect their information.

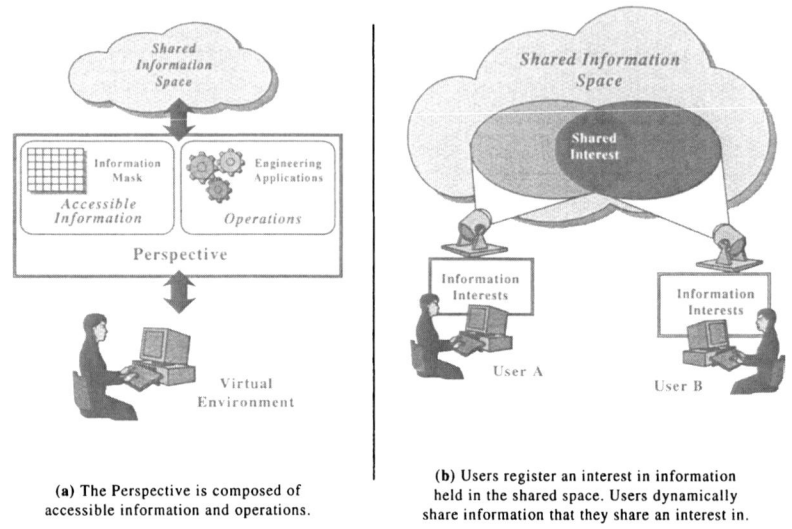

(a) The Perspective is composed of accessible information and operations.

(b) Users register an interest in information held in the shared space. Users dynamically share information that they share an interest in.

Figure 2   The Fundamental Concepts Employed by the DVE Environment

Internally the DVE environment maintains either a *passive* or *interactive* interest in an particular area of the shared information space for each user. A *passive* interest is automatically registered in all information that the user accesses. An *interactive* interested is registered in any information that the user attempts to modify or manipulate in any way. By distinguishing between passive and interactive interests in this way, locking techniques can be employed by the environment to eliminate any chance of inconsistency in the shared information space. The concepts and mechanisms of interest registration are illustrated in figure 2(b).

A Detailed View Of The Architecture

A layered view of the detailed DVE architecture is illustrated in Figure 3. At the heart of this architecture is the shared information space called the *Product Data Sharing System* (PDSS). This system makes extensive use of a library of sharable objects that can be instantiated and populated with product information. These objects and the mechanisms employed by the PDSS for sharing these objects are discussed in the following sections. The users access the shared information through a graphical user interface that visualises the information in a virtual environment. They may then directly interact with the information and perform operations supported by a set of engineering applications. The following sections discuss this interface and the underlying mechanisms that allow a user to configure their interface to suit their required perspective. Finally this section also discuess the integration of constraint based 3D manipulation within the architecture.

The Product Data Sharing System

This section discusses the structure of the sharable objects that are instantiated from an object library when required. It also descibes the distribution and sharing mechanisms used within PDSS.

The shared product information must be accessible by many different applications and therefore a neutral, usage independent representation for the information is important. A *product data model* is an integrated set of data schemata that describe such a standard format and content for storing product data. An instance of the product data model will contain data regarding a specific product and is called a *product model.* The most important issue here is that the information in a product model is in a neutral format which is totally independent of the way in which it is used. This makes a product data model an ideal format for the storage of product information within the objects used by the DVE environment.

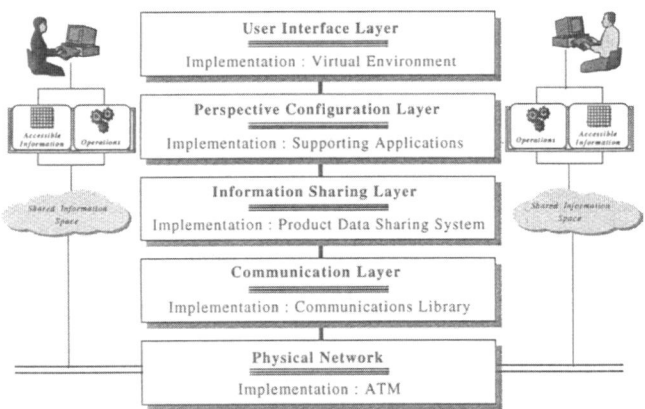

Figure 3   A Layered View of the DVE Environment

The objects generated and used by the DVE system, encapsulate product information and methods for validating the correctness of that information. This is held in an area of the object called the *payload.* Each object in the library consists of a payload (the information that will be shared) and useful methods for encoding and decoding the payload to assist in sharing the object. These methods are derived from a basic object type by all of the objects as illustrated in figure 4.

Given objects that encapsulate product information structured this way, it is feasible for this information to be exchanged, without any translation, with a product model database, if a common product data model is used to structure the information in both. Therefore DVE environment users can add information to the environment from a persistent product model database during the meeting. If the users also wish to keep any changes made as a result of decisions made in the meeting, then the modified information can be transferred back to the product model database at the end of the meeting.

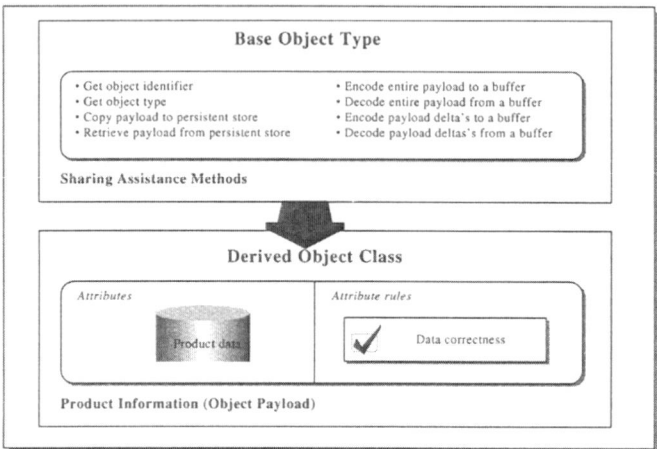

Figure 4   The Structure of a Typical Shared Object

At the heart of the PDSS is an *Object Manager*. The manager controls the instantiation, population, shared access and destruction of the objects. It does not distinguish between user interfaces and engineering applications but treats all entities that wish to share product information as *clients*. The object manager will distribute copies of the objects to its client on request. Each client has a *PDSS wrapper* that allows the client to register an interest in some product information using *registration* services and then view and edit the information using a standard set of *enquiry* and *modification* services. The wrappers will also inform the interface or application of any changes to its' local information made by another client, through a set of *notification* services. The PDSS wrappers effectively hide the sharing mechanisms and communicate with the object manager through a standard PDSS protocol which is currently built on top of a network and platform independent communications library. The internal components of the PDSS are illustrated in figure 5.

The object manager and PDSS wrappers all maintain local copies of the objects. The responsibility for maintaining consistency in the contents of the local copies is shared by the wrappers and the object manager. The PDSS protocol implemented in both the wrappers and object manager defines the interaction that is required to retrieve objects and maintain consistency in them. The main protocol services are summarised as :

- **Initialisation** : Each application and user interface must become a *client* of the object manager to gain access to the shared product information. During the initial exchange the each client will give the object manager a copy of its interest mask. The object manager will then send the client a high level list of components and assemblies that are currently available either, in the environment as virtual prototypes already, or available to retrieve from the product database.

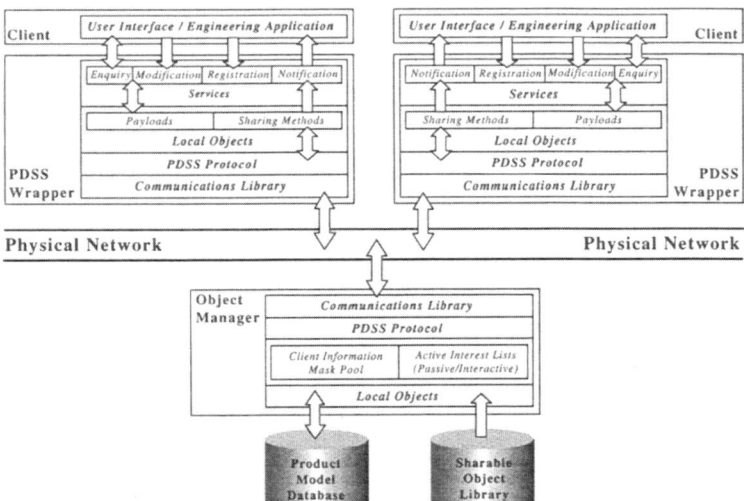

Figure 5   The Sharing Mechanisms

- **Registering interest** : When a user registers an interest in a component or assembly the object manager will first check whether the product information that represents it has already been added to the environment by another user. If it has not then the manager will retrieve the product information from the database and instantiate and populate the required objects to hold the data. The object manager applies the users information mask to the set of objects and gathers together the objects that will be meaningful to the client. These will then be dispatched to the client, where a local copy will be maintained in the PDSS wrapper.

- **Changing the contents of an object** : If a client modifies the product information in any way then the resulting changes or *delta's* in the payloads of the shared objects are propagated to the object manager by the clients' PDSS wrapper. The manager will propagate the deltas on to any other clients that are interested in the modified objects and then update its own local copy. Each interested client will receive the payload deltas, apply them to its own local object copies, and inform the client of the change using a notification service.

The Configurable Virtual Environment and Supporting Applications

The DVE system supports collaborative working between its users through 3D virtual environment interfaces. Each interface provides a view into the shared product information space managed by the PDSS. The interface visualises components and assemblies as 3D solid models and allows a user to manipulate the models in 3D and alter their viewing position by moving around in the environment.

The current implementation of the user interface uses non-immersive technology and does not represent the user in the environment in any way. Therefore to allow users to indicate regions or objects of interest in the environment to other users during a discussion, the interface provides the necessary tools for creating, manipulating, highlighting and labelling shared 3D pointers. Each user can construct a different perspective or personal interface for the virtual environment by configuring the basic interface to support a variety of engineering applications. Multiple perspectives of the same shared information can be supported by the DVE environment because the information is in a neutral and standard format which is accessible by all, but independent of any applications used by a client of the PDSS.

The user configures the virtual environment by choosing from a list of applications those that they wish to be supported by their virtual environment interface. Each application that chosen by a user relates to an engineering application that is integrated into the DVE architecture and invoked as a background process by the DVE environment. The user interface provides an *application toolbar* from which the user may select and adjust operations, modes and ranged values for each application selected. An operation is represented as a push button within the toolbar and may require parameters to be selected within the virtual environment. Examples of operation parameters that can be selected in the current environment include solids, faces, edges, vertices, or points on a face, edge or points in open space. Given the correct type and number of parameters, the operation can be requested by the user. A user selects the modes of an application using toggle buttons that can be switched on or off. Ranged values such as sizes, scales, and tolerances, are adjusted using sliders. Since the current implementation of the user interface is non-immersive it uses standard push buttons, toggle buttons, and sliders provided by the windowing environment to build the application toolbar. A picture of the toolbar (right) and virtual environment (left) taken from the current implementation is shown if figure 6.

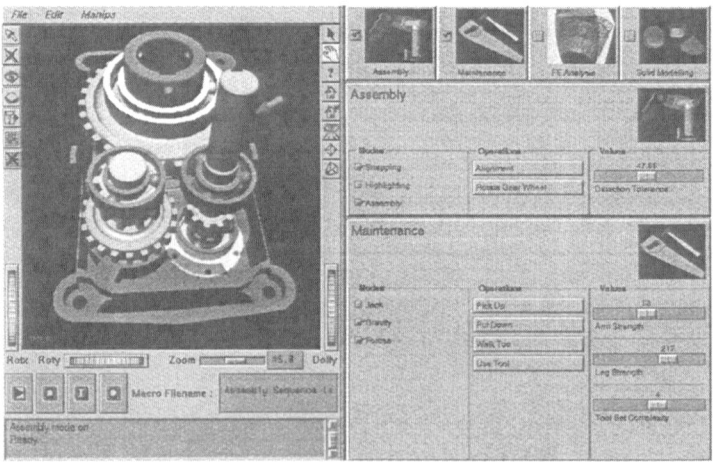

Figure 6   The Configurable Virtual Environment Interface

Figure 7 illustrates how the PDSS is used to support an individuals perspective within the virtual environment interfaces with support from engineering applications. When a user chooses a particular engineering application to support their perspective from the application toolbar, the DVE system invokes the appropriate background application to perform the operations required. The user interface and applications exist as different clients of the PDSS and use a special set of shared objects to communicate with each other. These objects are called *transient objects* because the lifetime of the information they contain is only as long as of the time the user spends within the DVE environment. The user interface informs each selected application of the users interests and requests operations, mode changes, and adjustments to ranged values using a *transient object* that is shared by the user interface and the application itself. Using this technique, an application can be shared between many users, if required, by simply sharing the applications' transient object.

The DVE architecture supports the integration of a diverse range of engineering applications. To integrate an application a configuration file must be created that describes how to execute the application and the operations, modes and ranged values that will be offered to the user in the toolbar. In addition to this file, the application must be made aware of the PDSS wrapper services so that it can access and manipulate the information held in the shared objects.

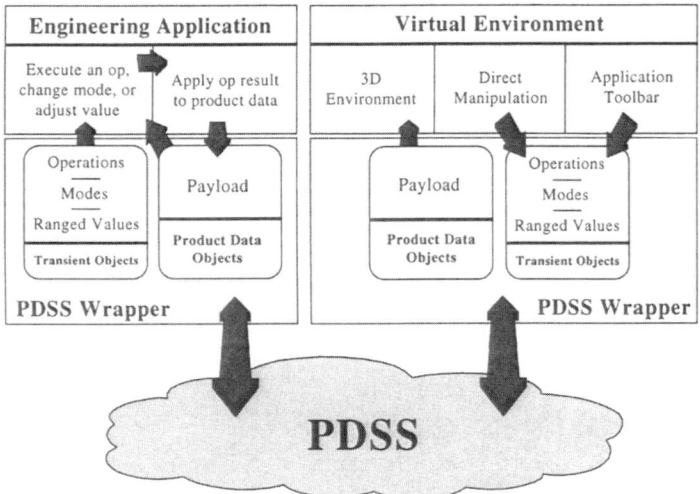

Figure 7 Virtual Environment and Supporting Application Interoperation.

Interactive Constraint Based 3D Manipulation

The management of geometric constraints within the shared information space is supported by a *Constraint Manager*. Constraint based 3D manipulation techniques, previously developed at Leeds (Fa *et al* 1993), are supported within the distributed virtual environment through the integration of this Constraint Manager within the DVE architecture.

The Constraint Manager automatically registers an interest in the information that the users are accessing and monitors the relative positions of the geometric solids as they are manipulating within the virtual environment. *Automatic constraint recognition techniques* are used within the Constraint Manager to recognise geometric constraints between geometric entities from the users' 3D manipulations. Constraints such as against, coincidence, cylindrical fit, gear fit and screw fit are automatically recognised using this technique. Once the geometric constraints are recognised, a technique referred to as *allowable motion* is used to manipulate the under constrained models without invalidating the previously satisfied constraints. For example, when the user position a block on top of a larger block through an against constraint, the allowable motion of the block is derived to be translation and rotation with respect to the bottom block. In the case of a gear fit between two gears, the rotation of one gear is transmitted to the second gear through a coupled rotation. The combination of automatic constraint recognition and allowable motion techniques supports the accurate positioning of solid models in 3D space using 3D input devices, such as a dataglove or spaceball, to assemble complex solid models. Once the objects are assembled, the kinematic behaviour of assemblies are automatically simulated using the allowable motions of the solid models. The Constraint Manager therefore supports the highly interactive virtual environment in which the users can carry out assembly modelling and kinematic simulation of virtual prototypes in a realistic manner.

IMPLEMENTATION AND RESULTS

The current prototype version of the DVE architecture runs on a network of Indy workstations over an ATM network. The architecture has been implemented using C++ and the current virtual environment interface is based on Inventor 2. The shared object library used in PDSS is generated by a compiler that has been developed by the authors to convert an Express schema into a set of C++ classes. The initial implementation uses part of the product data schema that has been developed within a project called MOSES, by the department of Mechanical Engineering at Leeds University (Henson 1994).

The current prototype version of the DVE architecture supports an assembly modelling perspective through the Constraint Manager. It enables engineers to register an interest in assembly models and perform assembly and disassembly operations within the shared information maintained within PDSS. A video conferencing system is run along-side with the virtual environment to support the communication between the members of the product development team. Several case studies are currently being used to test the feasibility of the environment for supporting collaboration among team members including a complex speed reducing gear box.

CONCLUSIONS AND FUTURE WORK

In this paper we have discussed an architecture for supporting multiperspective collaboration over complex interrelated product information for geographically dispersed virtual teams in concurrent engineering. The current implementation of the DVE environment supports assembly modelling and kinematic simulation of virtual prototypes. When used in conjunction with other collaborative tools, the DVE environment allows geometric and assembly problems to be explained clearly and assists a virtual team in reaching a common understanding of problems quickly. The environment will then also allow the users to discuss and test precise solutions in real time. The architecture is currently being evaluated and refined using case studies.

Work is now underway within the Keyworth Institute to integrate further engineering applications into the system to support a wider range of perspectives. These applications include the JACK (Badler *et al* 1993) human factor modeller and a solid modelling kernel based on Parasolid. JACK will allow us to demonstrate a maintenance engineers perspective more effectively by allowing a user to consider the ergonomic and spatial issues involved in real human maintenance, in more detail. The integration of a solid modelling kernel will allow parametric variation of the virtual prototypes in the DVE environment.

A more powerful communication architecture is being developed to support the DVE architecture which will deliver Quality of Service over ATM, synchronisation and directory services (X.500) supporting access to distributed data and users. Experiments involving the scalability of the architecture will study the replication of components of the PDSS and the management of multiple, but related meetings. We are also planning to integrate multimedia information sharing. This will allow us to integrate other collaborative tools into the system and support multimedia annotation of the product information, which can then be stored as interactive minutes for a meeting held in the DVE system.

ACKNOWLEDGEMENTS

The authors would like to thank the members of the DVE project and the Keyworth Institute of Manufacturing and Information Systems Engineering for supporting this work. Thanks also go to Brian Henson and Martin Ashworth within the Department of Mechanical Engineering.

REFERENCES

K. J. Cleetus. (1993) The Virtual Team Framework. In *Concurrent Engineering : Tool and Technologies for Mechanical Systems Design*. NATO ASI Series, Springer-Verlag.

A. Wong and D. Sriram (1993) Shared : An Information Model for Cooperative Product Development. Technical report, Intelligent Engineering Systems Laboratory, MIT, Department of Civil Engineering, MIT, Cambridge, MA, USA, July.

M. R. Cutkosky *et al* (1993) Pact : An Experiment in Integrating Concurrent Engineering systems. In *IEEE Computer*, pages 28-37, January.

B. Prasad, R. S. Morenc, and R. M. Rangan (1993) Information Management for Concurrent Engineering : Research issues. in *ISPE Concurrent Engineering : Research and Applications*, pages 3-20.

D. P. Clausing (1993) World Class Concurrent Engineering. *In Concurrent Engineering : Tool and Technologies for Mechanical Systems Design*. NATO ASI Series, Springer-Verlag.

Toye *et al* (1993) Share : A Methodology and Environment for Collaborative Product Development. In proceedings of the *IEEE Workshop on Enabling Technologies : Infrastructure for Collaborative Engineering*. IEEE Press.

G.M. Bayliss, A. Bowyer, R.I. Taylor, and P.J. Willis (1994) Virtual Manufacturing. In *CSG 94*. Information Geometrics, April.

Steve Benford, John Bowers, Stephen Gray, David Leevers, Tom Rodden, Michael Rygoland, and Vaughan Stanger (1994) The Virtuosi Project. In *VR94*, February.

D. Hancock (1993) 'Prototyping' the Hubble Fix. In *IEEE Spectrum, Special Issues on Virtual Reality, tools, trends and applications*, pages 34-39, October.

M. Fa, T. Fernando, and P. M. Dew (1993) A Virtual Environment for Interactive Constraint-based Solid Modelling. *Eurographics93*, September.

N. I. Badler, C. B. Philips, and B. L. Webber (1993) *Simulating Humans: Computer Graphics, Animation and Control.* Oxford University Press, edition 1, June.

B. Henson (1994) An Assembly Data Model. Technical Report, moses-misc-10, MOSES Group, Department of Mechanical Engineering, University of Leeds, Leeds, LS2 9JT, April.

BIOGRAPHY

John Maxfield is a PhD research scholar working on the Distributed Virtual Engineering Project within the Keyworth Institute of Manufacturing and Information Systems Engineering. His research interests include computer supported collaborative working over high speed networks, object oriented software engineering and distributed virtual environments for engineering applications. John received a first class honours degree in Computer Science from the University of Leeds in 1992.

Dr Terrence Fernando is a lecturer in the School of Computer Studies at the University of Leeds. His research interests include interactive constraint-based solid modelling and distributed virtual environments for concurrent engineering. He leads these research activities within the Virtual Environment Research Group. Terrence received his PhD in Computer Science from the University of Manchester Institute of Science and Technology (UMIST) in 1987.

Professor Peter Dew is the Professor of Computer Science at the University of Leeds and Deputy Director of the Keyworth Institute of Manufacturing and Information Systems Engineering. His main research interests include virtual working environments for virtual organisations and scalable, parallel and distributed computing. Peter is head of the division of computer science where he leads groups in virtual working environments and scalable systems and algorithms

14

Towards a Virtual Prototyping Environment

U. Jasnoch, H. Kress, J. Rix
Fraunhofer Institute for Computer Graphics
Wilhelminenstr. 7
64283 Darmstadt, Germany
Tel. : +49 6151 155 (245 | 212 | 221)
Fax: +49 6151 155 299
Email : (jasnoch | kress | rix)@igd.fhg.de

Abstract

The paper will show and discuss the impacts and opportunities of a Virtual Prototyping Environment. The ability to rapidly prototype a proposed design in a Virtual Environment is becoming a key contributor towards fulfilling the business requirements embodied in a short time-to-market, in cost-effective and high quality manufacturing, and in easy support and maintenance. The major goal of the proposed Virtual Prototyping Environment is the integration of existing technologies in CAD modeling, Computer-Supported Cooperative Work, User Interface design, knowledge-based reasoning, process management and documentation, and virtual reality to offer a distributed desktop, computer-based environment to support the concurrent engineering method as it is applied to the product development process. The paper gives hints to the underlying product data model and the related organization in the data management system. In addition, it describes the first realized prototype and the resulting experiences.

Keywords

Virtual Prototyping, Software Environments, Product Model, CSCW

1 INTRODUCTION

The ability to rapidly prototype a proposed design is becoming a key contributor towards fulfilling the business requirements embodied in a short time-to-market, in cost-effective and high quality manufacturing, and in easy support and maintenance. Reorganizing the design and development process along the lines implied by the concept, *concurrent engineering*, means that advanced information technologies must be taken advantage of. However, most of today's generic, commercial, off-the-shelf technologies have to be extended and adapted to the needs of the product development process. In this context the term Virtual Prototype has been formed, defined by [HKF-93] as „a computer based simulation of a prototype system or sub-system with a degree of functional realism that is comparable to that of a physical prototype". A Virtual Prototyping Environment provides the basis to take the greatest advantage of this technique.

The major goal of the environment is the integration of existing technologies in CAD modeling, Computer Supported Cooperative Work (CSCW), User Interface (UI) design, knowledge-based reasoning, process management and documentation, and virtual reality to offer a distributed desktop, computer-based environment to support concurrent engineering as it is applied to the product development process. The product development process in virtual enterprises leads to virtual prototypes, which are no longer present at a single location. Instead, different parts and aspects of the prototype exist at the different locations of the virtual enterprise and may even change during the product development process.

The opportunities offered by Virtual Prototyping are several:

• to make use of the computer generated object in an electronic form for further communication, like sending it as a file over email or as an object in the distributed object-oriented data base, or using it in an electronic conferencing environment without requiring a physical model or hard-copy printouts.
• to overcome time differences and local distances by CSCW techniques like file transfer, email, object sharing, conferencing with multimedia communication integrated in the end-users desktop station, directly used in the end-users application environment.
• to allow direct cooperation within one application like cooperative editing, collaborative design, or expert sign-off. It would also allow shared "fly-through" with marketing or manufacturing personnel, top management, or key customers.

Another goal of the environment in the area of CAD is to provide new presentation techniques, incorporating multi-media audio and video techniques. In addition, new interaction techniques and user interfaces for 3D modeling and for interacting with the virtual prototype as presented in the virtual environment should be provided. The integration of new presentations of logical devices and new physical devices will be also necessary for manipulating the objects in the virtual prototype system.

This paper is organized as follows. The next chapter describes one basic aspect of a virtual environment, the underlying product model. Afterwards, the proposed environment will be described. A brief description of the first prototype is given in the following chapter. The paper ends with a conclusion.

2   THE UNDERLYING PRODUCT MODEL

Regarding the design and the manufacture of a product as a complex chain of different processes, continuous and homogeneous support by an environment will help to improve the product development process in the direction of enterprise integration based on an underlying product model. A product consists of other products and therefore represents an assembly of numerous different parts normally designed by different engineering groups using different applications. To be able to work together, applications must read, write and exchange their data in a well defined way, whereby each data item has to have a well defined unique meaning.

For a Virtual Prototyping Environment we propose to use an underlying, STEP-based product model according to the methodology developed by TC184/SC4 of the International Standardization Organization (ISO). STEP is an acronym for "Standard for the Exchange and the Representation of Product Model Data" and characterizes the emerging standard for the exchange of product model data [ISO-1]. By means of the STEP product data model and rules for the handling of these models it will be possible to describe product data in a unified way and to share product data between different systems without any loss of information. As STEP is not only focusing on implementations like physical file exchange [ISO-21], the conceptual model should be the basis for product database implementations or application programming interfaces. The STEP product model should whenever possible cover all properties by which a product is characterized during its life cycle. This requires the representation of all data, which are generated in the Virtual Prototyping System during all phases from design, construction, manufacturing, assembly, installation, quality control, utilization, maintenance to dismantling, destruction and recycling of a product. The information model ensures that data remain consistent during the different life cycle phases.

As in STEP, the Virtual Prototyping data management approach uses a product model concept, which is based on coherent partial models. Partial models are well defined subsets of concepts, which are combined based on some commonalties (e.g. geometry or product structure) as indicated in figure 1.

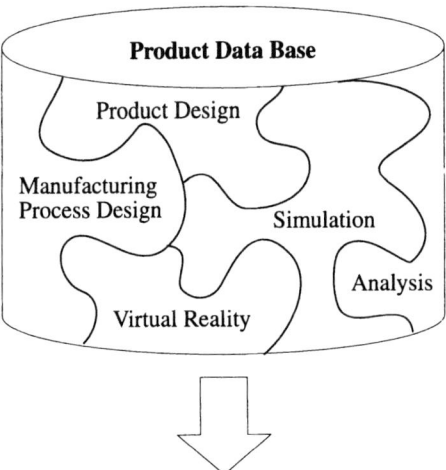

Manufacturing

Figure 1: The Product Data Base and the Partial Models.

The integrated product models cover all aspects of the applications integrated in the Virtual Prototyping Environment. According to the models, the applications are grouped into domains. The underlying product data model consists of the following views:

- Product Design
  This view covers most of the aspects needed for CAD-Systems. It could deal with geometry-based data as well as with feature-based one. Here, the initial parts of the product are described.
- Simulation and Analysis
  The Simulation and Analysis view describes the additional data needed for the analysis domain, e.g. FEM-Analysis, and the simulation domain. In addition to the tested data, the results are stored, to provide the basis for comparing different versions of a product.
- Virtual Reality
  This view deals with all the additional aspects for the presentation of the product data, e.g. texturing information or light models.
- Manufacturing Process Design
  In addition to the simulation and analysis view, the view covers the aspects of the pre-manufacturing process. Based on the data of the other views, additional data for manufacturing, e.g. NC-data, is defined.

These different views and partial models constitute the whole product data base, covering most of the product relevant data. Based on this created, simulated, and presented data, the real manufacturing process of the product could start.

STEP defines not only the off-line data exchange, it also defines the methods and interfaces for the on-line data access [ISO-22]. The document describes two different kinds of access methods:

- the generic interface methods (in terms of implementation: late binding)
- and the specific interface methods (in terms of implementation: early binding).

Using these methods definitions is a basis for the transparent access to the product data model for integrated applications. It ensures the correct semantics and usage of the different data items and provides the possibility of a homogeneous data flow through the different applications.

Besides the schema point of view to the product data base, the object and methods point of view to the data base is also of importance. Figure 2 presents a partial view on the resulting

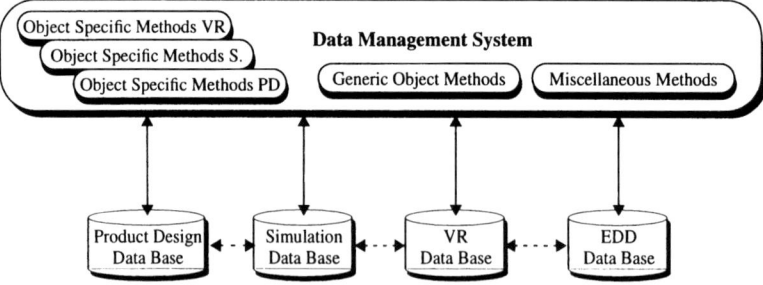

Figure 2: The Data Management System.

data management schema, where some of the schemas are skipped. Basically, this figure shows from the interface point of view that there exist globally usable miscellaneous methods, e.g. those for transaction management, generic object methods applicable for all objects, and schema and object specific methods directly attached to those objects. By providing such interfaces to the product data, uniform access and interchangeability is guaranteed. For logical and implementation reasons, the different views are separated into different data bases. By using a modern distributed object-oriented data base, the access to objects located in another data base is hidden and managed by the underlying data base. These relations are indicated by the dotted arrows.

One major benefit of using an object-oriented data base is the granularity of data access and locking. As the smallest available data item is an object, representing e.g. geometric data like a b-spline, the applications have the possibility to work and therefore lock only with the data which is really needed. Previously, a large amount of data was locked without taking care of the real data needed by the application. By providing this granularity with advanced locking and data management concepts, the parallel working is supported and the turn-around cycles are shortened. In addition, concepts as version and configuration management could be better supported based on the objects. These concepts are especially important in the early design phases.

3 THE ENVIRONMENT

To meet the goals, the Virtual Prototyping Environment consists of a framework and integrated applications, shown in figure 3. In this figure, only some of the existing application domains

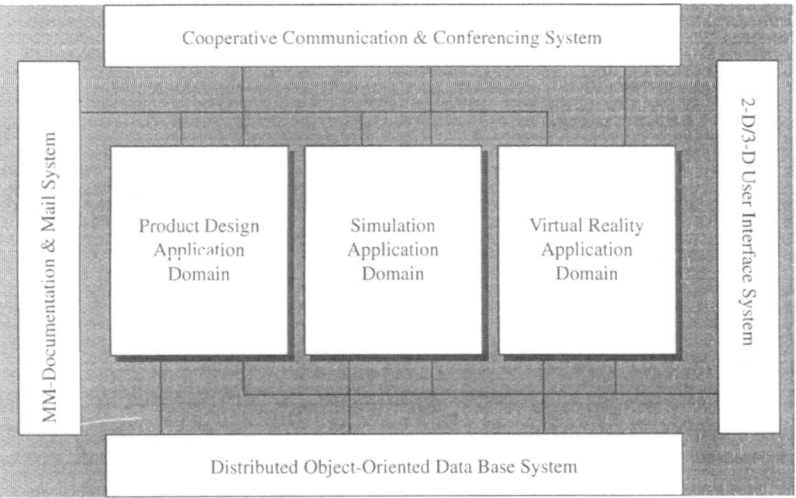

Figure 3: The Virtual Prototype Environment.

are mentioned.

The task of the environment is to offer libraries for the application development process and to provide a set of services for the applications during runtime in the areas of:

- communication and conferencing
  This service provides on the one hand an asynchronous broadcast and point to point message service for communication and on the other hand a synchronous communication service for conferencing. The asynchronous service is used by heterogeneous applications to exchange messages, thus informing each other about a special topic. Here it is important, that the applications are independent and each application decides separately which messages are important for the application and the kind of reaction. The synchronous service is used to support applications in the area of CSCW. Here, the basis services like group management in a conferencing session are provided.
- multi-media documentation and mail
  Integrated documentation becomes an important part of the product development process. The framework provides support for the integrated documentation in the area of dealing and exchanging multi-media objects and to convert and integrate different formats.
- 2-D/3-D user interface
  The user interface system provides services necessary to be able to develop homogeneous 2-D and 3-D user interfaces. Homogeneous user interfaces at the integrated applications raise the acceptance of the end users and help them working in the Virtual Prototyping Environment. Together with the communication and conferencing system, these services enable the development of wrappers to transfer single-user systems into conferencing applications.
- distributed object-oriented data management
  The data management component offers all services for dealing with objects in a distributed environment. The basic goals for the component are to hide the location of an object for the application and to enable a uniform working with transsistent and persistent objects. Advanced concepts like working data bases for user session, enabling concurrent access to objects, and check-in/check-out facilities are also topics of this service. To be open for the future, the data base system provides a STEP-based data modeling facility and the object provides an interface in conformance with the SDAI specification [ISO-22].

The integrated applications cover all aspects of virtual prototyping and are grouped in application domains. The virtual prototyping environment consists of the following domains:

- product design
  This domain consists of several CAD-systems for the design of the product. These systems are either geometry-based or feature-based. The purpose of this domain is to create the initial parts of the product.
- simulation
  The simulation domain embraces applications for the physical simulation and also for the analysis of the product.
- virtual reality
  The modeled products are put into the virtual reality domain to be presented in an advanced virtual environment. The virtual reality domain provides also the environment for the visualization of physical simulations or of analysis.
- manufacturing process design
  The domain covers all applications for the computation of the manufacturing process. Here,

based on the designed and simulated product data, the data for the real manufacturing process are computed.

While these application domains cover mainly all aspects of virtual prototyping, the environment is not limited to these. It could be extended into other directions. Not mentioned in the figure are the domain neutral applications. The purpose of these applications, e.g. the cockpit or the session manager, is to provide functionalities to the user which are needed to deal with the environment, but they are not related to a certain application area.

4 REALIZATION

The Fraunhofer-Institute for Computer Graphics (IGD) and the Fraunhofer Center for Research in Computer Graphics (CRCG) initiated a project for setting up a Virtual Prototyping Environment (see also [HaJa-94], [JKR-94], and [JKSU-94]). In the current state of the prototype, we use also several commercial tools and systems for the environment and in the application domains. The prototype works in a heterogeneous hardware environment, where different applications run on different platforms.

The first prototype supports the messaging between the applications by a communication system. This is based on ToolTalk (SunSoft) and supports messaging in local area networks. The set of used messages is oriented towards those defined in the CAD Framework Initiative (CFI) [CFI-92]. For new applications, we use Tcl/TK (University of Berkeley) for the user-interface. The data management bases on the commercial object-oriented data base Versant (USA). For the underlying product data model, we use the application protocol (AP) Configuration Controlled Design [ISO-203] as the basis. For aspects, which are not covered by this AP, we define extensions for dealing with this data.

For the administration of the environment and serving as the entry point, we implement an

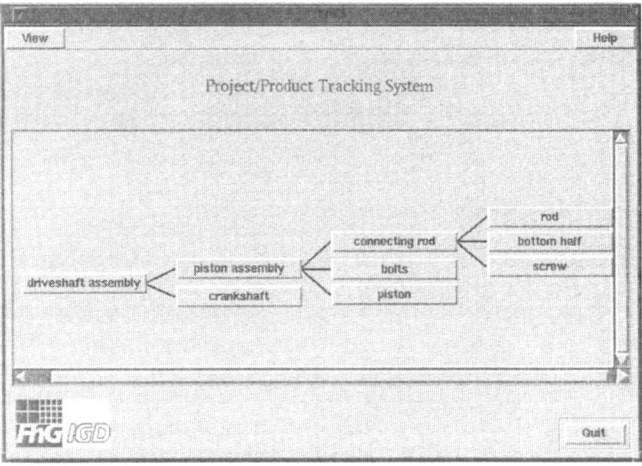

Figure 4: The interface of the Product/Project Tracking System.

application called cockpit. It serves for the designer as the entrance into the Virtual Prototyping

Environment and presents him, besides other information, the list of projects he is participating in. To simplify the work of a designer and to provide control to the data, the entry to a product for the designer is realized via a Product/Project Tracking System (PPTS). The interface is shown in figure 4. Here, the product structure is visualized to the designer, in this case a drive-shaft assembly. By pressing the buttons the designer could initiate different actions. The first action is the retrieval of additional information about one part, like identification number, material information, or the responsible person.

Another action could be the design of a part, which initiates the start of an integrated CAD-System. In our first prototype, we use CATIA (Dassault Systemes, France) as a commercial system.

The designer could also start a discussion about a specific aspect of the product. The environment supports the designer by offering him a list of possible discussion partners, the start of the discussion tool 3D CAD Viewer (figure 5), and the preparation of the data. The CSCW func-

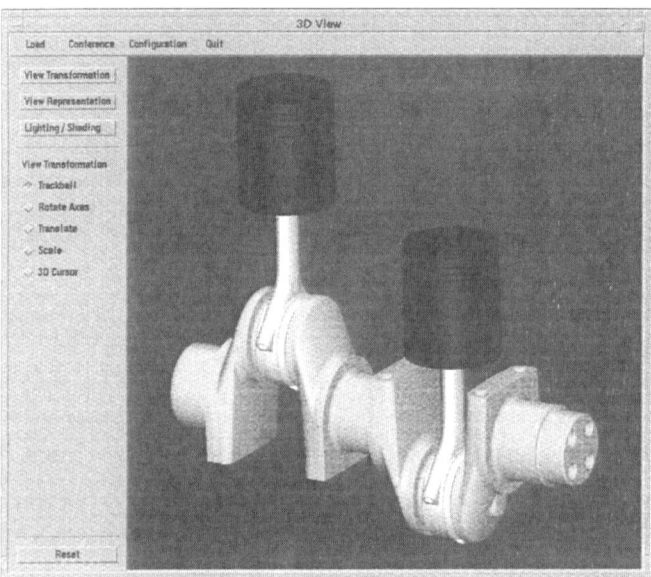

Figure 5: The 3D CAD Viewer showing the driveshaft assembly.

tionality of the 3D CAD Viewer includes shared viewing and annotation capabilities. Shared viewing means that view modifications made by one user can be seen by all partners of the conference. Annotations to the shape e.g. specific markers can be made by all partners simultaneously to support the discussion. The 3D CAD Viewer provides also enhanced visualization capabilities for 3D geometrical data in combination with additional organizational data. This means that not only the geometry will be visualized, but also the approval status or change requests for the product parts. The 3D CAD Viewer consists of two major parts, the data reader and the interactive viewer. The data reader reads the data from the database. The interactive viewer allows high quality visualization of the geometrical shape. The user can modify viewing, lighting and shading parameters interactively.

Currently, the advanced visualization of the product data with Virtual Design [AFM-93] is not fully integrated. Nevertheless, Virtual Design is able to read and visualize the exported product data. For the driveshaft assembly, Virtual Design offers a stereo projection of the data and a behavior animation of the assembly.

5 CONCLUSION

The paper described the benefits, concepts, architecture, and components of a Virtual Prototyping Environment. The described environment contains several major features. One feature is the availability of a common, distributed data repository, which stores the relevant data of the design process according to the Application Protocol 203 *Configuration Controlled Design* of the ISO 10303 Standard STEP. The access methods to the objects in the data repository are based on the *Standard Data Access Interface* specification of ISO 10303. The adoption of standards is a central objective for a Virtual Prototyping Environment and ensures the openness of the environment for future developments.

Features of the first prototype are browsing and editing applications for the product structure and the project context. For the design task, a commercial CAD-System is integrated. To enable synchronous communication among not co-located persons, CSCW tools as components of the Virtual Prototyping Environment were introduced.

The presented environment represents an open, extendable framework for the integration of engineering applications. Future extensions to the environment can address the underlying information model as well as the integration of applications which support specific tasks in the process chain of product development, such as kinematic or FEM analysis. The underlying data model can be extended by integrating additional STEP schemata. Thus, the whole data model of the prototype will develop in the direction of an integrated product model.

The described Virtual Prototyping environment supports concurrent design and integrates CAD systems in a heterogeneous computing environment. Nevertheless, the concept is open for the support of the whole engineering process as well as other application areas, e.g. Electronic Design Automation. Considering other approaches in the manufacturing area described by Kimura [KIM-93] who presented a total manufacturing integration by a virtual manufacturing concept the idea of Virtual Environments will be adapted to the whole enterprise.

6 REFERENZES

[AFM-93]   P. Astheimer, W. Felger, S. Mueller: *Virtual Design: A Generic VR System for Industrial Applications*; in: Computer & Graphics, Vol. 17; No. 6; pp. 671-677; 1993.

[CFI-92]   CAD Framework Initiative: *Intertool Communication Architecture*; CFI Pilot Release Document CFI-92-P7; 1992.

[KIM-93]   F. Kimura: *Virtual Manufacturing as a Basis for Concurrent Engineering*; IFIP TC5 Conference Towards World Class Manufacturing; September 12-16, 1993; Arizona, USA.

[HaJa-94]   S. Haas, U. Jasnoch: *Cooperative Working on Virtual Prototypes*; in Proc. of the IFIP 5.10 Workshop on Virtual Prototyping; Providence, USA; 1994.

[HKF-93]   E. J. Haug, J. G. Kuhl, F. F. Tsai: *Virtual Prototyping for Mechanical System Concurrent Engineering*; in Concurrent Engineering: Tools and Technologies for Mechanical System Design, Ed. E. J. Haug; pp. 851-879; Springer 1993.

[ISO-1]   ISO/IS 10303-1: *Industrial automation systems - Product data representation and exchange - Part 1: Overview and fundamental principles*; International Organization for Standardization; Geneva (Switzerland); 1994.

[ISO-21]   ISO/IS 10303-21: *Industrial automation systems - Product data representation and exchange - Part 21: Implementation methods: Clear text encoding of the exchange structure*; International Organization for Standardization; Geneva (Switzerland), 1994.

[ISO-22]   ISO/CD 10303-22: *Industrial automation systems - Product data representation and exchange - Part 22: Implementation methods: Standard Data Access Interface Specification*; ISO TC184/SC4 N225; National Institute of Standards and Technology; Gaithersburg (MD, USA); November 1993.

[ISO-203]   ISO/IS 10303-203: *Industrial automation systems - Product data representation and exchange - Part 203: Application protocol: Configuration controlled design*; International Organization for Standardization; Geneva (Switzerland); 1994.

[JKR-94]   U. Jasnoch, H. Kress, J. Rix: *Integrating Applications into a Virtual Prototyping Environment*; in Proc. of the IFIP 5.10 Workshop on Virtual Prototyping; Providence; 1994.

[JKSU-94]   U. Jasnoch, H. Kress, K. Schroeder, M. Ungerer: *CoConut: Computer Support for Concurrent Design Using STEP*; in Proc. of the IEEE Third Workshop on Enabling Technologies: Infrastructure for Collaborative Enterprises; Morgantown, 1994.

7 BIOGRAPHY

Uwe Jasnoch received his university diploma in Computer Science from the Technical University of Darmstadt in 1989. Then, he was a software engineer with Philips Kommunikations Industrie AG for one year. Afterwards, Uwe Jasnoch was a researcher with the Interactive Graphics Systems Group at the Technical University of Darmstadt, where he was involved in several R&D projects. Since 1992, he has been a researcher with the Fraunhofer Institute for Computer Graphics in the Industrial Applications Department. His main research topics are data modeling, open environments, and consistency management.

Holger Kress is a researcher in the Industrial Applications Department of the Fraunhofer Institute for Computer Graphics since 1991. He is currently involved in research projects in the area of product modeling, groupware, and CAD frameworks. He received a masters degree in mechanical engineering from the Technical University of Darmstadt in 1991. His research interests include concurrent engineering, product modeling, and computer supported cooperative work.

Dr. Joachim Rix is head of the department for Industrial Applications of the Fraunhofer Institute for Computer Graphics (IGD) in Darmstadt, Germany. From 1991 to 1992 he was Associate Manager of the Fraunhofer Computer Graphics Research Group (today; Fraunhofer Center for Research in Computer Graphics, Inc. (CRCG) in Providence, RI). Mr. Rix received his Diploma and Ph.D. in Computer Science from the University of Darmstadt. His topics of interest are in Computer Graphics, CSCW, CAD, and Product Modelling. This includes the integration and use of computer graphics with its presentation and interaction techniques in industrial applications, like CAD, CAM, Concurrent Engineering , and Groupwork Computing. Mr. Rix is member of the standards Committee of ISO/IEC JTC1/SC24 „Computer Graphics" and was Rapporteur of the study group „PREMO" (Presentation Environments for Multimedia Objects). Since 1985 he is member of the national and international committees (DIN NAM 96.4, ISO TC184/SC4) developping STEP (Standard for the Exchange and Representation of Product Model Data). Since 1994 he holds the position of a deputy convenor of its WG 3 „Product Modeling". Since 1981 Joachim Rix is member of the Eugrographics Association.

15

Network Requirements for Large Distributed Virtual Environments

R. B. Araujo[1] and Dr. C. Kirner[2]
Computer Science Department
Federal University of São Carlos
Via Washington Luiz, Km.235
13565-905 São Carlos - SP BRAZIL
Tel.: +55-162-74823 Fax: +55-162-712081
drea/dcki@power.ufscar.br

Abstract

This paper presents the Network requirements of a special class of telepresence applications: Large Distributed Virtual Environments. The characteristics of large virtual environments are introduced and their network requirements analysed from the network bandwidth and latency point of view. When multimedia information is integrated to these systems, demands over the network are even more stringent.

Keywords

Large distributed virtual environments, multimedia information, network latency, network bandwidth, multicasting.

1 INTRODUCTION

Virtual Environment - VE systems have emerged as a breakthrough in human-computer interaction, promising a revolution in areas ranging from entertainment to medicine. A VE is defined here as a system which simulates an environment so that a user can feel immersed in

---

[1] PhD Student in the Integrated Systems Laboratory of the University of São Paulo - LSI EPUSP.
[2] Visiting Professor in the Computer Science Department, University of Colorado at Colorado Springs (UCCS), CO, USA, with grant from CNPq, Brazil, Proc. 201443/93-7.

it and interacts with it in a direct and natural way. VE applications have different characteristics, placing different demands over their supporting systems. Some of these applications involve a few participants in a shared virtual world whose 3D geometric representation is small enough to run in every node of the system, as for instance, a Computer Supported Cooperative Work - CSCW application. Other applications may use the virtual world for direct interference into the real world, such as telerobotics, which makes strict real time demands upon the supporting system.

This paper is concerned with the network requirements of yet another class of VE applications: large multi-party distributed virtual environment systems, which is characterized by the participation of thousands of users in an environment such as a virtual city, a battle field for military training, star war-like games and so on. All these applications involve a large number of participants in areas ranging from a few to thousands of miles, what makes difficult to keep the whole virtual world representation in every node of the system. Moreover, in order to add more realism to the environment, the VE is populated with simulated objects, what will bring in further demands over a communication network.

This paper is organized as follows: Section 2 presents the characteristics of large scale VEs. Section 3 describes the communication issues involved in a VE. Section 4 analyses the network requirements for large VEs. The integration of multimedia information to VE systems is discussed in section 5. The design of a distributed virtual environment is presented in section 6, followed by conclusions, acknowledgements and references.

2 CHARACTERISTICS OF LARGE DISTRIBUTED VE SYSTEMS

According to Sheperd (Sheperd, 1993), VE systems are classified in two broad classes: *telepresence,* where a common synthetic environment is shared among several people, as an extension to the CSCW concept, and *teleoperation,* where robots act upon a subject, be it a living body, a product being manufactured etc. In *telepresence* systems, maintenance of illusion is essential otherwise the sense of presence would fall in disbelief. In order to maintain illusion, the system has to provide for response time within an acceptable threshold. Besides, a participant actions have to be reflected immediate and simultaneously to all participants, and environments have to be realistic.

This implies in the following demands from the system, as summarized by VanDam (vanDam, 1993): rapid update rates and minimal lag; handling of multiple input devices; simulation of a potentially large number of interacting objects with complex behavior.

These requirements demand not only powerful tools for building the virtual world and its graphical representation. They also demand high performance graphic processors, support for the complexities and required realism of the environments to be rendered, as well as accurate and confortable 3D input and output devices, adequate operating systems and networks that can guarantee the delivery of the information in time.

The above requirements are even more stringent to a special group of applications within the telepresence class: *multi-user distributed large virtual environments.* These applications present the following characteristics:
• large population of participants and simulated objects.
• large areas, ranging from a few to thousands of miles.

- participants are geographically distributed.
- multidimensional environments.
- interaction among participants and objects in real time.
- heterogeneous computer and network resources.

Few large scale VE systems were implemented. Some of the best known systems are SIMNET (Calvin, 1993) and NPSNET (Zyda, 1992) which continue to develop towards a scale of thousands of participants. The building of large VEs is quite difficult, integrating several technological areas. This paper deals with the network requirements for large scale VEs, discussing their communication aspects and integration with multimedia information.

3. COMMUNICATION ISSUES IN A VE SYSTEM

3.1 The population of a VE

A virtual environment is populated with objects which are classified here as static, dynamic user-driven and dynamic simulation-driven. Large part of the world population is composed of static objects, i.e., objects that are fixed such as buildings, mountains, trees, parks, furniture etc. Simulation-driven objects can range from those programmed to act according to a pre-defined pattern (scripts) up to highly complex objects driven by expert systems, such as humanoids, robots etc. These latter objects can be developed by different companies and plugged into the VE. User actions are reflected on the VE through the user-driven objects. These may be represented as abstract elements (geometric forms), parts of a human body (hands, feet etc.), mobile vehicles, gloves etc.

3.2 The Basic Model of a VE System

A basic model of a VE system comprises the following modules: user application (APPL), virtual world simulator (VWS), 3D geometric database (3DDB), 3D rendering (visual, acoustic, tactile), and I/O devices.

The APPL is defined by the application designer who decides about the objects characteristics, the simulation-driven objects behavior, how all objects interact, if the synthetic environment is to be shared among users or if every user will have his/her own view of the VE (Isdale, 1993), and so on. A user should be able to customize the application by adding, dynamically, new objects or changing the characteristics and/or behavior of existing objects.

The simulation (VWS), is supposed to support the application and can be made in two levels: general simulation and object simulation (Araujo, 1994a). General simulation runs no stop and is responsible for the positioning of all objects in the area under its control as well as collision detection and resolution, concurrency disputes, objects translation etc. Object simulation is responsible for controlling the animation details of each dynamic object. The position of dynamic objects are calculated following physical laws. Information on dynamic objects position is sent only when unpredictable behavior occurs - as the dead reckoning concept used in SIMNET (Calvin, 1993). In a user-driven car, for instance, if the user keeps speed and orientation steady, the repositioning of the car does not need to be sent to all its replicas across the system, unless the user decides to change any of these parameters.

User actions are captured from sensors connected to them (gloves, head and body movement trackers etc.) or devices pointed towards them (cameras). The VWS translates   this information into graphical and other commands which are sent to the 3D renderers for the repositioning process. The graphic renderer generates the final image to be presented to the user and the audio renderer "displays" audio which coincides with the location of the graphical objects. A user's actions, read from sensors, can trigger feedback actions reflected on other user(s) through actuators.

3.3 Distributed Nature of Large VEs

In a system where thousands of users can take part in the application, and where large areas are involved, distribution seems inherent. The geometrical information of an extensive area being synthesized, such as a city, may need to be divided in sub-areas which are distributed across the network. Environment servers - ES are responsible for the simulation and storage of sub-areas and their respective data set, as shown in Figure 1.

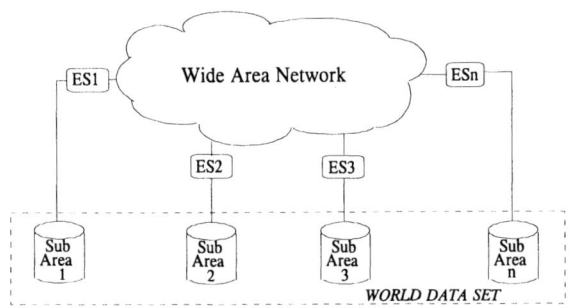

Figure 1 A Distributed VE System.

For visualization purposes, the world is stored in the environment servers in various degrees of resolution. Only objects within the user's view volume are rendered in full resolution.

The architecture to support such a system can vary from a peer-to-peer approach, client-server or a hybrid architecture. Sub-areas swapping, due to objects movements across their boundaries, may cause the loading of sub-area data sets located or maintained remotely. A hybrid communication architecture of a distributed VE system is presented in Araujo (Araujo, 1994a).

3.4 Communication in a Distributed VE

Considering a client-server architecture, where a sub-area, or part of a sub-area (area of interest) is downloaded to a client machine, replicas of the dynamic objects living in that area are also downloaded to the client machine. When there is more than one client machine for a same area of interest, i.e., there are other participants sharing the same virtual space, replicas of that area are downloaded to as many client machines as there are.

An action triggered by a participant, in a client machine, must have an immediate response. Moreover, this action must be broadcast to all client machines containing the same area of interest, so that all participants have a consistent view of the VE they are taking part of. The communication among objects in a VE involves basically the communication between an object and its replicas (communication type 1) and among dynamic objects (communication type 2). Figure 2 shows the flow of information in the two types of communication mentioned:

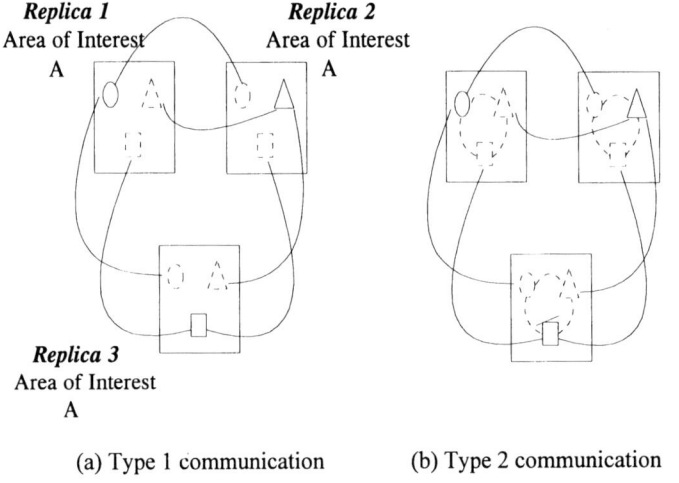

(a) Type 1 communication   (b) Type 2 communication

Figure 2 Flow of information in the communication among objects in a VE

In figure 2, the squares represent three replicas of a same area of interest. The solid geometric forms inside each square represent either a user-driven object, whose user resides in the client machine housing that replicated area of interest, or a master copy of a simulation-driven object. The solid lines reflect the flow of messages from a dynamic object to its replicas, which are represented as dashed geometric forms. Consider, for instance, the solid circle as a dynamic object and the dashed circles as a dynamically modeled representation of that object in other two machines. Every movement performed by the object represented as the solid circle, which can not be predicted by the code running in the modeled representation (depicted as dashed circles), is sent to its replicas through a repositioning message.

For the communication among dynamic objects (for instance, when a participant, through a user-driven object, sings to other participants of the same area of interest), a simplification of the concept of auras (Benford, 1994) is used. Every dynamic object is assigned an aura per medium. An aura is an area surrounding an object whose range varies according to the medium reaching area it represents - the ranging area for audio may be different from the ranging area for video, for smell etc. An audio message sent, for instance, from user-driven object A, will be sent to all its replicas. When that message reaches the replicas, only the objects within the audio aura of user-driven object A replicas are affected by the message. The auras can be established dynamically by the communicating objects. Figure 2 (b) reflects the communication among dynamic objects. The dashed lines represent the communication among objects which is in fact realized through the solid lines.

Besides the communication among objects described above, communication is also realized for: synchronization between client machines and environment servers, objects movement from one sub-area to another, message exchanges in the initial loading of the areas of interest from the environment server to the client machine, concurrency disputes results broadcast by the server to its clients etc.

3.5 Communication Among Objects in Related VE Systems

The DIVE System (Carlsson, 1993) adopts a peer-to-peer approach for the interaction among objects. BrickNet (Singh, 1994) is implemented as a client-server architecture where objects, kept in different clients, are updated through different servers which together maintain information on all objects. It uses unreliable User Datagram Protocol - UDP protocol for the communication. In SIMNET (Calvin, 1993), every participant system has a copy of the whole VE. Objects communicate among themselves by broadcasting events through the Distributed Interactive Simulation protocol - DIS. The data units of this protocol - PDUs, related to the state of the objects, are transmitted periodically even when they present no change, as no centralized information about the objects states is maintained, leading to a waste of bandwidth. NPSNET also makes use of DIS but over the unreliable IP multicasting network - MBone (Macedonia, 1994a). It associates spatial, temporal and functionally related entity classes with multicasting groups (Macedonia, 1995). Araujo (Araujo, 1994b) makes a more detailed analysis of the communication approaches being developed by these systems.

4 NETWORK REQUIREMENTS FOR LARGE DISTRIBUTED VES

As far as networks are concerned, latency is perhaps the most important performance parameter for real-time communication systems. For VE applications, there is a time dependency as maintenance of illusion is crucial to the success of the system - the whole operation of handling user's input, computation, synchronization, information distribution, rendering and presentation of the rendered image, must occur in less than 0.1 seconds, which is the time a user takes to notice lag.

As all users need to have an immediate response to their actions and their actions must be seen by all users sharing the same area of interest, system success will depend not only on the processing power for the 3D graphics rendering and the simulation itself, which can be extremely heavy, but also on the network capability to deliver this information in time. This turns distributed VE systems into real-time systems and, as such, low network latency is an important requirement.

Causes for Delay

A VE has many sources of delay. The network delay itself is a major source of lag. The total network delay is contributed by:
- the propagation delay which is related to the distance and the signal propagation speed;
- the transmission delay which is related to the transmission speed (10 Mbps for the Ethernet, 100 Mbps for the FDDI etc.) and the message length;
- the protocol processing delay which is related to the processing speed of the switching nodes and end stations;
- the store-and-forward delay which is related to the number of hops, the transmission delay, and the number of packets for a segmented message;
- the queuing delay which is related to the congestion level at a switch node.

Considering a packet switched network, Fluckiger (Fluckiger, 1991), in a study based on CERN experience, measured round trips delays over TCP/IP lines, linking CERN labs to many European and American sites. It was shown that round trip propagation only took a few milliseconds even for intercontinental connections (less than 25 ms for a one way trip between CERN and Cornell over a 1.544 Mbps T1 line).

Attention must be drawn to the difference between high bandwidth and high speed. Fluckiger showed that round-trip delay over 10 megabits Ethernet is 1.5 ms against 1.8ms for 100 megabit FDDI network - this delay is caused by the FDDI protocol processing overhead.

Considering that large delays are caused by the handling of input/output information[1], simulation processing, network information processing, rendering and presentation, every time unit saved is precious. NPSNET-IV (Macedonia, 1994b) uses a four processor machine for increasing frame rates where simulation, rendering, drawing and network traffic handling are run in different processors.

As technology advances, faster network intermediate equipment are employed, making internal and end system delays smaller. On the other hand, better machine architectures are needed to exploit high speed networks, as even super computers can present bottlenecks to a gigabit network.

Optimizing Bandwidth Requirements

Supposing that every participating host has an update rate of 10 frames per second and that one message is generated every 0.1 second. Considering a message of 100 bytes (a DIS Entity State PDU has at least 140 bytes (Pratt, 1993)), the following traffic is generated by each user:
10 x 100 x 8 = 8 Kbits per second.

---

[1] 10 ms is the transmission delay for mice, trackers etc.; 20-100 ms for visual output; 1-500 ms for sound and directional sound - values suggested by (Ellis, 1994).

In an application with 100.000 users, a traffic of 800 Mbits/s would be generated. Macedonia (Macedonia, 1995) wrote that network simulations using multicasting showed a traffic reduction of 90% and that dead reckoning decreases traffic by 50%. By applying these percentages on our example, the final generated traffic falls from 800 Mbps to 40 Mbps. It has to be observed that the network traffic is highly dependent on the number of dynamic objects and their behavior complexity.

From the numbers derived above and considering the emergence of the broad band highways, at first sight, VE systems does not seem to place a high demand for bandwidth upon the network. However, multicasting, although being successfully developed, presents some difficulties such as: Existence of a fixed number of multicasting addresses to serve VE applications (large scale VEs may demand large numbers of group addresses). As the use of these addresses can be highly dynamic, address collision may happen if a global addressing scheme is not used (Schooler, 1992); Time to disconnect from a multicast stream is non-deterministic, causing host resources to continue to be consumed longer than necessary (Moran, 1992); Experiments over the IP multicasting network between the Naval PostGraduate School and SRI showed delays between 100 to 1000 ms (Macedonia, 1994b), what can be very high even for rates of 10 frames per second. Also, it is not widely used nor easy to be installed (Macedonia, 1994a).

As for dead reckoning, in a large population of dynamic objects, with complex and unpredictable behavior, it may not apply. There is also the "closet syndrome" mentioned by Pratt (Pratt, 1993) where, regardless the size of the closet, it will eventually fill up (referring to the bandwidth utilization of high speed networks such as FDDI and ATM in the support of multimedia and VE applications).

ATM as the Paradigm to Support Distributed VE Systems
Asynchronous Transfer Mode - ATM networks arise as a promising alternative to the support of large VE systems because of its low latency, bandwidth on demand and support for real-time traffic. ATM is a cell switching technology where data is carried in cells of 53 bytes. However, ATM networks are virtual circuit networks in nature what require communication channels to be set up before the ATM cells can be sent. The initial virtual circuit setup time can be several hundred milliseconds. To reduce the setup delay, pre-arranged circuit setup can be performed before the VE system starts.

Hayter (Hayter, 1991) proposed the Desk Area Network (DAN), an internal architecture which uses an ATM switch to provide interconnections between the machine components and from these components to an ATM LAN and/or ATM WAN so that seamless connections are achieved, what means lower latency. Several other initiatives exist for a more effective use of high speed networks.

5 INTEGRATING MULTIMEDIA TO VE SYSTEMS

Multimedia information will certainly be a desired complement to VE systems. In a virtual city, for instance, a participant can enter a cinema to see a movie or, through hypermedia, select objects and have multimedia information being displayed on demand about those objects, within the VE environment.

Many interactive multimedia applications are considered as soft real-time applications, which are defined as time dependent and tolerant to some amount of packet loss (Aras, 1994). Due to the inherent interactive characteristic of VE systems, they also present this time dependency and a certain tolerance to losses. However, it is important that some types of messages exchanged between environment servers and client machines be delivered reliably, for instance, messages conveying concurrency dispute results, messages exchanged among environment servers notifying object movements from one sub-area to another etc. Packet losses can be avoided but require time-out and re transmission mechanisms, what augments latency.

Network support for soft real-time systems has been extensively discussed in the literature, and is summarized as follows: interactive multimedia applications, providing services of voice and video, may need a certain guaranteed Quality of Service - QOS before being admitted to the network. The determination of the QOS depends on the characterization of the application traffic and this can be difficult to determine for both audio and video sources: they can vary over the time, and, in the case of video, be strongly dependent on the video coding algorithm employed. VE traffic is not easy to be characterized either. It can be very bursty, as shown in (Macedonia, 1994b), and have a model highly dependent on the complexity of the dynamic objects.

The integration of multimedia data in a VE system will raise further issues, such as: association of a particular medium channel to an object in the VE - for a virtual TV or radio broadcasting to user-driven objects nearby (Frécon, 1992); coordination between the presentation of the multimedia information embedded or associated to an object in a VE and the presentation of this rendered object to the users; matching of possibly different network architectures to support multimedia and VE systems - for instance, a VE network architecture may adopt a peer-to-peer approach (as in DIVE) against a client-server model for the multimedia data.

6 THE DESIGN OF A LARGE VE SYSTEM AND RELATED WORK

The first version of a distributed virtual environment system, involving a network of workstations and PCs is being designed and implemented at UCCS and UFSCar[2]. The system comprises a number of environment servers responsible for storage and control of VE sub-areas (Araujo, 1994a). Other machines connected to the network are used as client machines to the environment servers for both participation and visualization.

The emphasis here is on the distributed structure and network issues in the support of VEs with large populations and extensive areas with heterogeneous end user equipment. Systems like MASSIVE (Benford, 1994), is more concerned with the interaction aspects among users with heterogeneous equipment who communicate over an ad-hoc mixture of media. NPSNET (Zyda, 1992), one of the only really large scale VE systems implemented, do not address, as yet, support for heterogeneous equipment, as well as BrickNet (Singh, 1994), what is important if thousands of users are expected to populate a virtual world. Other systems like

---

2 UCCS - University of Colorado at Colorado Springs; UFSCar - Universidade Federal de São Carlos.

DIVE (Carlsson, 1993) are now addressing simulation-like applications as the ones being considered here.

7 CONCLUSIONS

With the time dependency of VE applications, VE system success will depend not only on the processing power for the 3D graphics rendering and simulation, which can be extremely heavy, but also on the network capability to deliver this information in time.

This time dependency turns distributed virtual environments into real-time communication systems so that, a network to successfully support these systems must have the following characteristics: low latency with seamless connections, meaning low propagation time, low resource requirements within the network, low processing overhead in the handling of packets within the network and at the end system, high bandwidth, support for multicasting and good performance for a large number of connections. Networks as described above, allied to end systems that effectively use these high speed networks, are the candidates for supporting interactive large VE applications, specially when multimedia traffic is integrated to the VE system. ATM networks raise as promising alternative because of its low latency, bandwidth on demand and support for isochronous traffic.

A distributed VE system is being developed for PCs and workstations. The emphasis is on the distributed structure and network issues in the support to large populations with heterogeneous end user equipment and network resources.

8 ACKNOWLEDGEMENTS

We are particularly grateful to Dr. Edward Chow of the Computer Science Department at UCCS for his significant comments on this work. Thanks also to Professor Sebesta who made available to us the resources of the CS Department at UCCS.

9 REFERENCES

Aras, Ç. M. et alii (1994)   Real-Time Communication in Packet Switched Networks Proceedings of the IEEE, Vol.82, No.1, January 1994, pp.122-139.

Araujo, R. B. and Kirner, C. (1994a) Distributed Large Scale Virtual Environments Discussion, Proposals and Design, paper not published.

Araujo, R. B. and Kirner, C. (1994b) An Architecture to Support Large Scale Virtual Environments, to be presented at the Brazilian Symposium on Computer Networks, 22- 26 May, Belo Horizonte, MG, 1995 (in Portuguese).

Benford, S.et alii (1994) Managing Mutual Awareness in Collaborative Virtual Environments, Proceedings of the VRST'94 Conference, 23-26 August, 1994, Singapore, pp. 223-236.

Calvin, J. et alii (1993) The SIMNET Virtual World Architecture, in IEEE Virtual Reality Annual International Symposium, September 18-22, 1993, Seattle, WA, pp. 450-455.

Carlsson, C. and Hagsand, O. (1993) DIVE - a Multi-user Virtual Reality System, in IEEE Virtual Reality Annual International Symposium, September 18-22, 1993, Seattle, WA, pp. 394-400.

Ellis, S. R. (1994) What are Virtual Environments, IEEE Computer Graphics & Applications, January 1994, pp. 17-22.

Fluckiger, F. (1991) From Megabit to Gigabit: Possible Transition Scenarios, in Computer Networks and ISDN Systems, Vol. 23, 1991, pp. 129-138.

Frécon, E., Eriksson, H. and Carlsson, C. (1992) Audio and Video Communication in Distributed Virtual Environments, Proceedings of the 5th MultiG Workshop, Royal Institute of Technology, Stocholm, December 1992.

Hayter, M. and McAuley, D. (1991) The Desk Area Network, in ACM Operating System Review, May, 1991.

Isdale, J. (1993) What is Virtual Reality? A Home-brew Introduction, electronic document in URL=ftp://sunee.uwaterloo.ca/pub/vr.

Macedonia, M. and Brutzman, D. P. (1994a) MBone Provides Audio and Video Across the Internet, IEEE Computer, April 1994, pp. 30-36.

Macedonia, M. et alii (1994b) NPSNET: A Network Software Architecture for Large Scale Virtual Environments, Presence Magazine, Vol.3 No.4, 1994.

Macedonia, M. et alii (1995) Exploiting Reality with Multicast Groups: A Network Architecture for Large Scale Virtual Environments, Submitted to the 1995 VRAIS Conference.

Moran, M. and Gusella, R. (1992) System Support for Efficient Dynamically-Configurable Multi-Party Interactive Multimedia Applications, Proceedings of the 3rd International Workshop on Network OS Support for Audio and Video, San Diego, November 1992, pp.143-156.

Pratt, D. (1993) A Software Architecture for the Construction and Management of Real-Time Virtual Worlds, PhD Dissertation, Naval PostGraduate School, Monterey, CA, June 1993.

Schooler, E. M. (1992) The Impact of Scaling on a Multimedia Connection Architecture, Proceedings of the 3rd International Workshop on Network OS Support for Audio and Video, San Diego, November 1992, pp.341-346.

Sheperd, B. J. (1993) Rationale and Strategy for VR Standards in IEEE Virtual Reality Annual International Symposium, September 18-22, Seattle, WA, 1993.

Singh, G. et alii (1994) BrickNet: A Software Toolkit for Network-Based Virtual Worlds, in Presence Magazine, Vol.3, No.1, 1994, pp.19-34.

van Dam, A. (1993) VR as a Forcing Function: Software Implications of a new Paradigm in IEEE 1993 Symposium on Research Frontiers in Virtual Reality, October 25-26, San Jose, CA, 1993.

Zyda. M. J. et alii (1992) NPSNET: Constructing a 3D Virtual World, Proceedings of the 1992 Symposium on Interactive 3D Graphics, Cambridge, MA, pp. 147-156.

10 BIOGRAPHY

REGINA BORGES DE ARAUJO received her B.S. degree in Computer Science from UFSCar-Brazil in 1983, and the M.S. in Data Communication Networks and Distributed

Systems from the University College London in 1986. She is now finishing her PhD in Electrical Engineering at LSI/EPUSP (due in October-1995). She has been a lecturer in the CS Department at Federal University of São Carlos since 1989. Her current research interests are in distributed systems and network support for multimedia systems and virtual reality. She is a member of ACM and SBC (Brazilian Computer Society).

CLAUDIO KIRNER received the B.S. degree in electrical engineering from EESC/USP-Brazil in 1973, the M.S. degree in electronic engineering from ITA - Brazil in 1978, and the Ph.D. degree in systems engineering and computer science from COPPE/UFRJ-Brazil in 1986. He has been an adjunct professor in the Computer Science Department at Federal University of São Carlos - Brazil since 1974. He is currently a visiting professor in the Computer Science Department, University of Colorado at Colorado Springs - USA. His research interests involve parallel and distributed computing, computer graphics, and virtual reality. He is a member of ACM, IEEE, and SBC (Brazilian Computer Society).

PART FOUR

Advances in Virtual Reality Technologies

16

Interacting with virtual reality

M. M. Wloka
Science and Technology Center for Computer Graphics and Scientific Visualization, Brown University Site
Department of Computer Science, Box 1910, Brown University, Providence, RI 02912, USA.
Tel.: (401) 863 7600
Fax: (401) 863 7657
Email: mmw@cs.brown.edu

Abstract

Interacting with virtual reality is fundamentally different from interacing with traditional desktop graphics. The three features that characterize virtual reality interaction are immersion, rich interaction and presence; I define these features. To achieve them, virtual reality system designers need to address many different issues. I discuss some of these issues, in particular multiple inputs, multiple outputs, multiple participants, dynamic virtual worlds, user interface paradigms and performance.

Keywords

Virtual reality, interaction, user interface, performance

1 INTRODUCTION

Even though virtual reality (VR) originates in traditional, interactive 3D graphics, it now is its own area of research: VR has its own specific problems and specialized solutions that are often different from those in interactive 3D computer graphics. Therefore, the applicability of even basic concepts of interactive 3D graphics is questioned. For example, 3D graphics architectures have used essentially the same 3D viewing pipelines (Foley and van Dam, 1982) for more than a decade; recently, Regan and Pose (1994) proposed a much different pipeline that better satisfies the needs of VR applications.

Accordingly, interacting with VR is different from interacting with traditional 3D graphics applications. Three features characterize VR interaction: immersion, rich interaction and presence.

1.1 Immersion

Immersion implies that the user only perceives the artificial, computer-generated world. Typically, immersion is achieved with headmounted displays and head- and hand-trackers. The advantage of immersing the user is that the space with which the user interacts is no longer limited to the dimensions of the display, since the display now only represents a dynamic window into a vast data space.

In comparison, conventional 2D graphics lets the user only interact within the confines of the display-monitor. The size and resolution of the monitor severely limit the number and size of the displayable objects. The common desktop metaphor is thus severely flawed; as Fred Brooks, Jr. once remarked, it implies a desktop the size of an airplane tray-table – even a small desk typically covers more than two square meters.

1.2 Rich interaction

Three factors enrich the interaction metaphor in virtual environments. First, instead of using indirect methods such as menu-selections or command-line interfaces, an object's interface often is the object itself, i.e. manipulation is direct. Second, multiple and simultaneous input devices (two gloves, head- and hand-trackers, voice recognition, etc.) are common. Third and finally, the input devices used in VR all have a high number of degrees of freedom.

1.3 Presence

Finally, presence – the user's feeling of really being 'there' – is perhaps the most distinguishing difference between VR interaction and interactive 3D computer graphics. If presence is achieved the user is able to process presented information not only deliberately, but also on a subconscious level. Such subconscious information assimilation is common in real life: for example, people often know where their spouse's keys or eyeglasses are, even though they do not actively look out for them.

1.4 Overview

To achieve the above described three features, a VR system needs to address various issues, including multiple inputs, multiple outputs, multiple participants, dynamic virtual worlds, user interface paradigms and performance. I discuss each of these issues in their own section.

The sections on user interface paradigms and performance issues are more detailed than the others. This is not to say that the other issues are less important or less challenging – they certainly are not – but rather reflects my own current research interests.

2 MULTIPLE INPUTS

A VR system integrates several distinct user input devices, for example, trackers (for head and hands), gloves and speech recognizing software. Integrating these devices is hard, because not only do they have to function simultaneously, but they also have to operate in concert.

The following scenario illustrates the concept. A user interacts with a virtual world using two gloves and a speech recognizer. To move an object she says 'Move that over there!' while simultaneously pointing with the right hand at a particular object and indicating its new position with the left hand.

This simple example clarifies the need for VR systems to correlate the various user input streams to disambiguate the user's commands. Research is underway to address these problems, for example, by time-stamping and abstracting the input streams (Tarlton, 1993).

3   MULTIPLE OUTPUTS

To improve a user's feeling of presence it is important to engage more than just the visual sense. Movie producers know that the audience perceives a movie's visuals as of higher quality if the soundtrack is of high quality. Similarly, VR systems are more convincing if they provide aural and haptic feedback in addition to visuals.

Aural displays, i.e. audio systems that allow the 3D placement of various sound sources and software systems that drive such a system are becoming common (Beaudouin-Lafon and Gaver, 1994).

On the other hand, haptic displays for virtual reality, i.e. force-feedback displays and displays that let users feel or touch their environment, are currently an active area of research.

Integrating multimodal outputs into a single system is hard for several reasons. First, the various devices need to display information in tight synchronization, since human tolerances of synchronization errors are quite small (Schaufler, 1992; Bulterman et al., 1991). Second, varying delays in the various output devices makes proper synchronization even harder. Worse, synchronization errors also result from varying distances between user and devices. (Modern movie soundtracks must take into account the distance sound travels in an average movie theater before reaching the audience.) Third and last, synchronization is not always beneficial. For example, in limited bandwidth, computer-supported cooperative work applications, it is preferable to sacrifice synchronization to enable low-latency, audio-only communication (Isaacs and Tang, 1993).

4   MULTIPLE PARTICIPANTS

Future virtual worlds will host multiple participants – not two or three, but rather multiple thousand (Zyda et al., 1992). Current solutions, for example, for synchronizing actions and interactions among the participants, are ill-suited to handle such numbers.

Hosting multiple thousand participants also implies use of multiprocessing and networking. Incorporating those into VR systems in turn demands solutions to difficult problems, for example, where and how to store the underlying database, how to handle the unpredictable delays common to networking and how to approach fault-tolerant computing.

5   DYNAMIC VIRTUAL WORLDS

To make virtual worlds believable they have to be dynamic. (A static virtual world feels empty and dead.) To support such a dynamic environment, time has to become a first-class notion (Zeleznik et al., 1991): instead of simulating dynamic behavior by reediting a static database, the VR database must inherently support dynamic behavior. Doing so has immediate benefits, for example, allowing high-level optimizations (Elliott et al., 1994).

In order to manage the added complexity, control structures similar to those used for managing spatial information are useful. So far, the following advanced models of time are being proposed: encapsulating and reusing dynamic behaviors (Kalra and Barr, 1992), constructing hierarchical time coordinate systems (Tarlton and Tarlton, 1992) and using synchronization primitives (Gibbs, 1991).

The idea of making time a first-class notion also applies to interactive behaviors: instead of simulating interactive behavior by reediting a static database, interactive behaviors are described and stored directly in the database. Thus, describing interactive behaviors becomes easier and the VR system can apply high-level optimizations to interactive behaviors as well.

6   USER INTERFACE PARADIGMS

6.1   Floating menus

The floating menu paradigm displays two-dimensional menus in the three-dimensional world of VR. These menus are either text-based, describing the available choices with words, or graphically based, using icons to convey the available choices.

Because menus enable the display of large amounts of command choices and because they are well accepted in the 2D community, they are often used in VR. However, since menus do not have a counterpart in the real world, they seem out of place in the fully three-dimensional virtual world. Furthermore, since menus are inherently two-dimensional, the added third dimension complicates (rather than simplifies) interacting with a menu.

6.2   Gestures

Gestures enable a user to give commands by making certain hand-gestures or assuming certain hand-postures. For example, assuming a pointing posture switches the user into a mode where she flies around the virtual world.

Hand-gestures and -postures are a powerful paradigm: it corresponds to the user performing magic spells. However, the two disadvantages of the gesture paradigm are a direct result of this correspondence to magic. First, people do not use magic spells to interact with the real world and thus, the gesture paradigm becomes a nonintuitive paradigm that requires training. Second, it is difficult for the user to remember the various gestures and their effects. (Real magicians usually have a thick spell book to look up spells.)

6.3   3D widgets

3D widgets are objects in the virtual world that present an intuitive, direct manipulation interface
to the user. For example, the rack widget (Snibbe et al., 1992) graphically represents a vise with
several handles. The handles move in three dimensions; their movements directly map to the
parameters of a deform operator applied to an object in the vise.

Using 3D widgets as an user interface paradigm has many advantages. First, since 3D widgets are direct manipulation interfaces, it is straightforward for a user to interact with a widget.
Second, 3D widgets coexist in the same space as 3D application objects; thus, they take advantage of the user's familiarity with the three-dimensional world. Third, the widgets are implemented with the same software as used for programming the application objects. Therefore, only
a single development environment is required, high bandwidth communication between application objects and interface widgets is possible and the machinery used for the application objects
(constraints management, physically-based simulation, etc.) is reusable for the widgets. Fourth
and finally, since widgets are essentially the same as application objects, they help to reset the
artificial separation between the application and the user interface that was introduced with the
desktop metaphor (Figure 1).

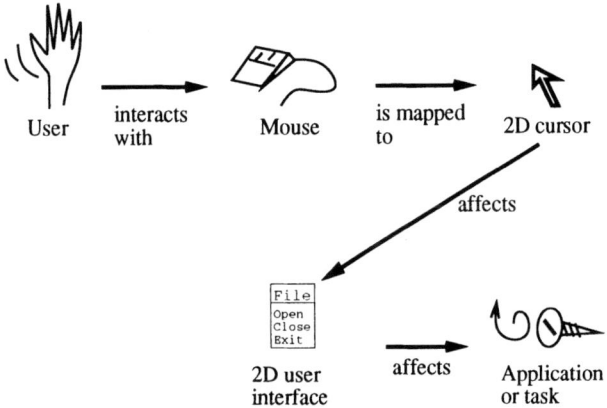

Figure 1   Interaction with traditional 2D applications is indirect.

Designing 3D widgets, however, is hard. First, concrete metaphors that map a task into a
tangible 3D widget are hard to define – it seems that the various attributes of a widget (for example, geometry and behavior) are task-dependent, as well as interdependent. Second, because
users are familiar with 3D space, they expect widgets to behave a certain way. In contrast, 2D
widget design is easier, because users do not have expectations about the behavior of a 2D object. Third, the number of primitives and attributes available to the widget designer is vast; it is
hard to control that many channels simultaneously. Fourth and last, it is easy to design a widget
that is visually cluttered, i.e. one that obscures itself or the object the user is modifying.

6.4   Discussion

Which of the above user interface paradigms or combination of user interface paradigms is better? This question is evaluable with respect to several criteria. There are the user-oriented criteria, for example, ease of learning, ease of use, speed and accuracy of performing specific tasks and 'attractiveness' (i.e. how much the user likes to work with the paradigm) and there are the more pragmatic criteria, for example, ease of implementation and resource utilization.

Instead of answering which paradigm or paradigm combination is better, I am answering a different, more essential question. Are any of the above paradigms well suited for VR? I believe the answer is no.

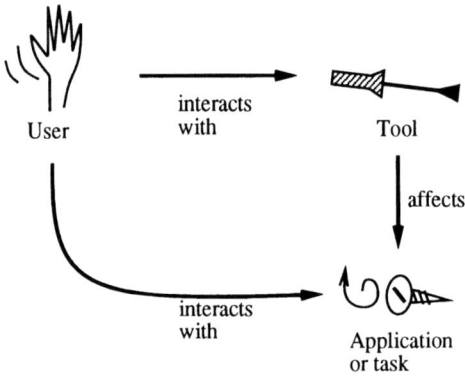

Figure 2   Interaction in the real world is direct, or at most one level removed, if a tool is used.

In particular, to provide a sense of presence, some form of haptic feedback seems essential. None of the above paradigms currently integrate haptic feedback, nor is it obvious how to extend them so that they could. Researchers are thus searching for different paradigms that inherently incorporate haptics. For example, Randy Pausch is taking the approach of building a specific user input device for every new application task (Bryson et al., 1994). While this approach is effective, it seems cumbersome to build new hardware for every software application.

In addition, all the above paradigms are less immediate than the human-object interactions in the real world. To manipulate an object, humans generally reach out and manipulate that object directly, or use a tool to manipulate the object (Figure 2).

As Figure 3 illustrates, 3D widgets are a more immediate paradigm than the typical desktop metaphor (Figure 1), but they are still less immediate than natural interactions.

I feel that the area of human-computer interaction paradigms for VR is largely unexplored. As a result many evolutionary and revolutionary improvements seem possible.

7   PERFORMANCE ISSUES

Performance of VR applications is often measured by the rate with which new frames are displayed. However, frame rate is only one of the important parameters that determine immersion:

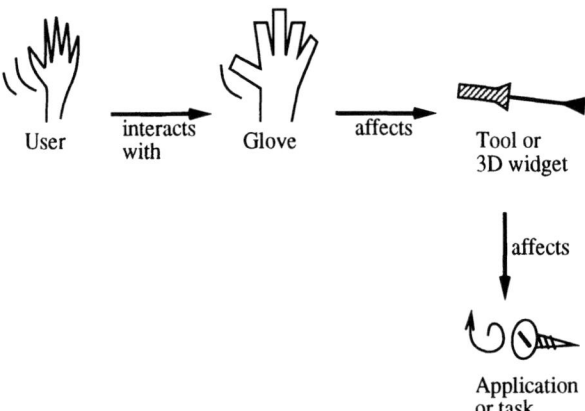

User — interacts with → Glove — affects → Tool or 3D widget

affects

Application or task

Figure 3 The 3D widget paradigm supports interaction that is more direct than 2D paradigms, but it is still less direct than interacting with the real world.

lag is equally important (Liu et al., 1993).

Lag is the time between when a user performs an action and when the application displays the result of that action. For example, when a user moves a 3D input device, various computation stages process and transform the 3D input data to make it visible on-screen, so that the user's movements are displayed with a finite delay. This process is illustrated in Figure 4.

Human beings are extremely sensitive to lag. For instance, depending on the task and surrounding environment, lag of as little as 100ms (less than a tenth of a second) degrades human performance (Liu et al., 1993; Held and Durlach, 1991). Even worse, if lag exceeds 300ms, humans start to dissociate their movements from the displayed effects, thus destroying any immersive effects (Held and Durlach, 1991).

Throughput measurements cannot substitute for lag measurements in assessing the interactivity of an application, since lag and throughput measure different quantities: lag measures how long a computation process delays data and throughput measures how frequently a computation process delivers a result. However, the two quantities are related. Reducing lag in a computation process (for example, by using a faster algorithm) proportionally increases the throughput of that process (but only if one is able to sample new input data quickly enough). Yet increasing throughput does not necessarily decrease lag. For example, using multiple processors in a pipelined configuration increases throughput yet maintains the same lag.

A VR system has several characteristic lag sources. The following two sections describe these sources and how to combat the lag, respectively. Additional information about lag in VR is described in (Wloka, 1995).

7.1 Lag sources in VR systems

User input device lag
The user input device in a VR application reports 3D position and orientation data. It is external to the host workstation and typically communicates data via the serial port. Total user input

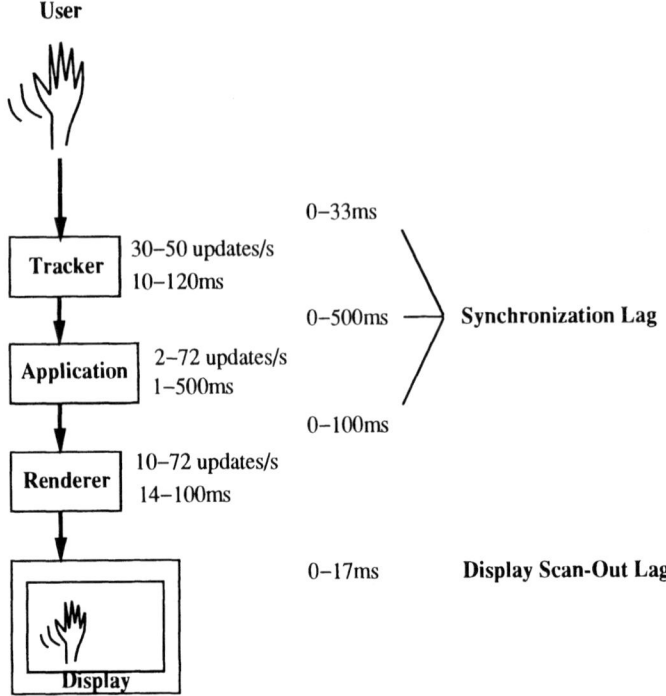

Figure 4 A typical VR application reacts to a user's actions with a finite delay caused by several characteristic components.

device lag includes signal generation and communication time.

Depending on the type of device and mode of operation (i.e. noise filtering on/off, different orientation reporting modes, etc.), lag ranges from 10ms to 120ms; throughput is between 30 and 50 samples per second (Adelstein et al., 1992).

Application-dependent processing lag

Once the user input device data arrives at the host workstation, the application processes it. Processing can be as simple as transforming the data from the device format to the rendering format, i.e. the application echoes the user input device position to the virtual environment. Other more complicated application processes are common, for example, interactive streamline computations for virtual windtunnels (Bryson and Levitt, 1991).

Processing lag is highly application-dependent and thus highly variable. The simple echoing scheme above is the lower bound; today's workstations perform these data transformations in less than one millisecond. The upper bound is harder to characterize. Keeping in mind, however, that the resulting VR system is supposedly immersive, I assume that the lag introduced by application processing does not exceed 500ms, since it is unlikely that applications with input processing requirements beyond 500ms can be made immersive. Therefore, application-dependent processing lag ranges from 1 to 500ms.*

Throughput of the application depends on the number of processors available. In the single-CPU case, the same processor computes the application and also feeds the renderer (see Section 7.1). Therefore, throughput is

$$\frac{1000}{(application\_lag + render\_lag)} \text{ times per second,} \tag{1}$$

with all lag times measured in milliseconds.

With at least two processors available, one is assigned to feed the renderer. Application throughput is thus at least $(1000/application\_lag)$ times per second, which translates to at least twice per second.

If more than two processors are available, the application task should first be parallelized so as to reduce lag. If thereafter application throughput is still worse than rendering throughput (see Section 7.1) and processors are idle, then several instances of the application should run on different user input data until application throughput is equal to or better than rendering throughput. Since rendering throughput is at most 72 times per second (see Section 7.1), application throughput is thus 2 to 72 times per second.

Rendering lag

Rendering lag is the time from sending data to the rendering hardware until the same data is displayed on the monitor. I assume double-buffering rendering hardware that does not use the CPU for rendering computations. Since double-buffering synchronizes the rendering hardware with the display refresh, the finite display refresh rate (typically 60-72Hz) causes a minimum

---

*Of course, application delays of 500ms are only permissible if head-tracking proceeds asynchronously and independently with considerably less lag and higher frame rates; otherwise immersion is not achievable.

rendering lag of 14ms. The maximum rendering lag derives from the minimum frame rate of 10 frames per second: 100ms. Rendering lag is highly scene- and viewpoint-dependent and thus is likely to vary during the runtime of an application.

The scan-out of the display, since it occurs with a frequency of 60 to 72Hz, causes additional lag. Depending on where the rendered data appears on the display and whether the display refreshes from top to bottom or vice versa, the data image may remain invisible for a further 0 to 17ms.

As in the application case, rendering throughput depends on the number of processors available. If only a single processor is available, rendering throughput equals application throughput, i.e. $1000/(application\_lag + render\_lag)$ times per second, since the single CPU computes the application and also feeds the renderer.

If at least two processors are available, assigning one of them exclusively to feed the renderer yields a rendering throughput of $1000/render\_lag$ times per second. Since I also assume that only a single graphics board renders into the framebuffer, the presence of additional processors cannot further influence rendering throughput.

Synchronization lag

Parallel VR applications process user input in several stages: the user input device processing stage, the application-dependent processing stage and the rendering stage.[*] Since these stages are independent, it is possible (and in fact likely) that, for example, the user input device deposits a new sample on the serial port shortly after the application reads the serial port. Thus, the application is busy processing the previous input before it reads the serial port again and starts to process the current input, so that user input data is delayed because it is waiting to be processed by a currently busy stage.

Synchronization lag is the total time data is waiting inbetween stages without being processed. Synchronization lag is thus inversely proportional to the throughput rates of the various stages. It also varies during the runtime of the application.

In the best case, synchronization lag is zero: each stage writes its output just before the next stage reads the data. The worst case is equally likely: each stage writes its output just after the next stage reads the data. Synchronization lag thus varies from 0 to a maximum of the sum of the inverse throughput rates of each stage. On average, synchronization lag is half that maximum, so that average synchronization lag varies depending on the throughput rates of the various stages, i.e. it varies from

$$\frac{\left(\dfrac{1000}{max\_throughput\_UID} + \dfrac{1000}{max\_throughput\_appl} + \dfrac{1000}{max\_throughput\_render}\right)}{2}ms$$

$$= \frac{(20 + 15 + 15)}{2}ms = 25ms \tag{2}$$

---

[*]Even in the single-CPU case, the user input device is separate and independent from the CPU computing the application and feeding the renderer. Thus, VR applications process user input in two stages when running on a single-CPU architecture.

to

$$\frac{\left(\dfrac{1000}{\text{min\_throughput\_UID}} + \dfrac{1000}{\text{min\_throughput\_appl}} + \dfrac{1000}{\text{min\_throughput\_render}}\right)}{2}\text{ms}$$

$$= \frac{(33 + 500 + 100)}{2}\text{ms} = 316.5\text{ms}. \tag{3}$$

While synchronization lag is easy to overlook, it contributes up to 50% of the total lag in a VR system.

7.2 Techniques to reduce lag

Prediction
Prediction methods extrapolate past user input data to future time points, thus reducing perceived lag (Friedmann et al., 1992). However, this extrapolation process introduces spatial inaccuracies that increase under the following three conditions (Friedmann et al., 1992): (1) the user input device throughput is too low; (2) prediction is too far in the future; and (3) the user input device acceleration is too high.

Yet prediction is the only available method that drastically reduces total perceived lag and in particular application-dependent processing lag (since I am discussing the general problem of transforming the user input nontrivially in the application-dependent processing stage). To minimize lag, the user input device stage projects the user input data to the time this data reaches the display. Thus, the user input device stage requires knowledge about the lag experienced by the predicted data in future stages. Prediction thus demands constant (or close to constant) application-dependent processing lag and rendering lag, as well as synchronization lag with as narrow a distribution as possible. In general, even prediction cannot eliminate perceived lag, because of variations in total end-to-end lag. Figure 5 illustrates these requirements.

Time-critical computing
It is not advisable to use time-critical computing (Wloka, 1993) directly to reduce lag. Since time-critical computing trades computation time for computation accuracy, saving maximum time by computing with the least accuracy produces gross visual errors while still not fully eliminating lag. The benefit gained is questionable.

I propose to use time-critical computing to assure constant or nearly constant application-dependent processing lag and rendering lag. This brings us one step closer to being able to use prediction, as shown in Figure 5.

Multiple processors
Multiple processors reduce lag in a VR application in several ways. If the application process is parallelized, application-dependent processing time is reduced directly. Pipelining the application or running several instances of it increases the throughput of the application and thus decreases the expected average synchronization lag of data waiting to be processed by the application. However, the most popular use of multiple processors is to assign at least one to each

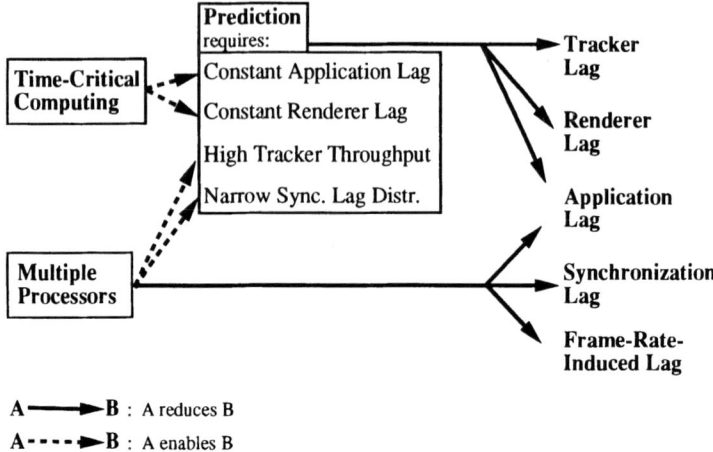

A ———▶ B : A reduces B

A - - - - ▶ B : A enables B

Figure 5   Interdependencies of the various lag-reduction techniques.

computation stage.

Using at least one CPU for each computation stage in a VR application, even in asynchronous communications mode (Shaw et al., 1992; Lewis et al., 1991), has four advantages. First, the user input device is independent from all other stages and thus runs with maximum throughput, allowing use of prediction (Figure 5). Second, rendering also proceeds at maximum throughput. Third, the distribution of synchronization lag is also narrower and thus better than in the single-CPU case. Fourth and finally, by allowing the user-input processor to communicate user input device data directly to the rendering stage, the application can be 'short-circuited:' a low-lag cursor echoing the user input device position is displayed directly in addition to the high-lag application-computed feedback.

8   CONCLUSIONS

Interacting with VR is currently difficult. Many problems in the areas of multiple inputs, multiple outputs, multiple participants, dynamic virtual worlds, user interface paradigms and performance issues exist and must be addressed to advance immersion, rich interaction and presence.

The most limiting factor in today's VR systems is the lack of performance. For now, decreasing lag to less than 100ms is more important than improving image fidelity, model geometry, or model behavior. Video games, such as Doom (Id Software, 1994), convincingly illustrate this observation: Doom achieves presence even without immersive hardware.

9 ACKNOWLEDGEMENTS

I thank my advisor Andries van Dam for his continuing support and José Teixeira for inviting me to the IFIP TC 5 WG 5.10 Workshop on Virtual Environments.

This work was sponsored in part by NSF/ARPA Science and Technology Center for Computer Graphics and Scientific Visualization, Sun, Autodesk, Taco Inc., Microsoft, NASA, NCR and ONR Grant N00014-91-J-4052, ARPA Order 8225.

10 REFERENCES

Adelstein, B.D., Johnston, E.R. and Ellis, S.R. (1992) A testbed for characterizing dynamic response of virtual environment spatial sensors. *1992 UIST Proceedings*, 15–22.

Beaudouin-Lafon, M. and Gaver, W.W. (1994) ENO: Synthesizing structured sound spaces. *1994 UIST Proceedings*, 49–57.

Bryson, S., Feiner, S., Pausch, R., Proffitt, D., Sowizral, H. and van Dam, A. (1994) Developing advanced virtual reality applications. *Course Notes of Course 02 at SIGGRAPH'94*.

Bryson, S. and Levitt, C. (1991) The virtual windtunnel: An environment for the exploration of three-dimensional unsteady flows. *Visualization '91*, 17–24.

Bulterman, D.C.A., van Rossum, G. and van Liere, R. (1991) A structure for transportable, dynamic multimedia documents. *Usenix - Summer 1991*.

Elliott, C., Schechter, G., Yeung, R. and Abi-Ezzi, S. (1994) TBAG: A high level framework for interactive, animated 3D graphics applications. *Computer Graphics (SIGGRAPH '94 Proceedings)*, **28**, 421–434.

Foley, J.D. and van Dam, A. (1982) *Fundamentals of Interactive Computer Graphics*. Addison-Wesley, Reading, MA.

Friedmann, M., Starner, T. and Pentland, A. (1992) Device synchronization using an optimal linear filter. *Computer Graphics (1992 Symposium on Interactive 3D Graphics)*, **25**(2), 57–62.

Gibbs, S. (1991) Composite multimedia and active objects. *Proceedings of OOPSLA'91*, 97–112.

Held, R. and Durlach, N. (1991) Telepresence, time delay and adaptation, in *Pictorial Communication in Virtual and Real Environments* (ed. S.R. Ellis), chapter 14. Taylor and Francis.

Id Software. (1994) Doom. Video game for IBM PC compatible computers.

Isaacs, E.A. and Tang, J.C. (1993) What video can and can't do for collaboration: A case study. *Computer Graphics (Multimedia '93 Proceedings)*, 199–206. ACM, Addison-Wesley.

Kalra, D. and Barr, A.H. (1992) Modeling with time and events in computer animation. *Computer Graphics Forum (EUROGRAPHICS'92)*, **11**(3), C45–C58.

Lewis, J.B., Koved, L. and Ling, D.T. (1991) Dialogue structures for virtual worlds. *Proceedings of CHI'91*, 131–136.

Liu, A., Tharp, G., French, L., Lai, S. and Stark, L. (1993) Some of what one needs to know about using headmounted displays to improve teleoperator performance. *IEEE Transactions on Robotics and Automation*, **9**(5), 638–648.

Regan, M. and Pose, R. (1994) Priority rendering with a virtual reality address recalculation pipeline. *Computer Graphics (SIGGRAPH '94 Proceedings)*, **28**, 155–162.

Schaufler, R. (1992) Realtime workstation performance for MIDI. *Usenix Winter '92.*

Shaw, C., Liang, J., Green, M. and Sun, Y. (1992) The decoupled simulation model for virtual reality systems. *Proceedings of CHI'92*, 321–328.

Snibbe, S.S., Herndon, K.P., Robbins, D.C., Conner, D.B. and van Dam, A. (1992) Using deformations to explore 3D widget design. *Computer Graphics (SIGGRAPH '92 Proceedings)*, **26(2)**, 351–352.

Tarlton, M.A. and Tarlton, N.P. (1992) A framework for dynamic visual applications. *Computer Graphics (1992 Symposium on Interactive 3D Graphics)*, **25(2)**, 161–164.

Tarlton, M.A. (1993) *A Declarative Representation System for Dynamic Visualization.* PhD thesis, University of Texas at Austin.

Wloka, M.M. (1993) PhD thesis proposal: Time-critical graphics. Technical Report CS-93-50, Brown University, Department of Computer Science, Providence, RI.

Wloka, M.M. (1995) Lag in multiprocessor virtual reality. *PRESENCE: Teleoperators and Virtual Environments*, **4(1)**.

Zeleznik, R.C., Conner, D.B., Wloka, M.M., Aliaga, D.G., Huang, N.T., Hubbard, P.M., Knep, B., Kaufman, H., Hughes, J.F. and van Dam, A. (1991) An object-oriented framework for the integration of interactive animation techniques. *Computer Graphics (SIGGRAPH'91 Proceedings)*, **25(4)**, 105–112.

Zyda, M.J., Pratt, D.R., Monahan, J.G. and Wilson, K.P. (1992) NPSNET: Constructing a 3D virtual world. *Computer Graphics (1992 Symposium on Interactive 3D Graphics)*, **25(2)**, 147.

11   BIOGRAPHY

Matthias M. Wloka is currently a PhD student at Brown University, Providence, USA; he is working with Professor Andries van Dam and the Brown Computer Graphics Group. Mr. Wloka received his B.Sci. in Computer Science from Christian Albrechts University, Kiel, Germany in 1987 and his M.Sci. degree from Brown University in 1990. His research interests include interactive, real-time computer graphics, in particular computing object behavior and geometry time-critically.

17

Isaac: Building Simulations for Virtual Environments

James F. Cremer
University of Iowa
Computer Science Department, University of Iowa, Iowa City, IA, 52242
USA. email: `cremer@cs.uiowa.edu`

George Vaněček
Purdue University
Computer Science Department, Purdue University, West Lafayette, IN,
47907 USA. email: `vanecek@cs.purdue.edu`

Abstract

This paper describes the architecture and initial implementation of the *Isaac* system. Our general research goal is to develop simulation support for virtual environments. Existing virtual environments are often graphically rich, but behaviorally impoverished. On the other hand, existing physical system simulation software is not well-suited for use within virtual environments. The *Isaac* system is a distributed simulation server that integrates multibody dynamics, geometry, and control and is designed to support the needs of virtual environments.

Keywords

rigid-body dynamics, simulation, virtual environments, geometric modeling, contact, collisions, control.

1 INTRODUCTION

Existing virtual environments are often visually rich but behaviorally impoverished; users may be able to walk through geometrically complex worlds rendered using high-performance graphics hardware and software, but the worlds are typically populated with objects that do not behave as humans expect them to. Either users can only look at the objects, or the objects do not behave in physically satisfying ways — objects released from a hand don't fall, for example, or active non-user entities such as robots are purely scripted and not reactive. Such shortcomings can now be addressed by physically-based simulations. These simulations can greatly enrich virtual environments and will certainly become an integral part. Yet research on incorporating physical simulations into interactive virtual environments lags behind other developments. An approach combining the sound technical basis of mechanical engineering work in dynamics simulation with the interactivity and controllability of graphics and animation systems is needed.

The graphics, animation, and virtual environment communities have shown great interest in physically-based simulation since it provides a means to enhance the believability of their products. In the engineering community, an enormous amount of physical systems simulation research has been carried out. However, existing simulation tools were not designed specifically to support the requirements of virtual environments and, in fact, do not support them well. Dynamics simulation systems from mechanical engineering — e.g. DADS (Haug, 1989), ADAMS (Orlandea, 1987) — support analysis of mechanisms in a standard paradigm: formulate motion equations and kinematic constraints, and then numerically integrate them over time. They do not support control of complex high-degree-of-freedom objects, and do not integrate geometry and dynamics well enough to support n-body collision detection and two-body contact analysis on other than a rudimentary level. Work in the graphics and animation community has produced software that is somewhat more usable in virtual environments—for instance, they support interactivity and some collision detection techniques. However, many are not technically sophisticated or robust — they are not based on sound, accurate, efficient numerical techniques, and they will not scale to virtual environments of interesting size (e.g. multiple many-legged walking robots interacting in complex geometric environments).

In this paper, we introduce the architecture of *Isaac*, a distributed simulation server designed to provide efficient, robust, and flexible physically-based simulation within virtual environments. It is designed to support simulation of complex physical systems at interactive rates, to efficiently and robustly handle collision and contact phenomena, and to provide powerful, clean mechanisms for motion and scenario control. It will be possible to simulate virtual worlds populated with autonomous creatures under various modes of control. The control could range from simple kinematics-based scripting to semi-autonomous behavior and scenario control.

In the next section, we introduce the basic components of *Isaac*. Section 3 presents some historical context for *Isaac*, describing its roots in our earlier work on dynamics simulation and geometric modeling. Section 4 presents the *Isaac* system architecture, with sections covering the simulation core, dynamics, geometry, control, and distributed computation. The status of the *Isaac*, including the first implementation, is presented in Section 5.

2   ISAAC OVERVIEW

Designed to provide efficient, robust and flexible simulation support for virtual environments, *Isaac* comprises five basic components:

- a simulation core that contains state-of-the-art numerical methods and efficiently and robustly handles *on-line constraint changes*. In virtual environments collisions occur, contact relationships change, and motor control programs or high-level plans change state. In *Isaac*, these correspond to constraint changes in the underlying equations.
- a dynamics module that is responsible for formulating the motion equations that capture the basic behavior of physical objects and for interacting with geometry to handle collision and contact dynamics.

- a geometry module that efficiently and robustly supports n-body collision detection and two-body contact analysis; also, a geometric database that will manage the global geometric information of a virtual environment to enable such operations as proximity queries and planning.
- a control module that supports high-level specification of motion control (including specification of low-level controllers such as PID controllers for, say, robot joints, as well as higher-level controllers coordinating a high-degree of freedom mechanism such as an anthropomorphic robot) as well as scenario and behavioral control (including coordinating of multiple agents, planning and control high-level agents behavior).
- a task management module that manages the distribution of computations across a set of *Isaac* server processes. The task manager oversees resource allocation, synchronizes computations as necessary, and manages interprocess communication.

Isaac integrates three components crucial to virtual environments—dynamics, geometry, and control. To ensure scalability and efficiency, it is designed for distributed computation.

3   BACKGROUND — *Isaac*'S FORERUNNERS

Our *Newton*(Cremer, 1989; Cremer and Stewart, 1989) system was one of the first attempts to integrate geometry and control with dynamics. One of the driving problems was the design of multifingered dextrous robot manipulators. A system that is to be used to evaluate hand designs must support not only dynamics but also geometry, to analyze contact during manipulations, and control, to test controllability of the hand. It was successful as an test for collision, contact, and control research. However, the contact and collisions module never reached an acceptably robust and efficient level, primarily because the interplay between dynamics and geometry had not been carefully studied before the development of *Newton* and was not well understood.

Until 1989, *Newton* dealt with geometry only as parameterized primitives. Later, Vaněček integrated the *ProtoSolid* polyhedral geometric modeling system with *Newton*, providing the system a broad range of convex and nonconvex polyhedral objects. The integration was difficult because *ProtoSolid* was originally designed to support mechanical design, not dynamics simulation. Specialized support for collision detection had to be added. This was accomplished in part by using Binary Space Partition (BSP) Trees (Thibault and Naylor, 1987). With the BSP tree support *Newton* could perform simulations with any polyhedra, but only for simple contacts. Complex contact models such as the simulation of the tumbling rings, shown in Figure 1, failed due to insufficient contact information obtainable from the BSP support. Consequently, Vaněček generalized the trees to multi-dimensional structures and later to the Brep-Index, and is incorporating this in a system called Proxima (Sun, Van Vleet, and Vaněček, 1992).

The control portion of *Newton* was its most successful component. Paradigms for *programming* the control of high-degree-of-freedom mechanisms were developed and applied to graphics and animation as well as mechanical engineering and robotics. This work has been influential, for example, in the development of the scenario control subsystem of the Iowa Driving Simulator (Cremer and Kearney, 1994). At Iowa, Hansen develop a programming framework that enabled users to create complex simulations involving multiple interacting robots from clear, concise, control programs (Hansen, 1993). At Purdue, Bouma was able to program control a four-legged walking robot in *Newton*(see Figure 2). At Cornell, Pai and others programmed bipeds and hopping machines for tasks including standing-up, sitting, walking, jumping and riding a bicycle (Pai, 1991; Stewart and Cremer, 1992; Kearney, Hansen, and Cremer, 1993).

Figure 1 *Newton* simulation of tumbling rings. This idea was obtained from an article in "Mathematical Games", Scientific American, 1965.

4 *Isaac* ARCHITECTURE

The basic *Isaac* architecture is shown in Figure 3. Each of the dynamics, geometry, and control modules interacts with the simulation core in terms of constraints. Dynamics formulates basic motion equations and kinematic constraints and hands them to the simulation core. During simulation, the dynamics, control, and geometry modules modify the initial equation set by adding and removing equations as warranted by the occurrence of events.

4.1 Simulation core

A crucial feature of a simulation system for virtual environments is the ability to handle changes: collisions occur; contacts form, remain for a while, and break; motion control algorithms change state; active agents change their goals based on sensed information; and so on. In *Isaac* such changes are signaled by *events*. Handling of events generally consists of changing the set of equations representing object behavior. For example, two initially not-in-contact mechanisms may be modeled by two independent sets of equations (motion equations, kinematic constraints, and perhaps some control equations). If the mechanisms come into contact (and don't immediately break contact - i.e. that don't just bounce away from each other) an equation representing a new kinematic constraint will be added. This equation couples the two previously independent sets of equations. At some later time the contact might break; the equation set would then be modified again.

The design of *Isaac*'s simulation core has two major goals; namely,

Figure 2 Controlling GUMBY, a four-legged simulated walking robot.

- support efficient constraint changes, and
- support modularity and "constraint programming" style of module interaction.

From our work with *Newton*, we found that it was especially convenient to view the module interaction in "constraint programming" terms. The *Isaac* architecture makes this explicit. At the lowest level of the system lies the *simulation core*. It is ultimately responsible for solving a set of equations and advancing the simulation through time. The set of equations that the core solves can be viewed as a set of constraints that the other modules — dynamics, geometry, and control — manipulate. When events occur these modules may add, remove, or modify constraints. These higher-level modules are provided with a "constraint programming" view of the simulation. They interact with the simulation core through a simple well-defined constraint manipulation interface. Note that while the interface may be simple to define (e.g. containing a small number of constraint set manipulation routines) it is not a trivial matter to implement it well. The simulation core will contain a variety of equation solving methods. For example, for some problems standard DAE solvers like MEXX (Lubich et al., 1992) will be appropriate. For others, especially those involving a significant number of collisions and contact changes, a more specialized integrator such as that outlined in Section 4.2 will be necessary. For efficiency purposes, the constraint programming interface routines will each have a number of implementations based on the various solvers. If instead, we had a generic set of interface routines that worked in terms of some common symbolic equation format, the simulation core would have to translate between that representation and a particular solver's representation at run-time. This would, in general, lead to unacceptable performance.

Within the simulation core lies the event manager. Various *Isaac* modules can define *events* by specifying how an event is to be *detected* and how it is to be *resolved*. Event detection may correspond to a

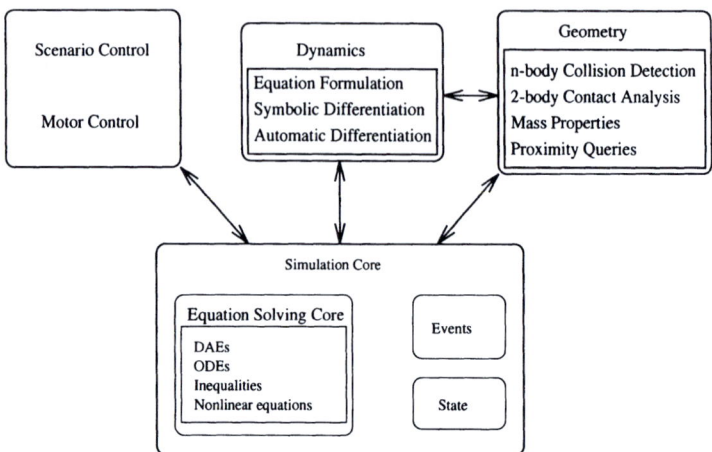

Figure 3 The *Isaac* architecture.

function value passing through zero or to detection of geometric interpenetration. Event resolution may involve formulating a set of equations representing handling of impact, adding or removing equations corresponding to contact constraint changes, or simply changing gain values within a controller. A number of issues complicate event handling; event occurrence time must be *isolated* efficiently, and continuation of the simulation after an event must be done with maximal efficiency and accuracy.

4.2   Dynamics

The dynamics module of *Isaac* is responsible for formulating a set of motion equations and for providing them to the simulation core. It is also responsible for formulating equations (and related mathematical information such as Jacobians) corresponding to kinematic constraints. For mechanisms involving only *permanent* kinematic constraints, those corresponding to standard physical joints such as revolute joints, a variety of standard dynamics formulations will be used. The basic formulation uses a maximal set of Cartesian coordinates, in the style of Haug/DADS (Haug, 1989) and Cremer/Newton (Cremer, 1989). Such formulations are particularly amenable to specifying and implementing constraint changes. As development of *Isaac* proceeds and as efficiency considerations require, other formulations, such as the recursive formulations developed by Haug and colleagues at Iowa (Tsai and Haug, 1991; Bae and Haug 1987), will be introduced.

When contact constraints, called *temporary constraints*, are present, the dynamics module interacts with the geometry module to formulate the appropriate set of equations*. As outlined below, a contact

---

*Throughout the paper, we use the term *equations* to include inequalities as well as true equations.

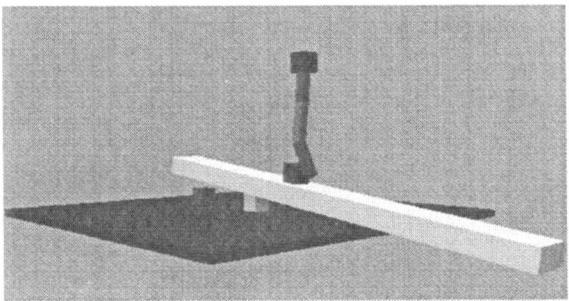

Figure 4 A simple simulation involving collisions and sliding contact.

constraint is modeled using two sets of inequalities: one for the dynamics portion of the constraint and one for the geometric portion of the constraint. As an example, Figure 4 shows a set of blocks in temporary contact.

Integrating Geometry and Dynamics

To achieve efficient and robust simulations, the dynamics and geometry components of the system must be integrated with great care. In particular, the roles of geometry and dynamics components in contact analysis must be well-defined. Consider, for example, a block sliding down an inclined table. Suppose, for simplicity, that the block is oriented so that just one of its corners is in contact with the table. The point-on-face contact is modeled using (1) an inequality that constrains the point to be on or above the plane of the table, (2) a force condition that says that the contact remains only so long as the reaction force is non-tensile, and (3) conditions indicating that the contact remains only as long as the point remains within the geometric bounds of the table top (the bounds of the face). We distinguish two ways in which the contact can break. One involves a force pulling up on the block such that it breaks contact by lifting off the face. This is a dynamics event — it arises because the non-compressive reaction force condition cannot be consistently maintained. The second type of contact breakage involves the block sliding off the end of the table. This is a geometric event corresponding to violation of the conditions about face bounds.

In early versions of *Newton*, these types of events were not carefully discriminated. In the given example, the geometry module would check at each time instant to see if the objects were in contact or not. If the objects had been in contact at one time instant, but the geometry module determined that at the next time instant they were not in contact, an event would be generated and the contact constraint would be removed. This is, in fact, the wrong thing to do in many cases. Numerical integration can only maintain contact within some prescribed tolerance. It is difficult, at best, to maintain the complete consistency between the dynamics and geometry modules' tolerances that would be required for a geometry-based decision in this situation to be guaranteed to be correct. Precise consistency is, however, not necessary for handling dynamics-type contact events. Unless the contact has reached the face boundary, it can only be broken by the inability to maintain the force condition. Thus, it does not matter if the contact exists or not from the point of view of the geometry module; dynamics can and should make the decision. On the other hand, the second type of contact breakage, that involving the block sliding off the end of the table, does correspond to a geometric event. When it is geometrically determined that the point has reached the face boundaries, a geometric event is signaled and the contact analysis routines analyze the situation and update the equations with new contact constraints.

To most efficiently support our model of dynamics-geometry integration, novel numerical integration techniques are required. Historically, multibody dynamics simulation programs have relied on ordinary differential equation (ODE) integrators with additional code wrapped around them to allow them to accurately handle differential-algebraic equations systems (DAEs). Recently, integrators designed especially for differential-algebraic equations have become available. Currently, Cremer and F. Potra are are working to develop to develop a new DAE integrator that is designed especially to efficiently and robustly handle changing constraint sets that include inequalities.

4.3 Geometry

Geometric support for simulation in virtual environments must include:

1. the representation of the geometry of the environment which takes into account moving objects, fixed objects such as the floor, and proximity queries,
2. the determination of mass properties of the movable solid objects,
3. fast n-body collision prediction,
4. fast two-body collision detection, and
5. fast and robust two-body contact analysis.

Representing the Geometric Environment

Currently, to limit complexity, we assume that environments consist of nondeformable objects that do not interpenetrate. We partition the objects into two categories: movable objects and fixed, immovable objects (e.g. the ground or walls). Objects that move are currently modeled using planar polyhedra. We plan to extend the representations to include free-form surfaces. Fixed objects can be modeled using relatively large, oriented lamina (i.e., surfaces) and not closed volumes. In terms of the dynamics, the objects that do not move need not have their mass properties computed or motion equations formulated; when convenient, all fixed objects can be combined geometrically into a single complex object.

Mass Properties

For objects to move, the dynamics module must formulate the motion equations using the mass properties of the objects. This consists of computing the inertia matrix which encodes the moments of inertia of the object. Since the objects are assumed rigid, the inertia matrix does not change during a simulation and can be precomputed. This is a well-understood problem for which efficient boundary-based algorithms exist.

Collision Detection

Isaac relies on automatic detection of collisions between objects. Note that two objects that are already in contact, such as a book on a table, may also collide, such as when the book falls over. To illustrate this, consider Object s_1 of Figure 5 sliding on top of Object s_2; at some later time, s_1 collides with the vertical inside wall of s_2. This subsequent collision is handled as a contact detection and analysis problem. From

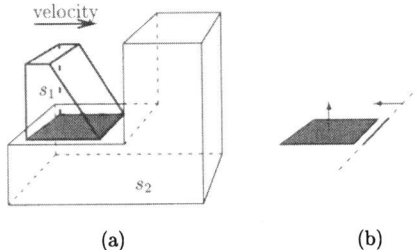

Figure 5 *An example where a two objects in temporary-contact will also collide. (a) show the objects at time t, and (b) shows the two regions at time t + Δt, one for the temporary-contact, the other for the collision.*

the geometry alone it is not possible to distinguish a contact from a collision. Thus once two objects come into contact, the collisions and contact have to be analyzed simultaneously.

Since all collision events stop time to change the equation motion velocities, and this can happen quite frequently, the detection algorithm must be very fast. For objects that are far apart, the exact geometry of the boundary is not important. For this reason and for computational benefits, one may approximate objects that are far apart by their convex hull and then check their proximity using the fastest known algorithm for convex object, the Lin and Canny's algorithm (Lin and Canny, 1991). However, we have to handle n objects simultaneously. Even this fast two-body collision detection algorithm requires $O(n^2)$ time if it performs pair-wise collision checks, and this is prohibitive if large number of moving objects are simulated.

There are a number of algorithms that address the n-body problem. For instance, Lin, Manocha and Canny give a simple extension of the Lin and Canny's algorithm for convex objects by estimating the possible time of collision (Lin, Manocha, and Canny, 1993). A similar extension was proposed by Dworkin and Zeltzer at MIT; it uses simple time-parameterized object trajectories to predict possible intersections (Dworkin and Zeltzer, 1993). In both cases, the predicted times are placed into a time-prioritized queue. The simulation then continues until the time of the first event on the queue, at which time the two objects indicated by the event are checked. The simulation then continues until the next event on the queue. These approaches assume that the number of collisions in any time interval is small, the trajectories of objects are known a priori and that objects are relatively far away from each other. These assumptions are not always reasonable.

In our case, however, we use a totally different approach mainly due to the assumption that objects can be of any shape and that they can be in prolonged contact. We are developing an n-body collision detection algorithm that integrates three components: a Locally-resolvable Boundary Representation (LRBep) (Gonzalez-Ochoa and Vaněček, 1994), an Extended Binary Space Partitioning (EBSP) tree, and lazy evaluation. LRB-reps are basically a generalization of bounding-volumes. The faces of a B-rep are locally abstracted to form a single multi-resolution structure that can support a locally resolvable wrapper. A wrapper is a subset of the LRB-rep's faces which enclose the original object but which can be locally refined to higher levels of detail. Each object can thus be approximated by a wrapper. As two objects come close, their wrappers may intersect. However, these wrapper can be locally refined at the intersection to expose more detail. To attain the n-body collision check, the wrappers of all objects are

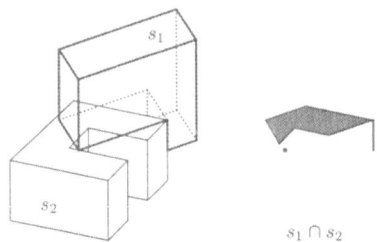

Figure 6 *Two objects in contact, and their three contact regions resulting from a set-theoretic intersection.*

inserted into a global BSP tree. The insertion is only partial in that a wrapper is not completely processed by the BSP tree. Only the parts that conflict with other wrappers are processed and only at local regions of intersections are the wrappers refined to higher levels of detail.

Contact Analysis

Contact analysis is a process that provides a detailed description of the two-body contact regions. Note that the property that objects cannot interpenetrate is not supported by the representations of the objects. Regardless of the representation—BRep, CSG or Octrees, for instance—there is no inherent support for disallowing their interpenetration. The property must be supported computationally. The dynamics module does this by adding constraint equations to the set of motion equations which reduces the degrees of motion freedom. The contact regions are converted to equations that describe where forces are applied to the geometric limits of the region to keep the two objects from interpenetrating.

Bouma and Vaněček have shown that the contact analysis requires a full set-theoretic intersection of the two objects to determine the contact regions, followed by the analysis of the contact regions (Bouma and Vanecek, 1993). For example, Figure 6 shows two objects in contact and the set of contact regions obtained from the set-theoretic intersection.

This can be done easily in $O(N^2)$ time where N is the number of vertices, edges and faces in both objects. For large N, the cost can be prohibitive. To speed up this analysis, Vaněček has developed two techniques: Brep-index (Vaněček, 1991) and back-face culling (Vaněček, 1994). Basically, the analysis begins by culling the vertices, edges and faces that are know a priori to be moving away from the other object and therefore cannot be contact and then classifying the unculled vertices, edges and faces of one object against the Brep-index of the second object. For efficiency, we check the Brep of the smaller object against the Brep-index of the larger object. The topological entities that lie on the boundary of the other object are retained and combined into contact regions. For robustness, only the Brep of one object is checked against the other. This alleviates classic robustness problems found in geometric modeling systems.

The Brep-Index

Contact analysis is based on analyzing contact regions; thus, it is boundary based. However, the classification that obtains the contact regions is inherently spatial, not boundary based. For this reason, Vaněček developed a spatial representation of an object that recursively subdivides space into open halfspaces called a multi-dimensional space partitioning (MSP) tree. It is a direct extension of the binary-space partition (BSP) tree. The MSP structure allows for a fast search that quickly converges to the region containing the query point, line segment or polygon. To gain the benefit of both the efficient spatial search and the detail of the boundary, Vaněček combined the MSP tree and the BRep to yield a single unified representation for objects. This representation is called the Brep-index (Vaněček, 1991b).

Back-Face Culling

In order to find contact regions, we can classify all the faces, edges and vertices of one of the objects against the Brep-index of the other object. Although this reduces the cost of classification from quadratic to subquadratic, we can reduce the cost even more. Vaněček has applied the well-known back-face culling idea of computer graphics to that of the contact analysis, basing decisions relative velocity instead of on view direction. A face may be culled (i.e., it is known a priori not to collide) when the relative-velocity vectors of the points on the face are all pointing in the opposite direction of the normal. Combining the back-face culling technique with the brep-index yields a very efficient technique for detecting collisions for objects in close proximity and for determining contact regions.

Proxima

Proxima is a set of C++ routines provided as a library, and intended to provide the geometric support described in this section (Sun, Van Vleet, and Vaněček, 1992). The primary representation for objects is the BRep, with the MSP tree and the Brep-index as secondary representations. Through these representations, *Proxima* provides a wealth of low-level geometric operations to query the boundary, classify entities, and to obtain mass properties. Initial implementations of the contact analysis and collision detection components of *Proxima* are complete, but significant further development is required.

On the Geometric Complexity

On first inspection, it may appear that the geometric support is unnecessarily complex. This complexity is, never-the-less, inherent in the need to have a well integrated spatial and boundary representation, and the need to support consistent and robust operations in interactive times. Although there are other possible choices for the data structures and algorithms, we feel that we've chosen the best of the current state of the art.

4.4   Control

The ability to control, direct, and choreograph the activities and behaviors of complex active entities is an essential ingredient of a virtual environment system. Here, we make a somewhat hazy distinction between *motion control* and *scenario control*. Motion control consists of specifying and implementing the control of mechanisms in physical terms — i.e. motion control typically consists of specifying joint torques, forces, and accelerations, or constraints on such quantities. Motion control can be quite complex and can involve significant programming in terms of control events that dictate when control parameters or constraints should change. We include in motion control such basic control mechanisms as PD and PID joint controllers. Less clearly in the realm of motion control are programs that control an anthropomorphic robot to walk.

Scenario control consists of higher level controlled activity of simulated entities. It can include AI-style planning activities, the results of which activate appropriate motion control programs. It also includes the coordinating, directing, and choreographing of the activities of multiple simulated entities in accordance with the goals of the scenario author. Virtual environments will have to be flexible and provide a means for a person (either the VE designer/builder, an experimenter, or even the user) to mold the scenario to fit their needs. One person will want four robots behaving and interacting with the user in a particular way, while another will want some different number of robots doing substantially different things.

Isaac is designed to support both motion control and scenario control. As described in the Section 4, the control module will interact with the simulation core in a constraint-programming style. At the lowest level control programs correspond to time-varying sets of constraints, with control events determining the constraint set modification times. At the user-level control programs will be specified in a framework based on Cremer, Kearney, and Hansen's previous work (Kearney, Hansen, and Cremer, 1993; Hansen, 1993; Hansen, Kearney, and Cremer, 1994) on control for mechanical simulation and on related work by others (Harel, 1987; Brooks, 1989; Reynolds, 1987). The framework is based on a notion of concurrent, hierarchical state machines that is currently being developed by Cremer and Kearney for behavior modeling and scenario control in Iowa Driving Simulator (Cremer and Kearney, 1994; Ahmad et al., 1994). We also intend to integrate our work on control some of the Bates' work (Bates, 1992; Kelso, Weyrauch, and Bates, 1993) on "believable agents".

4.5   Distributed computation with *Isaac*

The geometry, dynamics, control, and simulation core components of *Isaac* are being developed with state-of-the-art efficiency in mind. However, a system such as *Isaac* will naturally benefit from a distributed system organization. Thus, at the top level *Isaac* consists of a set of *Isaac* simulation-server processes managed by a *task management* process. Each *Isaac* process is a self-contained simulation server; the entire computation *could* be done within any single process. However, it is the combined responsibility of server processes and the task manager to distribute computations across multiple *Isaac* processes. For example, at a fairly simple level, there can be one *Isaac* process for each independent (e.g. kinematically independent) mechanism. Each process acquires updated state information about other objects as needed. In this model, each process has a local cache containing the complete simulation state, but it is only responsible and authorized to manipulate the components of the state corresponding to its assigned object. Such information replication and implied state communication is reasonable within *Isaac* because the anticipated maximum number of entities is relatively small (i.e. hundreds of moving objects).

5   IMPLEMENTATION STATUS

Isaac is being developed jointly by the University of Iowa and Purdue University. Iowa is responsible for the design of the overall system architecture, and for the dynamics and control components of the system. Purdue is responsible for addressing the substantial geometric demands of *Isaac*.

Isaac ran its first simulations in July 1994. The first version was able to simulate multiple fairly simple articulated objects (e.g. pendulums) in real-time (30 frames/second or better) on a network of workstations. Individual workstations simulate one or more mechanisms, depending on their complexity.

We developed a deterministic time, distributed virtual environment system that allowed users to sit at graphics workstations and interactively examine the simulation. Visualization clients can run on all workstations in a local-area network, each displaying a unique view into the synchronized environment. *Isaac* is being developed in C++ on Silicon Graphics workstations. Two visualization clients have been developed — one using openGL and a simpler X windows-only version. A tool that allows users to create complex mechanism and manipulate them into desired initial configurations is being developed using SGI's Open Inventor. Future versions of *Isaac* will be publically available.

6 CONCLUDING REMARKS

The development of high-performance computer graphics hardware and software, head-mounted displays, and a range of position tracking, haptic, and multi-degree of freedom interaction devices, has enabled the creation of visually rich interactive virtual environments. One of the most important, but least developed, facets of virtual environments research is the support for *behavior*. To provide a realistic, satisfying experience, users must be able to interact with objects they encounter in virtual environments, and have those objects behave according to a physically-based model. Users must be able to pick up, carry, push, throw, swing various things they encounter in then environment. Furthermore, a successful environment oftens requires more complex, active, semi-autonomous entities — objects, such as robots or humanoids, that have a substantial control component integrated with the physically-based model.

Our system, *Isaac*, is being designed to address the simulation needs of virtual environments. By creating a distributed simulation server that integrates efficient dynamics with geometric computation and control programming facilities, *Isaac* will make it possible to populate virtual environments with complex objects exhibiting believable and interesting physically-based behavior.

7 ACKNOWLEDGEMENTS

This work is supported by Office of Naval Research grant N00014-94-1-0576. Iowa's Michael Booth and Paul Henning developed much of the first *Isaac* implementation.

REFERENCES

Ahmad, O., Cremer, J., Hansen, S., Kearney, J. and Willemsen, P. (1994) Hierarchical, concurrent state machines for behavior modeling and scenario control. In *Proceedings of the 1994 Conference on AI, Simulation, and Planning in High Autonomy Systems*, Gainesville, December, pp. 36–42.

Bae, D. S. and Haug, E. J. (1987) A recursive formulation for constrained mechanical systems, part II — closed loop. *Mechanics of Structures and Machines*, **15**(4).

Joseph Bates. (1992) Virtual reality, art, and entertainment. *PRESENCE: Teleoperators and Virtual Environments*, **1**(1), pp. 133–138.

Bouma, W. and Vaněček, G. (1993) Modeling contacts in a physically based simulation. In *Proceedings of 2nd ACM Symposium on Solid Modeling and Applications*, Montreal, May 1993, pp. 409–418. (revised version to appear in Computer Aided Design).

Brooks, R. A. (1989) A robot that walks: Emergent behaviors from a carefully evolved network. In *Proceedings of the 1989 IEEE International Conference on Robotics and Automation*, May, pp. 692–696.

Cremer, J. (1989) *An Architecture for General Purpose Physical System Simulation — Integrating Geometry, Dynamics, and Control*. PhD thesis, Cornell University, May 1989.

Cremer, J. and Kearney, J. (1994) Scenario authoring for virtual environments. In *Proceedings of the IMAGE VII Conference*, Tucson, AZ, June, pp. 141–149.

Cremer, J. and Stewart, A. J. (1989) The architecture of Newton, a general-purpose dynamics simulator. In *Proceedings of the 1989 IEEE International Conference on Robotics and Automation*, May, pp. 1806–1811.

Dworkin, P. and Zeltzer, D. (1993) A new model for efficient dynamic simulation. In *Proceedings of the 4th Eurographics Workshop on Animation and Simulation*, Barcelona, September 1993, pp. 135–147.

Gonzalez-Ochoa, C. and Vaněček, G. (1994) Locally resolvable b-reps (revised january 1995). Technical Report TR-94-076, Dept. of Computer Science, Purdue University, November 1994.

Hansen, S. (1993) *Conceptual Control Programming for Physical System Simulation*. PhD thesis, Computer Science Department, University of Iowa, May 1993.

Hansen, S., Kearney, J. and Cremer, J. (1994) Motion control through communicating, hierarchical state machines. In *Proceedings of the 5th Eurographics Workshop on Animation and Simulation*, Oslo, September 1994.

Harel, D. (1987) Statecharts: A visual formalism for complex systems. *Science of Computer Programming*, 8(3), June 1987, pp. 231–274.

Haug, E. (1989) *Computer Aided Kinematics and Dynamics of Mechanical Systems, Volume I: Basic Methods*, Allyn and Bacon.

Kearney, J., Hansen, S. and Cremer, J. (1993) Programming mechanical simulations. *The Journal of Visualization and Computer Animation*, 4(2), April-June 1993, pp. 113–129.

Kelso, M., Weyhrauch, P. and Bates, J. (1993) Dramatic presence. *PRESENCE: Teleoperators and Virtual Environments*, 2(1).

Lin, M. and Canny, J. (1991) Efficient algorithms for incremental distance computation. In *Proceedings of the 1991 IEEE International Conference on Robotics and Automation*.

Lin, M., Manocha, D. and Canny, J. (1993) Fast collision detection between geometric models. Technical Report TR93-004, Department of Computer Science, University of North Carolina at Chapel Hill, January 1993.

Lubich, Ch., Nowak, U., Pöle, U. and Engstler, Ch. (1992) Mexx — numerical software for the integration of constrained mechanical systems. Technical Report Technical Report SC-92-12, Konrad-Zuse-Zentrum für Informationstechnik, Berlin.

Orlandea, N. (1987) Adams – theory and practice. *Vehicle Systems Dynamics*, **16**, pp. 121–166.

Pai, D. (1991) Least constraint: A framework for the control of complex mechanical systems. In *Proceedings of the American Control Conference*, Boston, MA, American Automatic Control Council, pp. 1615–1621.

Reynolds, C. (1987) Flocks, herds, and schools: A distributed behavioral model. In *Computer Graphics (SIGGRAPH 1987)*, ACM, July, pp. 25–34.

Stewart, A. J. and Cremer, J. (1992) Animation of 3d human locomotion: Climbing stairs and descending stairs. In *Proceedings of the 3rd Eurographics Workshop on Animation and Simulation*, Cambridge, England, September 1992.

Sun, G., Van Vleet, P. and Vaněček, G. (1992) Proxima, a polyhedral distance and classification support in C++. Technical Report CER-92-41, Dept. of Computer Science, Purdue University, November 1992.

Thibault, W. and Naylor, B. (1987) Set operations on polyhedra using binary space partitioning trees. In *Computer Graphics (SIGGRAPH 1987)*, ACM, July, pp. 153–162.

Tsai, F. and Haug, E. (1991) Real-time multibody system dynamic simulation, part I — modified recusive formulation and topological analysis. *Mechanics of Structures and Machines*, **19**(1).

Vaněček, G. (1991) Brep-index: A multidimensional space partitioning tree (revised). *International Journal of Computational Geometry and Applications*, **1**(3), September 1991, pp. 243–262.

Vaněček, G. (1991b) Brep-index: A multi-dimensional space partitioning tree. In *Proceedings of 1st ACM Symposium on Solid Modeling and Applications*, Austin, June 1991, pp. 35–44.

Vaněček, G. (1994) Back-face culling applied to collision detection of polyhedra. *Journal of Visualization and Computer Animation*, **5**(1), January 1994, pp. 55–63.

AUTHOR BIOGRAPHIES

James F. Cremer is an Assistant Professor of Computer Science at the University of Iowa. He received his Ph.D. from Cornell University in 1989. He remained at Cornell as a Research Associate from 1989–1992 and led research and development of the Newton dynamics simulator and, with Rick Palmer, initiated the SimLab project. Since moving to Iowa in 1992, he has been active in research in environments for constructing simulators (SimLab and related projects), scenario control for real-time vehicle simulation (with Iowa's Center for Computer-Aided Design and the Iowa Driving Simulator), and techniques for efficiently handling constraint changes related to contact, collision, and control in simulation-based virtual environments (project Isaac).

George Vanecek is an Assistant Professor in the Department of Computer Sciences at Purdue University. His work includes geometric modeling, computer graphics, animation, simulation, virtual environments, and recently, enterprise integration. In 1992, trying to solve the collision problem, he formulated a representation of polyhedra that unified a spatial representation, a multi-dimensional space partition trees, and a boundary representation, called the Brep-Index. The work later led to a model of contact in his work on contact-analysis of objects. Currently he is working on the geometric support for simulations of complex virtual environments and on virtual enterprise integration on the World-Wide Web. Vanecek received his Ph.D. from the Department of Computer Science at University of Maryland, College Park.

18

Task performance using 3D displays

T. N. Bardsley and I. Sexton
Imaging and Displays Research Group
Department of Computer Science
De Montfort University
The Gateway, Leicester LE1 9BH
ENGLAND
Tel.: (-533) 577498   Fax: (-533) 541891
e-mail: tnb/sexton@dmu.ac.uk

Abstract

This paper presents the results of a recent Ph.D. research programme encompassing the design, development, and evaluation of a variety of stereoscopic display systems. Emphasis is placed upon four distinct methodologies: anaglyphic display, a commercial frame sequential display using a switchable polarisation rotator and passive spectacles, a frame sequential display using active spectacles, and an autostereoscopic (no glasses) system, developed in our laboratory which uses a flat panel display with a specially produced lenticular faceplate.

The operation and technical limitations of the various display types is discussed and a robust assessment methodology based upon visual search and spatial tracking tasks is presented. The limitations of the four display types are discussed to provide an indication of the ergonomic factors likely to contribute to a differential display performance.

The data analysis and inferences drawn from this cross-display evaluation are presented and a hypothesis is formulated which describes the causal relationship between stereoscopic display methodology and isolated aspects of observer performance. A subjective ranking of 3D display 'quality' was solicited from participants in the evaluation and this is compared with the display ordering derived from the experimental study.

Keywords

3D displays, stereoscopic, autostereoscopic, computer Graphics

INTRODUCTION

Recent technological advances and the demands of more sophisticated methods of interaction (virtual reality is one particular example) have engendered an unprecedented interest in the techniques and applications of 3D imagery.

Although the head mounted display is probably the preferred display apparatus for true (immersive) virtual reality systems these are often costly and cumbersome. Because of this, and in view of rising concern regarding the health and safety issues surrounding such displays, alternative non immersive or desktop VR systems are often preferable. Although many commercially available examples are monoscopic implementations, the arguments supporting the use of stereoscopic displays are quite compelling.

The Imaging and Displays Research Group at De Montfort University is involved in various aspects of three dimensional and stereoscopic display research. A central theme of this research has been the development of autostereoscopic display systems; particularly display types which utilise view selecting mechanisms in the form of parallax barriers and lenticular screens. Autostereoscopic displays are quite different in their implementation to the more conventional stereoscopic systems and as a consequence the subjective impression is also quite different.

The use of stereoscopic and 3D displays has become well established in niche application areas where it is possible to clearly demonstrate an improvement in operator performance, however this is usually achieved by comparing a particular stereoscopic system to a monoscopic counterpart (Pepper *et al.*, 1997). In areas where the use of 3D displays has become established, the evidence to support the superior performance is often incontrovertible but few researchers have the opportunity to make direct comparisons of different 3D systems. Because of this, the choice of display methodology is often quite arbitrary.

There are many ways of producing a monoscopic display and many more ways of producing a stereoscopic system. This is to be expected as monoscopic displays are often components of stereoscopic systems. Each display type exhibits a set of artefacts; some of which are unique to the implementation. These typically encompass limited screen resolution (influencing depth quantization level), crosstalk, retinal rivalry, and perceptual flicker, all of which ultimately impinge upon operator performance. It is important for potential users of such displays to recognise these limitations in order to determine the applicability of a particular methodology.

Two of the displays in this investigation are frame sequential, wherein the two views comprising the stereo pair are temporally multiplexed. The system which uses active spectacles decodes the image by means of liquid crystal shutter devices arranged in lieu of the lenses in a pair of spectacles. These shutters can selectively block an image from either eye. The system which uses passive spectacles has a large liquid crystal device (in this case a Tektronix SGS410 Stereoscopic Modulator) interposed between the monitor and the viewer. This device encodes alternate frames with either a clockwise or anticlockwise polarisation. These are subsequently decoded by polarising spectacles worn by the viewer.

The anaglyph display rarely requires explanation. The left and right views are encoded using different colours on the display and are decoded by means of coloured (usually red and green) spectacles worn by the viewer.

The autostereoscopic display differs fundamentally from the others in this evaluation. The left and right views are spatially multiplexed and the viewer is not required to wear any special spectacles. Appropriate left and right views are presented to the viewers eyes by means of a lenticular sheet at the display surface. In this particular case the display device is a colour (TFT) liquid crystal panel.

It is possible to make a purely objective assessment of a 3D display; invariably such systems depend upon the channelling of appropriate views to each eye of the observer. Although the techniques used to achieve this objective differ markedly between display strategies it is a common requirement across display types. Consequently, a revealing measure may be obtained by determining how effectively the appropriate images are channelled to each of the observers eyes. Unfortunately this approach takes no account of the fundamental ergonomic differences between display types, and it was considered that more appropriate metrics should engage the concept of *fitness for purpose* whereby task related performance could be determined. A comparative assessment of the four participating display types was performed on this basis.

Two separate experimental procedures were developed to enable a quantitative assessment of display capability in supporting both static and dynamic graphical interaction. The first experiment comprised a *visual search* task which required observers to correctly identify pairs of visual stimuli which were uniquely correlated in depth. The second experiment utilised real-time animated graphics to evaluate observer performance during a continuous *spatial tracking* task.

EXPERIMENTAL OBJECTIVES

A controlled study was devised to examine the efficacy of each stereoscopic display technique (anaglyph, lenticular, LC polariser and LC shutter glasses) during a quantitative assessments of operator task performance. The hypothesis underlying the experiment being that each stereoscopic display methodology will have a specific and measurable effect on critical aspects of human performance. Objectives of the display evaluation were to determine the potential of each display technique as the basis for implementing a stereoscopic graphics workstation, and to establish a performance envelope for the recently developed lenticular prototype. A subsidiary goal was to identify a set of universally applicable performance metrics to serve as a basis for future stereoscopic display assessment.

Previous psycho-optical investigations have revealed certain aspects of human performance which are sensitive to the provision of binocular stereopsis. Human factors experiments, originally intended to explore the superiority of stereoscopic presentation over conventional monoscopic display formats, have shown marked improvements in spatial positioning accuracy [Pepper *et al.*, 1977; Kim *et al.*, 1987; Reinhart, 1991; Takemura *et al.*, 1989; Beaton, 1990), enhanced speed and accuracy during depth correlation judgements (Reinhart *et al.*, 1990; Zenyuh *et al.*, 1988; Miller and Beaton, 1991; Yeh and Silverstein 1990) and a significant reduction in task completion times (Pepper *et al.*, 1977; Cole *et al.*, 1990). In the light of these observations, it seems reasonable to assume that the relative magnitude of such performance increases will be contingent upon the efficiency with which individual display methodologies implement their stereoscopic effect.

Two separate task scenarios were devised to investigate differences in stereoscopic display performance during the presentation of both static and dynamic visual stimuli. The requirement for interaction with dynamic stereoscopic imagery is universally recognised

(Butts and McAllister, 1988), and an investigation of display performance in this context is easily justified. A static visual search task was included to investigate each display's disposition for one form of visual stimulus over the other.

A *target tracking* task was conducted to monitor the *spatial positioning error* associated with the continuous pursuit of a dynamic visual stimulus as it manoeuvres within the confines of a 3D display viewing volume. A composite measure of tracking performance was obtained by computing the *mean Euclidean distance error* during tracking operations. Six additional error metrics were calculated based upon the *mean* and *absolute* tracking errors associated with each of the independent X, Y and Z display axes. Measures of absolute deviation form the target trajectory were used to provide a useful indication of the error magnitude attributable to each display axis, while the polarity of mean tracking scores was intended to reveal any systematic bias in positional deviation.

A second *target acquisition* task was designed to examine the potential of each display for facilitating three-dimensional *visual search* and *depth correlation* judgements. The response measures associated with this particular task were *decision response time* and *accuracy* during depth correlation judgements. The degree of accuracy was determined in two ways. Firstly by computing the percentage of correct depth correlations for each display and secondly, by recording the degree of error when objects were incorrectly matched in depth.

Both the tasks and the visual stimuli used to depict them, were designed to draw upon the perceptual and motor skills necessary for comprehending and interacting with three-dimensional graphical environments. As the purpose of the study was to provide a measure of stereoscopic effect, all monocular cues with the exception of motion parallax, size constancy, linear perspective and object interposition were eliminated. Adopting a subset of the available monocular cues was intended to isolate the influence of binocular stereopsis, while maintaining important binocular/monocular cue consistency.

Both visual search and interactive cursor positioning operations can be characterised by their high levels of task demand (Reinhart, 1991; Zenyuh *et al.*, 1988). This is fortunate as both the workload and the degree of stress experienced during task performance are considered to be influential factors which determine the benefit and efficacy of retinal disparity cues (Miller and Beaton, 1991). The high cognitive load associated with both the target acquisition and tracking tasks, help to ensure the adopted human performance metrics are well sensitised to small differences in display quality, and therefore constitute good indicators of stereoscopic display performance.

A subjective assessment of display quality and performance was obtained by interviewing participants at the end of the experimental procedure. Subjects were asked to rank the four displays in order of preference based on their confidence level while performing the experimental tasks. In the event that subjects could not differentiate the displays using this criteria, they were instructed to consider additional factors such as viewing comfort, visual fatigue and any disturbing visual characteristics the displays may have.

METHOD

Subject population

Thirty two potential participants were screened for satisfactory stereopsis using a 16 depth level random-dot stereogram (RDS) test stimulus. The absence of extrastereoscopic cues in the RDS image make it an unfakeable test for stereopsis (Reinhart, 1991) and this permits early rejection of subjects with stereoanomalous vision. All stereodeficient candidates were eliminated from the study. Twenty nine subjects (23 male and 6 female) passed this initial screening procedure by virtue of being able to discriminate the smallest values of positive and negative parallax embodied within the RDS test image. At twenty nine, the number of subjects participating in the evaluation study was reassuringly large in comparison to the numbers commonly employed in similar investigative procedures (Kim *et al.*, 1987; Williams and Parrish, 1990), consequently both the experimental power and the prospects for generating statistically significant results were greatly improved.

All test subjects taking part in the evaluation study were unpaid volunteers drawn from the staff and students at De Montfort University and from enthusiastic members of the public. Subjects were aged between 17 and 48 years old; a mean age of 25½ characterising the group. Previous exposure to stereoscopic imaging techniques varied widely between subjects. Some individuals had experienced the full gamut of cinematic, holographic and virtual reality based entertainment systems, while others had little or no prior exposure to stereoscopic media. None of the participants reported recent or extensive exposure to any of the stereographic displays involved in the evaluation study.

Environmental conditions and apparatus

The experimental trials took place over a three week period during which time every effort was made to ensure consistent environmental conditions for all participating subjects. Ambient lighting conditions were carefully controlled to optimise the performance of all the displays involved in the experiment. Particular attention was given to the location and brightness of individual light sources to avoid introducing first and second order reflections from CRT faceplates and screen mounted optical components.

The red and green colour components used to represent anaglyphic stereo pairs were optimised to yield screen colours which closely matched the chromatic filter characteristics of the anaglyph spectacles. Maximising transmission for corresponding colours and reducing the leakage of unwanted colour components helped to improve filter extinction ratios and reduced the detrimental effects of image crosstalk.

A choice of either full-frame or clip-on glasses were supplied for use with the Tektronix polarizing display. This allowed subjects with corrected vision to retain their prescription lenses and participate more effectively in the experimental study.

Care was taken to avoid unfairly compromising any of the displays involved in the investigation, however, none of the steps taken to ensure optimal display viewing conditions were considered to be unusual or contrived. After all, careful regulation of the viewing environment and the provision of well implemented stereoscopic display techniques are prerequisites for a robust evaluation strategy (Merritt, 1983).

The Spaceball Technologies Spaceball™ was used to provide subject input responses during spatial tracking operations. The Spaceball was chosen in preference to a glove-based positioning device, as the users arm is fully supported during use, and this reduces the influence of arm fatigue on tracking performance. The Spaceball comprises a rubberised sphere about the size of a tennis ball mounted on a stable platform. The Spaceball is sensitive to the fingertip pressures and torsional forces applied to it, resolving these simultaneously into X, Y and Z translations and rotational components. Although the Spaceball is capable of simultaneously encoding six degrees of freedom (S6DOF), rotational effects were disabled during the tracking task. The Spaceball was programmed to provide output responses that were proportional to the magnitude of applied translational forces. This meant that cursor velocity could be freely controlled during pursuit of the target stimulus.

Target tracking

Task structure
The three-axis tracking task was designed to measure subject performance during the pursuit of a target symbol which moved in a random and continuous fashion throughout the confines of a three-dimensional viewing volume. The hypothesis under investigation being that choice of stereo display methodology will significantly influence the accuracy with which subjects perform spatial tracking operations.

Subjects were required to manipulate a three-dimensional cursor using the Spaceball positioning device which afforded simultaneous control over all three cursor axes. Participants were instructed to follow the target symbol as closely as possible in order to minimise their X, Y and Z-axis tracking error.

A continuous target motion was chosen in preference to a static stimulus to increase the cognitive load associated with the spatial positioning task. Other investigators have employed static target stimuli which results in an oversimplification of the cursor alignment task by allowing a method of limits approach to cursor alignment (Reinhart, 1991; Takemura *et al.*, 1989; Drascic, 1991). Although use of a static target stimulus would have enabled the assessment of cursor positioning times, the associated task demand was considered too low to constitute a good measure of interactive stereoscopic display capability.

Visual stimulus
Visual representation of the manual pursuit tracking task comprised a wire-frame perspective cube inscribed within the boundaries of the stereoscopic display screen, a flat-shaded target symbol and a wire-frame *full-space jack* (FSJ) cursor. Each of these components was assigned a unique colour and rendered in order of decreasing depth on a black background to replicate the extrastereoscopic cue of object interposition. A pictorial representation of the target tracking task appears in figure1.

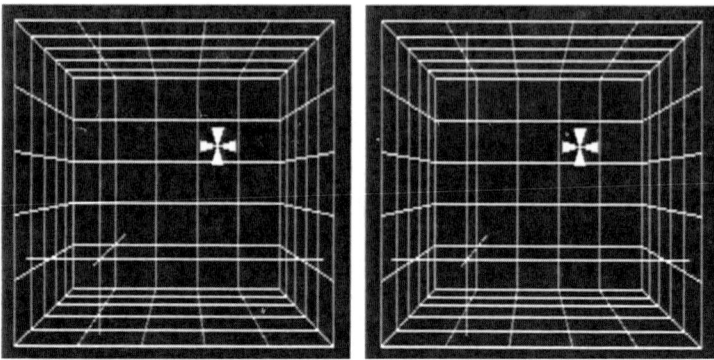

Figure 1   Stereogram depicting the target tracking stimulus (arranged for transverse viewing)

Both the simplistic perspective groundplane (Kim *et al.*, 1987; Yeh and Silverstein 1990; Grotch, 1983), and the more complex viewing volume reference cube (Butts and McAllister, 1988; Barham and McAllister, 1991; Beaton *et al.*, 1987), have been utilised extensively in the depiction of stereoscopic data and three-dimensional task scenarios. Such enhancements are routinely used to increase the perceived realism and facilitate relative spatial judgements in both 2½D and true 3D images. A view volume reference cube was provided in the target tracking task to delineate the extent of the three-dimensional workspace and indicate the intended limits of target and cursor motion.

Four criteria were used to determine an appropriate target shape for the pursuit tracking task. (1) The 'focus' of the target must remain visible regardless of its position and orientation within the viewing volume. (2) The precise focus of the target must be unambiguous and should be, at most, 1 pixel wide about all 3 display dimensions. (3) The target must undergo an appropriate change in angular subtense as it travels back and forth within the confines of the viewing volume. (4) Interposition cues must accurately reflect the relative depth of target and cursor stimuli. The planar Maltese Cross depicted in figure 1 was thought to be a reasonable compromise which satisfied all four design constraints. The planar nature of the stimulus gave it a zero depth extent thereby ensuring an unequivocal depth location. The focus located at the intersection of the cross axes was precisely one pixel wide and the target stimulus was made opaque to afford good object interposition characteristics.

The target trajectory for the three-axis manual tracking task was randomised to inhibit predictability and avoid track repetition. Target motion was composed of three independent single-axis trajectories, each comprised of a series of sinusoids having random frequency. Target position was updated during successive animation frames at a rate of 10 Hz. At no time was the target symbol permitted to exceed the bounds of the viewing volume.

A full-space jack (FSJ) cursor was chosen to indicate subject response during the target tracking operations. This cursor format was chosen as it embodies motion parallax, linear perspective and size constancy cues effectively, and can be rendered quickly. The FSJ is a modified two-dimensional cross hair cursor in which the depth axis is represented by a third vector lying perpendicular to the display surface. Each axis of the FSJ cursor is extended to meet the boundaries of the viewing volume to indicate absolute position within the 3D workspace. The cursor *hot-point*, from which all positioning errors are measured, lies at the intersection of these three vector components. Cursor position was updated between animation frames using force data supplied by the Spaceball; cursor velocity being directly proportional to the applied force on each Spaceball axis.

Target acquisition

Task structure

A target acquisition study was designed to reveal the influence of the anaglyphic, lenticular, polarizing and optical shutter based display techniques on subject performance during visual search and depth correlation operations. The underlying hypothesis being that choice of stereo display methodology will significantly effect both the speed of visual search and the accuracy of depth correlation judgements.

Subjects were presented with a sequence of static images, each containing numeric symbols scattered randomly throughout the confines of the display viewing volume. Subjects were required to identify verbally, which of these numeric symbols occupied the same depth plane as a single reference symbol lying elsewhere in the viewing volume. Only one numeric symbol was co-located in depth with the reference stimulus; the remainder serving as distracters, both to increase task demand during visual search operations and further confound estimates of cue/target depth correlation.

Visual stimulus

A number of factors are known to influence the speed and accuracy of relative stereoscopic depth judgements. The number of depth planes (Reinhart, 1991), the spatial proximity between cue and target stimuli (Yeh and Silverstein, 1989; Reeves and Tijus, 1990) and the number of distracter symbols present in the viewing context (Miller and Beaton, 1991) all influence the potency of binocular disparity cues.

Numeric symbols were utilised throughout the target acquisition study to enable participants to quickly and unambiguously identify their choice of co-located stimuli. Planar symbols with a parallel screen orientation were used to avoid generating stimuli that spanned several depth planes simultaneously, thereby confusing the depth matching process. A four point star symbol was chosen, quite arbitrarily, to represent the reference or *cue* stimulus. The precise viewing context of the target acquisition task is illustrated in figure 2.

All stimuli occupied the same visual angle regardless of their apparent depth within the viewing volume, and the spatial location of each symbol was carefully controlled to eliminate any character overlap. The elimination of retinal image size and object interposition cues was a necessary precaution to increase the cognitive load associated with visual search and depth correlation operations while to emphasising the impact of binocular stereopsis.

The depth axis was partitioned into 21 discrete depth planes encompassing the fusible extent of the display viewing volume. Each depth plane was assigned a unique value of screen parallax. Crossed and uncrossed disparity levels for each of the 21 depth planes ranged from -10 to +10 display pixels; the zero parallax condition coinciding with the plane of the display screen. The location and proximity of co-located symbols were randomised for successive stimulus conditions to prevent systematic cue/target separations. All non-correlated distracter symbols were randomly distributed throughout the viewing volume.

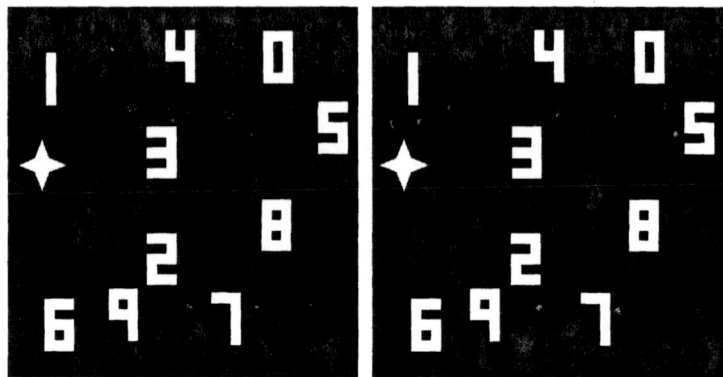

Figure 2 Stereogram of the target acquisition stimulus (arranged for transverse viewing)

Experimental procedure

All 29 test subjects undertook an initial training period to aid familiarisation with the Spaceball input device. This precaution was necessary to saturate the learning effects inherent in mastering an unfamiliar input device, and helped to ensure a consistent dexterity level for all subjects participating in the target tracking evaluation. During this training procedure, 2 subjects failed to meet the minimum performance criteria for steady-state tracking error and were subsequently eliminated from the target tracking study.

The remaining 27 subjects performed the spatial tracking and target acquisition tasks consecutively on each of the four stereoscopic displays. The order of exposure to each display was randomised between subjects to distribute the impact of learning and carry-over effects associated with task repetition.

Subjects performed three tracking runs of 60 seconds duration. Partitioning the task in this way helped to reduce the influence of fatigue on tracking performance. An initial warm-up period of between 10 and 15 seconds was provided before an audible tone announced the onset of data collection. Following the 60 second recording period, a second audible tone indicated the completion of each tracking run.

All 29 subjects who passed the initial RDS screening procedure completed 20 replications of the target acquisition task on each of the four stereoscopic display systems. Unique stimulus configurations were constructed for each task replication, and the display screen was blanked during the calculation and rendering of new target and distracter locations. Subjects indicated their choice of target symbol by calling out a number between zero and nine. This verbal response was immediately entered on a numeric keypad by the evaluation supervisor. This elaborate response procedure was required to maintain the attention localisation of each subject and to eliminate the influence of keypad search intervals on response time measures. During subject briefing, participants were instructed to make accuracy their prime concern while remaining mindful of the fact that their decision response times were also being monitored.

Dependent measures

All spatial tracking measures were computed from recorded differences in target and cursor trajectories. Uni-directionally, *absolute tracking errors* were calculated for X, Y and Z to establish the error magnitude associated with each display axis. *Mean errors* were also produced to highlight any systematic bias left-right, above-below or fore-aft of the target position.

Positional deviations in the X-Y plane were measured in world coordinates, while errors in depth were quantified in terms of the number of discrete depth planes separating the target and cursor location; each depth plane corresponding to a specific value of screen parallax. Tracking errors in the depth dimension were measured in discrete depth planes to compensate for apparent differences in display resolution between the X-Y plane and the depth axis. Small changes in an object's world-coordinate position are quickly reflected in X-Y screen location, however, such changes do not become apparent on the depth axis until their magnitude is sufficient to generate a variation in screen parallax. Consequently, measuring Z-axis tracking precision at a level which exceeds the display's capacity for representation makes little sense.

A composite measure of spatial tracking performance was obtained by averaging the *mean Euclidean distance* errors separating the target and cursor throughout the tracking interval. The average of all such instantaneous errors serving as a composite measure of tracking accuracy. Object-to-object distances were calculated thus,

$$Error \ = \ \sqrt{(X_t - X_c)^2 + (Y_t - Y_c)^2 + (Z_t - Z_c)^2} \ , \tag{1}$$

where t and c denote major axis coordinate positions of the target and cursor respectively.

Human performance during the target acquisition study was characterised by the length of time taken to search the viewing volume for a cue/target match, and by the accuracy attained in nominating co-located target symbols. Four experimental measures were used in quantifying subject performance.

Decision response time was measured from the instant each stimulus condition was presented to the moment subjects announced their response. Response times were measured in real-time clock increments giving an effective resolution of 1/18.2 sec.

The number of *search errors* expressed as the 'percentage of correct depth correlations', provided a useful indication of the 'hit-rate' attained by subjects when using each of the stereo displays. In each of the 20 task replications, there is a 1 in 10 probability that participants will select the correct target symbol by chance alone. Consequently hit rates in excess of 10% could be considered evidence of a positive display influence during depth correlation.

A measure of *absolute depth plane error* was generated to reflect the magnitude of search errors arising from incorrect cue/target associations. *Mean depth plane errors* were calculated to reveal any directional bias fore or aft of the cue stimulus during depth correlation judgements.

Experimental design

Response-time and correlation-error data obtained in the target acquisition experiment were collapsed across trial replications to provide averaged performance metrics for each subject using each of the four displays.

Data from the target tracking and target acquisition experiments were submitted to separate, doubly multivariate, analysis of variance (MANOVA) procedures for repeated measures. Separate analyses were used to avoid complications arising from differences in the sample sizes involved (27 and 29 subjects respectively). Four levels of the single within-subjects variable *display type* (anaglyphic, lenticular, polarizing and shutter-glasses) were used in a mixed effects design. Subjective ranking scores obtained during post experimental interview were assessed using a simple one-way analysis of variance procedure.

Examination of the descriptive statistics and normal plots obtained for each dependent variable, revealed all data to be normally distributed with good homogeneity of variance characteristics (Devore, 1990; Norusis, 1986). As all data appeared to satisfy the basic assumptions required for the proper application of analysis of variance procedures (Keppel, 1973), Greenhouse-Geisser probabilities were not used to adjust univariate test results. Post-hoc Student-Newman-Keuls multiple comparisons tests, conducted at the 0.05 significance level were used to identify critical differences between performance means on each display.

RESULTS

Only the most salient results to emerge from the target tracking and acquisition studies will be presented in the sections which follow. The interested reader is referred instead to the full thesis for a more detailed discussion (Bardsley, 1994).

Target tracking performance

All multivariate test criteria (Pillai's Trace $F[21,222]=5.6$, $p<0.001$, Wilks' Lambda $F[21,207]=7.7$, $p<0.001$ and Hotelling's Trace $F[21,212]=10.3$, $p<0.001$) suggest that the single within-subjects factor *display type* has a significant effect on all combined measures of tracking performance. Univariate F-tests conducted with (3,78) degrees of freedom were used to determine the effect significance in relation to individual measures of tracking performance.

Euclidean distance tracking error
The composite measure of tracking accuracy, Euclidean distance error, was significantly influenced by variations in display type ($F[3,78]=46.93$, $p<0.001$). A clear indication of this effect can be seen in figure 3 which portrays the mean Euclidean distance error scores associated with each display type.

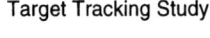

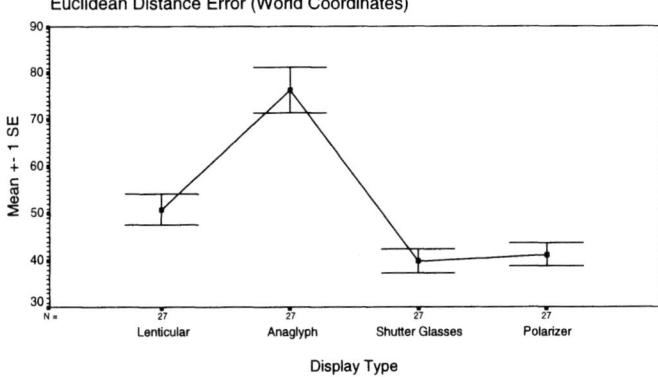

Figure 3 Euclidean distance tracking error

These apparent differences in display performance were investigated using a Student-Newman-Keuls multiple range test conducted at the 0.05 significance level. Results confirmed the world coordinate Euclidean distance error to be significantly greater for the anaglyphic display (76.3) than for the lenticular (50.8), shutter glasses (39.9) or polarizing (41.3) variants. No significant differences were found between either the lenticular, shutter glasses or polarizing displays, suggesting an equitable performance in respect of this composite measure of spatial tracking accuracy.

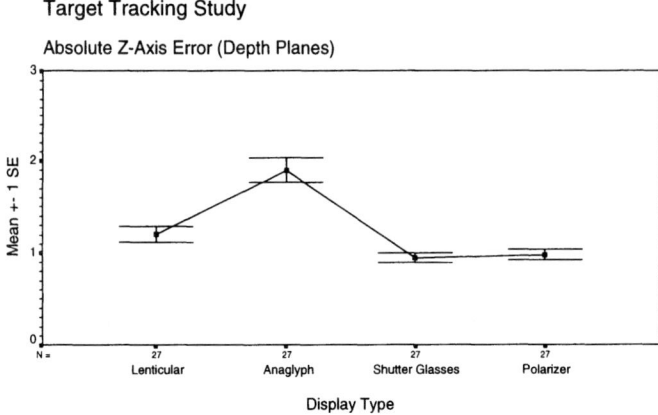

Figure 4 Average magnitude of depth axis tracking errors

Z-axis tracking error

In accordance with expectation, the largest impact on Euclidean distance error was made by the cursor/target displacements arising along the depth axis. It is generally acknowledged that performing accurate cursor positioning in depth is inherently more difficult than conventional manipulations in the X-Y plane (Beaton *et al.*, 1987). This observation seems to hold true regardless of the stereoscopic technique or the spatial positioning device employed, however stereoscopic display fidelity is expected to influence the magnitude of depth placement errors.

Figure 4 illustrates the absolute average tracking error measured in discrete depth planes along the Z-axis for each display. The most striking feature is the profile similarity of figures 3 and 4, establishing Z-axis deviation as the major component of Euclidean distance error. Not surprisingly, the type of stereoscopic display was found to have a significant influence on depth axis tracking error (F[3,78]=46.55, p<0.001). The only critical difference in performance means lay between the anaglyph display (1.9) and the lenticular (1.2), shutter glasses (0.96) and polarizing (0.99) variants. This result confirms the anaglyphic display as the worst performer during continuous spatial positioning operations, leaving the lenticular, shutter glasses and polarizing displays equal on merit.

A peculiar trend was detected during analysis of the mean z-axis error scores (measured in depth planes) for each subject participating in the tracking study. All data displayed a strong negative bias, indicating that subjects consistently positioned the stereoscopic cursor well in front of the target stimulus. Although the foreground positional bias was evident for all stereoscopic displays, the magnitude of the effect was display dependent (F[3,78]=56.91, p<0.001). Student-Newman-Keuls tests revealed the effect to be greatest for the anaglyph technique (-1.6), while for the first time, the lenticular display (-0.75) registered an effect size worse than either the shutter glasses (-0.34) or polarizing (-0.41) techniques.

Target Tracking Study

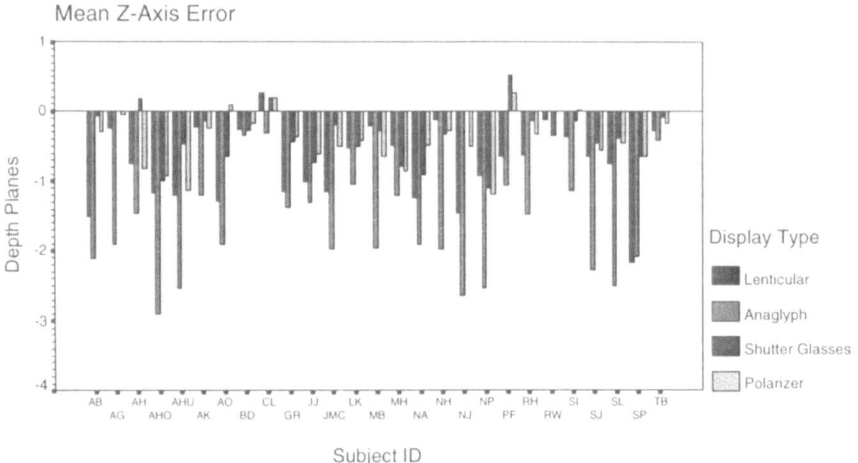

Figure 5 Mean Z-axis tracking errors

Figure 5 shows the mean z-axis deviation for each subject when using each of the four display types. The figure provides a clear indication of the consistent tendency to underestimate target depth.

Target acquisition performance

All multivariate test criteria (Pillai's Trace F[12,249]=3.7, p<0.001, Wilks' Lambda F[12,214]=3.8, p<0.001 and Hotelling's Trace F[12,239]=3.9, p<0.001) suggest that the single within-subjects factor *display type* has a significant effect on all combined measures of target acquisition performance. Univariate F-tests conducted with (3,84) degrees of freedom were used to determine the effect significance in relation to individual measures of target acquisition performance.

Decision response times
Surprisingly, the observed variation in decision responses times for each of the stereoscopic displays barely achieved statistical significance (F[3,84]=4.05, p=0.01), and Student-Newman-Keuls tests revealed that mean response times for the lenticular (7.64 seconds), anaglyph (7.34 seconds), shutter glasses (6.49 seconds) and polarizing (7.44 seconds) displays were not significantly different at the 0.05 level.

Percentage of correct depth correlations
The number of search errors committed by subjects when trying to identify co-located cue and target symbols was significantly influenced by stereoscopic display methodology used (F[3,84]=9.29, p<0.001). Figure 6 shows the average correlation performance or 'hit rate' attained on each display, expressed as a percentage of the total number of depth judgements performed.

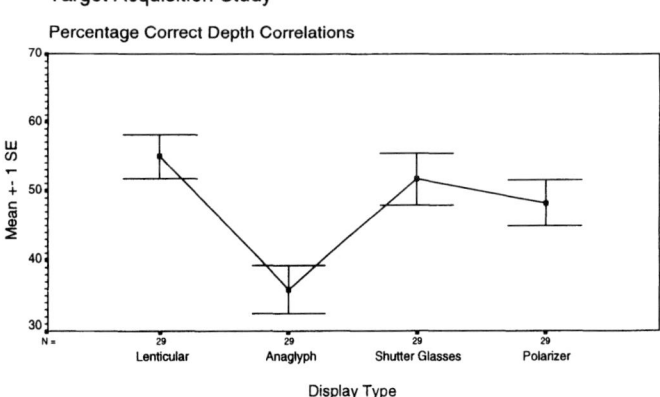

Figure 6   Percentage of correct depth correlations

The apparent superiority of the lenticular prototype was investigated using Student-Newman-Keuls multiple comparisons conducted at the 0.05 significance level. Test results failed to confirm any significant performance advantage for the lenticular display, indicating instead that the lenticular (55%), shutter glasses (52%) and polarizing (48%) variants performed equally well in respect of depth correlation accuracy. The anaglyph display was (predictably by now) the worst performer, with a mean percentage score of 35.

The hit rates attained by all subjects using each display were well above the anticipated 10% threshold. That is to say, each of the four display techniques substantially increased the number of correct depth correlation judgements beyond levels normally associated with random stimulus selection (2 in 20).

Depth correlation error

The absolute magnitude of errors arising from incorrect cue/target depth correlations was found to be dependent on display type ($F[3,84]=11.11$, $p<0.001$). The degree of absolute correlation error associated with each display type is shown in figure 7. On average, depth judgement errors are little more than 1 depth plane in magnitude for all of the displays participating in the study. Depth judgement errors reported in previous investigations are typically one unit pixel disparity in size (Drascic, 1991; Yeh and Silverstein, 1989). As depth planes and pixel disparities are synonymous in the current display context, the observed correlation errors are within expected limits.

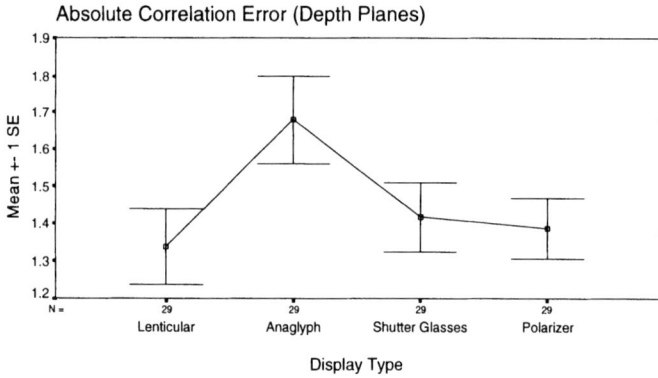

Figure 7 Average depth correlation error

The comparatively poor performance of the anaglyphic display (1.68) was confirmed during post hoc analysis. However, means comparisons also concluded that error magnitudes for the lenticular (1.34), shutter glasses (1.42) and polarizing (1.39) displays were not critically different. No significant bias to either the front or rear of the cue stimulus was detected, suggesting correlation performance to be independent of the relative depth placement of cue and target stimuli.

Subjective ranking scores

Following the conclusion of the target tracking and acquisition exercises, participants completed a subjective ranking scale to determine their display preference. Subjects were instructed to base their decision primarily upon the confidence level experienced while performing the tracking and acquisition tasks. If subjects were unable to differentiate between specific displays using this criteria, then additional considerations such as viewing comfort, fatigue, image quality and other extraneous factors were to be taken into account. A integer between 1 and 4 was uniquely assigned to each display format in order of increasing merit.

Display type was found to have a significant effect on subjective ranking scores (F[3,84]=25.84, p<0.001), the means of the scores awarded for each display appear in figure 8. Post hoc analysis failed to detect any significant differences in mean display rankings between the lenticular (2.97), shutter glasses (2.86) and polarizing (3.03) systems, however the anaglyph display was clearly isolated in terms of subject preference with a mean ranking of 1.17.

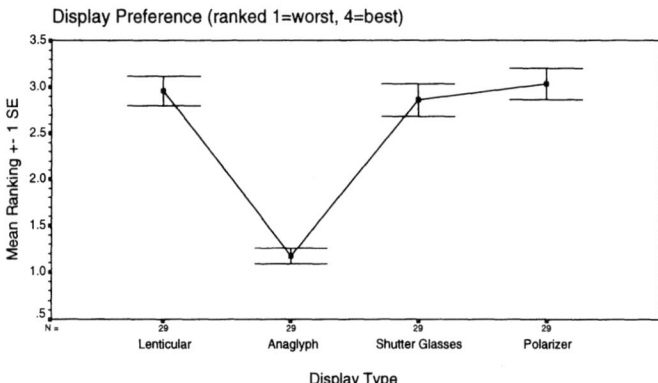

Figure 8 Average display ranking scores

DISCUSSION

The most salient result to emerge from the preceding analysis is the comparative performance of the lenticular, shutter glasses and polarizing display systems. Selected human performance measures indicate these three displays are inseparable in terms of their ability to support spatial tracking and visual search operations; a verdict supported by subjective ratings of display performance. The comparatively poor results obtained with the anaglyph display system suggests the evaluation strategy and response measures used were an effective tool for discriminating stereoscopic display performance.

It is tempting, at first, to attribute poor spatial tracking performance on the anaglyph display to difficulties in replicating the monocular interposition cues so vital to depth perception.

However, Barham (1991) suggests that stereopsis is used to achieve initial cursor placement only, after which interposition cues become dominant, allowing finer depth adjustments. The sheer magnitude of the Euclidean distance and Z-axis tracking errors associated with the anaglyph technique indicate that more complex perceptual difficulties exist when using this display. After all, the target acquisition task used a visual stimulus that was totally devoid of object interposition cues and this produced an equally damning verdict on anaglyph display performance in respect of depth correlation accuracy.

The tendency for subjects to underestimate object depth during the target tracking study gives cause for concern. Drascic (1991) has noted a similar phenomenon in his experiments to compare the positioning accuracy afforded by a stereovideo representation of a real world pointer and a computer generated stereographic pointer. A small but significant foreground bias was detected during the alignment of both the real and virtual pointer with a static target stimulus. Drascic offers no concrete explanation for the observed phenomenon, claiming the lack of an experimental precedent and the failure of researchers to report mean positioning errors is hindering the search for a solution. A cursory inspection of the subject profiles obtained during post experimental interview would indicate that an individual's propensity for underestimating object depth is inversely proportional to their stereo experience level. Consequently, a longer stereo acclimatisation period may well see a reduction in foreground positional bias during spatial tracking operations.

The lack of significant response time effects associated with the target acquisition task was disappointing, as it suggests that stereoscopic display discrimination can not take place on the basis of timed assessments of task performance. However, it appears that this observation is not without precedent. During an investigation of monocular and binocular cue saliency levels, Reinhart (1990) discovered that the inclusion of binocular disparity cues did not produce significantly faster response times for subjects performing simple relative depth judgements. Zenyuh (1988) also concluded that a stereoscopic display format did not significantly improve response times during visual search and object counting operations. Unfortunately neither investigator offers a conclusive explanation for their observation. There evidence to suggest that a degree of cognitive capture occurs during exposure to stereoscopic display techniques (Miller and Beaton, 1991), and this is likely to confound measurements of human performance. Response times can be expected to decrease as subjects gain proficiency during speeded depth judgements while progressing form one display to the next. Such variations can mask the impact of differential display performance on decision response times. Again, extensive stereo pre-training should be considered to exhaust such learning effects and heighten the influence of disparity cues.

In retrospect, the inclusion of a fifth, monoscopic display format, would have provided a useful experimental control throughout the display evaluation study. Performance metrics obtained on a monoscopic display format would establish a base-line performance reference to assist in the verification of any apparent stereoscopic display advantages.

CONCLUSIONS

The validity of comparing dissimilar display implementations was initially of some concern. All of the displays in the evaluation share common objectives - the portrayal of a spatial image - it might be argued that the manner in which these objectives are achieved differs sufficiently to preclude a direct comparison.

Simple differences in physical display attributes such as screen resolution for example, can have a profound influence upon display efficacy. A consequence of finite two dimensional display resolution is that the depth dimension is quantized; in the absence of suitable anti-aliasing techniques this quantization yields discrete depth planes corresponding to integral pixel disparities. Because screen pixel pitch varies between display implementations, the range and separation of the available depth planes will too. It may be argued that relative depth judgements are easier to perform when depth resolution is coarsely quantized, a perverse consequence of this being that a lower depth resolution will enhance depth correlation judgements. Clearly an objective display comparison must incorporate a comprehensive range of response measures and task scenarios.

Preliminary results obtained from this investigation, indicate that stereoscopic display techniques can indeed be differentiated on the basis of a quantitative assessment of human task performance. Unfortunately, the inability to discriminate between lenticular, shutter glasses and polarising display systems suggests that the simple tasks adopted during the evaluation were insufficient to resolve subtle differences in performance. Extending the nature and scope of the experimental tasks is expected to yield an even more decisive measure of display quality.

Finally, the authors would like to express their appreciation to the 29 observers who took part in this investigation.

BIBLIOGRAPHY

R. L. Pepper, R. E. Cole, and D. C. Smith, "Operator performance using conventional or stereo displays," *SPIE Three dimensional imaging*, vol. 120, pp. 92-99, 1977.

W. S. Kim, S. R. Ellis, and M. E. Tyler, "Quantitative evaluation of perspective and stereoscopic displays in three axis manual tracking tasks," *IEEE Transactions on systems man and cybernetics*, vol. Smc-17, pp. 61-72, 1987.

W. F. Reinhart, "Depth cueing for visual search and cursor positioning," *SPIE Stereoscopic displays and applications II*, vol. 1457, pp. 221-232, 1991.

H. Takemura, A. Tomono, and Y. Kobayashi, "A study of human-computer interaction via stereoscopic display," *Proceedings of III Conference on Human-Computer Interaction*, vol. 1, pp. 496-503, 1989.

R. J. Beaton, "Displaying information in depth," *SID Digest*, vol. 1990, pp. 355-358, 1990.

W. Reinhart F, R. Beaton J, and H. Snyder L, "Comparison of depth cues for relative depth judgements," *SPIE Stereoscopic displays and applications* , vol. 1256, pp. 12-21, 1990.

J. P. Zenyuh, J. M. Reising, S. Walchi, and D. Biers, "A comparison of a stereographic 3-D display versus a 2-D display," *Proceedings of the human factors society-32nd annual meeting*, vol. 1, pp. 53-57, 1988.

R. Miller H and R. Beaton J, "Some effects on depth-perception and course-prediction judgements in 2-D and 3-D displays," *SPIE Stereoscopic displays and applications II*, vol. 1457, pp. 248-258, 1991.

Yei-Y. Yeh and L. D. Silverstein, "Visual performance with monoscopic and stereoscopic presentations of identical three-dimensional visual tasks," *SID Digest*, vol. 1990, pp. 359-362, 1990.

R. E. Cole, J. O. Merritt, S. Fore, and P. Lester, "Remote manipulator tasks impossible without stereo TV," *SPIE Stereoscopic displays and applications*, vol. 1256, pp. 255-265, 1990.

R. W. Butts and D. F. McAllister, "Implementation of true 3D cursors in computer graphics," *SPIE Three dimensional imaging and remote sensing imaging*, vol. 902, 1988.

S. P. Williams and R. Parrish V, "New computational control techniques and increased understanding for stereo 3D displays," *SPIE Stereoscopic displays and applications*, vol. 1256, pp. 73-82, 1990.

J. O. Merritt, "Common problems in the evaluation of 3D displays," *SID Digest*, pp. 192-193, 1983.

D. Drascic, "Pointing accuracy of a virtual stereographic pointer in a real stereoscopic video world," *Proceedings ITEC '91 International symposium on 3D imagery*, pp. 623-626, 1991.

S. L. Grotch, "Three-dimensional and stereoscopic graphics for scientific data display and analysis," *IEEE CG&A*, vol. November, pp. 31-43, 1983.

P. T. Barham and D. F. McAllister, "A comparison of stereoscopic cursors for the interactive manipulation of B-Splines," *SPIE Stereoscopic displays and applications II*, vol. 1457, pp. 18-26, 1991.

R. J. Beaton, R. J. DeHoff, N. Weiman, and P. W. Hildebrandt, "An evaluation of input devices for 3D computer display workstations," *SPIE True 3d imaging techniques and display technologies*, vol. 761, pp. 94-101, 1987.

Yei-Y. Yeh and L. D. Silverstein, "Depth discrimination in stereoscopic displays," *SID Digest*, pp. 372-375, 1989.

A. Reeves and C. A. Tijus, "The pop-out effect in simple 3D visual matching task," *Proceedings of the 3rd cognitive symposium*, pp. 559-564, 1990.

J. L. Devore, *Probability and statistics for engineering and the sciences*, 3rd ed. California: Duxbury Press, 1990.

M. J. Norusis, *Advanced statistics SPSS/PC+ for the IBM PC/XT/AT*. Chicago: SPSS Inc., 1986.

G. Keppel, *Design and analysis: A researchers handbook*, 1st ed. New Jersey: Prentice-Hall Inc., 1973.

T. N. Bardsley, "Design and evaluation of an autostereoscopic computer graphics display," Ph.D. Thesis, De Montfort University Leicester, 1994.

19

Virtual Reality for Enhanced Computer Vision

Wilhelm Burger
Johannes Kepler University, Department of Systems Science
A-4040 Linz, Austria. email: `wilbur@cast.uni-linz.ac.at`

Matthew J. Barth
University of California, College of Engineering
Riverside, CA 92521-0425, U.S.A. email: `barth@ucrengr.ucr.edu`

Abstract

Computer Vision and Virtual Reality are research areas with almost opposite goals. However, synthetic environments are urgently needed to handle some of the engineering and testing problems for computer vision systems. On the other hand, vision technology will be helpful to simulate intelligent "creatures" within future virtual reality systems.

Keywords

Computer vision, virtual environment, robot vision, immersive perception.

1 INTRODUCTION

Vision systems for autonomous and reactive robotic applications need to perform reliably under a variety of conditions. Since these systems are quite complex, formal verification methods usually do not exist and extensive testing provides the only way to ascertain the required performance. In a typical robot vision design process, testing of individual system components is limited to pre-recorded images and image sequences that are processed in a passive fashion. Closed-loop testing can only be performed *after* the complete system has been integrated and deployed on the real robot. Frequently, the task environment or even the actual robot is not accessible at all during system development, *e.g.*, in many space applications. Moreover, systems equipped with adaptation and learning usually require many more training runs and additional feedback from the environment. Generally, the lack of test data (particularly with corresponding ground truth data) and realistic testing facilities constitutes a severe bottleneck in the vision system design process. Thus two urgent needs exist:

Figure 1 Synthetic image demonstrating the degree of realism and sophistication available from current simulators. This example includes physics-based object and surface models, shading, focusing effects, motion blur, and atmospheric conditions (image source: Evans & Sutherland).

1. A large number of test cases (perhaps two orders of magnitude more than what is available now) with access to ground truth information is necessary for tuning and validating vision systems.
2. It is necessary to have an *environment* that allows for the testing of complete vision systems *before* deployment, thus eliminating the high risk of system damage or mission failure.

While simulation has always played some role in image processing, computer vision, and control system engineering [2, 6, 8, 16], there has been *no* simulation environment available that could support the complexity of complete perception and control systems, and provide the required realism at the same time (Figure 1). In the following, we describe our work towards an "immersive" simulation environment for perception-based robotic applications. While the focus is on "computer vision" throughout the text, most arguments equally apply to other forms of sensory inputs, including lasers, radar, infrared, and (to some extent) sound. In the remaining parts of the paper we concentrate on two aspects of the CV/VR theme. First we describe our work towards a VR-based simulation environment for computer vision engineering applications, called the "Immersive Perception Simulator", and then we discuss possible applications of CV technology to simulate "perceptive" creatures within virtual environments.

2 WHY SIMULATION?

The foremost expectations from this approach are:
a. Simulation allows *extended testing* in arbitrary environments, under changing environmental conditions, and in extreme situations that may be rarely encountered in reality but may be

crucial for system operation. The simulations can be run *unsupervised*, at any time of day, under any weather conditions, and on several computers in parallel. The increased number of situations and test cases explored in this way should improve system reliability dramatically.

b. *Ground truth information*, essential for validating many vision tasks, is available at any time. In contrast, currently used pre-recorded image data come either with very limited ground truth data, if any.

c. *Adaptation and learning* at all vision system levels can be performed efficiently and autonomously. Large sets of training examples can be processed without human intervention in both supervised and unsupervised modes.

d. Realistic testing of individual modules in a complete system environment is possible at a very early phase, even when the real robot or other parts of the system (*e.g.*, sensors) are *unavailable*. The evaluation of system design alternatives is possible without actual deployment on the robot.

e. The *costs* for designing, building, and running the robot can be reduced. Simulation results will generally influence the robot's design, thus avoiding costly redesigns later in the process. Further, the transition between testing and deployment of the system can be streamlined. Usually, transferring a vision system from the lab onto the real robot requires a major logistic effort.

f. Processing *hardware requirements* and *real-time capability* can be estimated much earlier in the design process. Usually, hardware performance requirements are difficult to estimate before the complete system is running. Real-time behavior can be simulated even when the required high-speed hardware is still unavailable during system development.

Commercial flight simulators are among the systems that would meet many of these specifications. However, they are expensive, physically large, designed for a very specific purpose, and they have to cut corners to be fast enough (*i.e.*, by not implementing reflections, glare, shadows of moving objects, *etc.*). Although flight simulators have actually been used with computer vision systems (closed-loop experiments for automatic aircraft landing [6]), their sophisticated mechanical design (*i.e.*, hydraulics for emulating aircraft motion) is unnecessary for most perception tasks.

3 THE IMMERSIVE PERCEPTION SIMULATOR (IPS)

The IPS is a new software environment intended to support a wide range of simulation and evaluation tasks in computer vision. Its main purpose is to provide a simple, flexible, and cost-effective mechanism for testing and tuning perception-based reactive systems in a sufficiently realistic environment. The core of the IPS is based on public-domain software modules, which supports the intention to make this system freely available to the vision and robotics community. Flexibility is achieved by providing an abstract interface protocol that simplifies the communication with existing vision systems, *e.g.*, KHOROS, KBVision, and others.

3.1 IPS Architecture

The basic architecture of the IPS consists of three main components: (a) the perception-based robot system, (b) the virtual environment, and (c) the IPS interface (Figure 2). The vision and robot control system interacts with the virtual environment simulator in a closed-loop fashion,

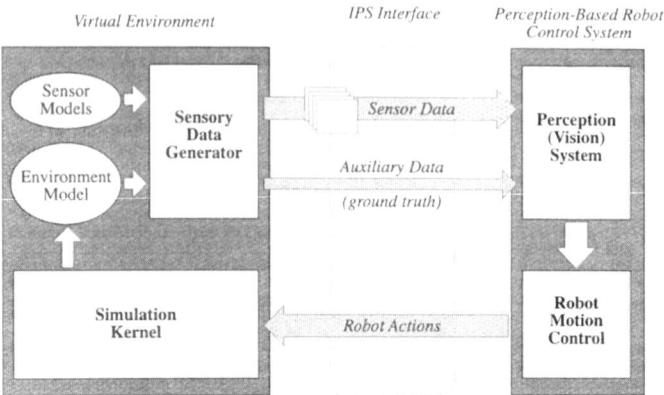

Figure 2 Immersive Perception Simulator (IPS) simplified architecture. The perception-based robot system interacts with the virtual environment through the interface layer in a closed-loop fashion, exchanging mainly sensory data, ground-truth data, and action commands.

where all communication is handled by the IPS interface. The main flow of data consists of *sensory* data generated by the simulator and *action commands* from the robot system. It should be noted that the robot system can be implemented in software, hardware, or as a combination of both. The necessary synchronization mechanisms are provided by the IPS interface, which is implemented on the UNIX process level.

3.2   The Virtual Environment

The virtual environment (Figure 2) contains an *environment model*, which is a description of the environment and the state of the robot at any point in time, that is maintained by what we call the *simulation kernel*. The environment model is used, in conjunction with the appropriate *sensor models*, to produce realistic images of the scene viewed by the robot's sensors. The structure of the actual simulator resembles that of contemporary "virtual reality" (VR) systems [9] and, in fact, we are borrowing several technological ingredients from VR, such as the modeling of object dynamics and collision detection. However, the main thrust of VR research today is on the implementation of sensors and manipulators for man/machine interaction and high simulation speed, which are not important in our case. On the other hand, the degree of physical realism that we require is not nearly available from any existing VR system. Thus the IPS could be considered an extremely realistic but "slow-motion" VR system.

3.3   Sensory Data Generation

While physical realism is the ultimate goal and requirement of the IPS, there is always a tradeoff between the amount of modeled detail and available computing resources. To make the simulator useful, rendering times must be kept within certain limits. They should be shortest at the beginning of the development cycle, when many alternatives still need to be evaluated, and can

Table 1 Comparison of image synthesis techniques.

Class	Method	Implem.	Features	Availability
1	polygon shading	SW/HW	flat-shaded polygons	
2	flat shading with z-buffer	SW/HW	depth values	
3	Goraud shading with z-buffer	SW/HW	smooth shading, simple fog, point light sources	SGI entry models
4	Phong shading with z-buffer	SW/HW	highlights	
5	texture mapping with z-buffer	SW/HW	surface textures, simple shadows	SGI high end, flight simulators
6	reflection mapping with z-buffer	SW/HW	reflections	SGI next generation
7	ray-tracing	SW	refraction, real camera model, area light sources with penumbra, realistic material models	common ray-tracers
8	ray-tracing + global illumination simulation	SW	indirect illumination	*Radiance*
9	ray-tracing + global illumination simulation + "participating" media	(SW)	realistic clouds, scattering	current research

be longer towards the end when the perception system has almost converged to its optimal performance. Fortunately, the amount of detail and realism required at the early stages will also be lower than towards the end, so that in general a reasonable tradeoff should be possible. The conclusion is that a *single* rendering approach cannot cover the whole range of applications and the different requirements encountered during the system development process. The solution in IPS is to provide several *different* rendering modules that can can be invoked selectively but use a single environment model.

Some of the available options for rendering visual scenes are listed in Table 1, ranging from simple polygon shaders on the low (but fast) end up to sophisticated (but slow) ray-tracing techniques. At the high end, *i.e.*, for generating the maximally realistic images, we are using an enhanced version of the *Radiance* rendering system, which is described below. The choice of the lower-end techniques depends upon the (vision) application and the available rendering hardware. On state-of-the-art graphics workstations (*e.g.*, HP, SGI, Sun), interfaces to hardware-accelerated Goraud shading, Phong shading, and texture mapping (classes 3—5) are going to be provided. On these platforms, limited availability of ground-truth data is provided by reading the z-buffer contents. Also, hardware-accelerated techniques do generally not allow to implement non-standard sensor models, but this may be tolerable at the early test stages.

3.4 The *Radiance* Rendering System

The main scene generation tool in the IPS is *Radiance* [15], a physics-based rendering system for producing photo-realistic images of complex scenes. Although initially developed for applications

Advances in virtual reality technologies

in architecture and lighting design, *Radiance* is currently the most widely used non-commercial tool for general photo-realistic image synthesis. *Radiance* considers both direct and indirect (global) light sources and uses a combination of deterministic and stochastic ray-tracing techniques to balance between speed and accuracy. It supports a wide variety of object shapes, materials, and textures, and accepts many different CAD input formats for describing the scene. Parallel (distributed) processing and limited animation are also supported. The most important *deficiencies* of *Radiance* for this application are related to the sensor model. *Radiance* uses a simple pinhole camera and ideal shutter model and we are currently extending the package to consider lens distortions, depth-of-field effects, and motion blur. Also, we are enhancing the system to improve outdoor scene generation by incorporating fractal terrain models, plant shapes, and enhanced textures.

Lens distortion: Real lens systems, wide-angle lenses in particular, are not entirely free from geometric distortions. As a consequence, straight lines in 3-D do not generally map onto straight lines in the 2-D image. However, this assumption is frequently made for vision algorithms that depend on calibrated imagery, such as binocular stereo algorithms using epi-polar planes or the Hough transform for finding straight lines and parametric curves.

Depth-of-field effects: In the pinhole camera model, every object in the scene is in sharp focus. This is not the case with a real (thick) lens, where images have a finite "depth of field" that depends primarily on the focal length, the aperture setting, the focus setting, and the distance of the object. This is mainly important for close-range viewing, such as in robotic applications, but can probably be ignored in many outdoor tasks. Practical solutions for simulating the depth-of-field effect exist [5, 11].

Motion blur: Image motion (and thus motion blur) is induced by a moving camera, a moving object, or a combination of both. Due to occlusion effects, moving shadows, *etc.*, motion blur is an extremely complex phenomenon and its simulation time-consuming [12]. Distributed ray-tracing, as proposed in [5], appears to be the only solution that allows arbitrary object shapes and motion paths.

Enhanced outdoor environments: Since *Radiance* is not primarily aimed at simulating outdoor scenes, we are enhancing the system to support the generation of fractal terrain models, plant shapes, and corresponding textures.

3.5   Ground-Truth Data

The acquisition of ground-truth data is usually expensive and tedious (*e.g.*, through separate theodolite or radar measurements [7, 14]), therefore the possibility to access highly reliable ground-truth data is a key motive for using simulation in many vision applications. Rendering systems like *Radiance* do not have this access capability built-in, but most of the internal data structures are already available and *Radiance* is currently being extended to provide this functionality. Each application requires specific ground-truth data, including local measures such as pointwise 3-D depth (z-value), local surface orientation, local surface curvature, surface color, material type, and object identity. Selective query mechanisms for these modalities are provided through the IPS interface protocol.

4   SELECTED VISION PROBLEMS

For many mainstream vision problems, such as 3-D scene reconstruction, shape-from-X techniques, sensor calibration, multi-sensor integration, motion analysis, obstacle avoidance, autonomous navigation, terrain interpretation, *etc.*, the huge amount of test data provided by an immersive simulation, with simultaneous access to ground-truth data, should allow to boost the accuracy and robustness of many algorithms. While space does not permit to expand on these issues here, the areas of active vision, adaptation and learning, and performance evaluation deserve some additional thoughts:

Active Vision: The idea of active vision is to control the sensor(s) in a goal-direction fashion, thus allowing, *e.g.*, to focus on specific parts of the scene, track moving objects, or to compensate for platform motion [1]. Since the resulting systems are inherently closed-loop, testing is usually only possible when the whole setup is complete and operational. Simulation allows to perform meaningful experiments at a much earlier phase and without finalizing the mechanical design. Typical hardware components for active vision are motorized pan/tilt heads, motorized lenses, stereo vergence mechanisms, and gyroscopes, all of which need to (and can) be properly modeled within the simulation system. The "Level of Detail" concept in graphics simulation [13], where the currently viewed parts of the scene are presented in more detail than others, is closely related to the ideas of active vision and multi-scale representations.

Adaptation and Learning: Current machine perception techniques, computer vision in particular, lack the required robustness, reliability, and flexibility to cope with the large variety of situations encountered in a real-world environment. Many existing techniques are brittle in the sense that even minor changes in the expected task environment (*e.g.*, different lighting conditions, geometrical distortions, changing vegetation, *etc.*) can strongly degrade the performance of the system or even make it fail completely. Although results have been disappointing in the past, there are good reasons to believe that adaptation and learning can increase the robustness and flexibility of current vision techniques. There are many forms of learning applicable for vision, including statistical parameter estimation, clustering, function approximation, structural learning, self-organization, and neural network training. Existing applications include low-level processing, feature selection and grouping, model acquisition from examples, map learning, and 3-D object recognition [3, 4]. The issue of automatic or interactive acquisition of knowledge for object recognition (acquisition of object models) is probably among the most urging and challenging problems in vision.

Task-specific adaptation of vision systems is best done by learning on examples, either in a passive (open-loop) or active (closed loop) fashion, where the learning system itself can determine the choice of training examples. Unfortunately, slow convergence and the need for large sets of training examples are typical characteristics of most common learning methods, making practical learning tedious and time-consuming. This is particularly true for any kind of supervised or interactive learning. The use of immersive simulation can provide several important solutions with respect to the learning problem in vision:

   a. Large sets of realistic examples can be created and processed with reasonable effort.

   b. Supervised learning is possible without human intervention, since ground-truth data and actual model information are available from the simulator.

   c. Closed-loop (or "exploratory") learning, where specific, critical training data are generated in response to the learning progress, can be performed if needed.

Of course, learning in this environment is not restricted to the vision system alone but is equally important for all other aspects of agent control. As a result, future vision and robot control system may have to spend just as many (and probably more) hours "inside the simulator" as human pilots before they are ready to perform their tasks reliably.

4.1 Performance Evaluation and Prediction

While measuring the performance of a *passive* perception system is, at least, difficult, evaluating a vision-based, closed-loop control system is almost impossible before the complete system is actually deployed. Pre-recorded test image sequences are appropriate for performance comparison of passive vision systems but are insufficient for testing *reactive* systems. In a simulation test suite, the system would not only be required to *interpret given images*, but to *solve a particular task* within the (virtual) environment. Effects of variations in sensor performance (such noise, jitter, vibration, dynamic range limitations, *etc.*) can be obtained without additional costs in the simulation process.

The *prediction* of complete system performance under real conditions is of equal importance, particularly for mission-critical applications. For real-time system development, it is often necessary to evaluate different hardware configurations, which are either physically available or only in the form of software emulators. Specific system components may be gradually replaced by real-time hardware components as during the development process. Tools that support this kind of complex and heterogeneous engineering process are emerging only now [10].

5 INTELLIGENT CREATURES AND "EMBEDDED PERCEPTION" IN VIRTUAL ENVIRONMENTS

Current VR systems can represent and simulate almost arbitrary objects, even if they do not obey the laws of physics, by using (simplifying) abstractions of their behavior and interactions. "Intelligent" creatures, however, cannot be modeled without describing their detailed behavior, *i.e.*, without actually implementing them in software.

There is no such abstraction for intelligence as there is for rigid body motion in mechanics. Moreover, intelligent creatures in a VR environment need to have means of perception, including visual, acoustic, and other sensors, raising the question, how these perceptual system components can be modeled and simulated within VR. Although simulating the descision-making process of the individual creatures may not involve much intelligence at all, the complexity of their *perceptual* processing is (as we know) considerable. For example, in order to simulate the behavior and decision of the pedestrians shown in Figure 3, one would need to know what they see and hear in the given situation. Naturally, they cannot perceive the entire scene at any point in time. Which parts of the environment are visible to each of them, how much detail is seen, is the motion of the approaching car registered, are distances sufficiently estimated, is the situation too complex to allow predictable reactions? Computer vision is concerned with finding answers to at least some of these problems for the purpose of building intelligent machines. The same technology could be useful to simulate this behavior in VR, and thus we can expect that computer vision may indeed become an inherent component of "intelligent" VR systems some day in the future.

This opens interesting perspectives with respect to enhancing the capabilities of VR. Once the robot's perception and control system are integrated with the VR system, the simulated

Figure 3 "Intelligent creatures" in a virtual environment cannot be implemented without simulating *perception* at the same time. For example, a realistic driving simulator will include pedestrians, whose actions will depend on what they *see* and *hear*. This requires the implementation of *embedded perception systems* within VR, likely based on the technology provided by *Computer Vision* (image source: *California Driver Handbook 1992*).

robot actually becomes part of the VR itself. As a consequence, we will be able to simulate and participate in the interactions between multiple "intelligent agents", as *e.g.*, driving a car in urban traffic.

6 SUMMARY

In the past, experiments on synthetic imagery have generally not been considered conclusive for a computer vision system's performance under real operating conditions. We believe that state-of-the-art simulation and computer graphics can provide the degree of detail and realism required to make synthetic experiments just as important as experiments on real data. The underlying assumption is that, if a vision system *does* perform well on large sets of sufficiently realistic synthetic imagery, there is a good chance it will show similar performance in reality. More importantly, if the system *fails* even on synthetic data, there will be little hope for reliable operation in practice.

The need to test computer vision systems without sufficient test data or in closed-loop applications has led to many *ad hoc* simulation setups that are limited to very specific tasks. A generic simulation environment, such as the IPS described here, should reduce this expensive "reinventing-the-wheel" syndrome and make comprehensive testing more practical. In addition to providing extensive training sets for vision system validation, immersive simulation technology facilitates the implementation and testing of complex reactive systems, for which sensory data cannot be obtained off-line. This is of particular importance for the design of future perception-based systems, such as planetary rovers and similar critical systems.

As a more futuristic perspective, we have argued that computer vision offers some of the technological ingredients needed to implement intelligent creatures within virtual environments themselves, in particular their perception component. Both directions, VR in a solid engineering application for CV on one side, and CV as an enhancement to VR on the other side, should lead to an interesting cross-fertilization between these two areas.

7 BIBLIOGRAPHY

[1] Y. Aloimonos. Purposive and qualitative vision. In *Proc. DARPA Image Understanding Workshop*, pages 816–825, 1990.

[2] M. Asada. Map building for a mobile robot from sensory data. *IEEE Trans. Systems, Man, and Cybernetics*, 37(6):1326–1336, 1990.

[3] B. Bhanu and T. Poggio. Special section on learning in computer vision. *IEEE Trans. on Pattern Analysis and Machine Intelligence*, 16(9):865–919, September 1994.

[4] K.W. Bowyer, L.O. Hall, P. Langley, B. Bhanu, and B. Draper. Report of the AAAI fall symposium on machine learning and computer vision: What, why and how? In *Proc. DARPA Image Understanding Workshop*, pages 727–731, 1994.

[5] R.L. Cook. Stochastic sampling and distributed ray tracing. In A.S. Glassner, editor, *An Introduction to Ray Tracing*, pages 161–199. Academic Press, 1989.

[6] E.D. Dickmanns and F.R. Schell. Autonomous landing of airplanes by dynamic machine vision. In *Proc. IEEE Workshop on Applications of Computer Vision*, pages 172–179, Palm Springs, CA, December 1992.

[7] R. Dutta, R. Manmatha, L R. Williams, and E.M. Riseman. A data set for quantitative motion analysis. In *Proc. Conf. on Computer Vision and Pattern Recognition*, pages 159–164, June 1989.

[8] K.E. Olin and D.Y. Tseng. Autonomous cross-country navigation: an integrated perception and planning system. *IEEE Expert*, pages 16–30, August 1991.

[9] K. Pimentel and K. Teixeira. *Virtual Reality: Through the New Looking Glass*. Windcrest Books, 1993.

[10] J. Pino, S. Ha, E. Lee, and J.T. Buck. Ptolemy: A framework for simulating and prototyping heterogeneous systems. Technical report, University of California, Berkeley, August 1992. Department of Electrical Engineering and Computer Science.

[11] M. Potmesil and I. Chakravarty. Synthetic image generation with a lens and aperture camera model. *ACM Trans. Graphics*, 1(2):85–108, April 1982.

[12] M. Potmesil and I. Chakravarty. Modeling motion blur in computer-generated images. *Computer Graphics*, 17(3):389–399, July 1983.

[13] G. Schaufler and W. Stürzlinger. Generating multiple levels of detail for polygonal geometry models. In *Proc. Virtual Environments 94*, pages 727–731, Monte Carlo, January 1995.

[14] B. Sridhar, R. Suorsa, P. Smith, and B. Hussien. Vision-based obstacle detection for rotorcraft flight. *Journal of Robotic Systems*, 9(6):709–727, September 1992.

[15] G.J. Ward. The RADIANCE lighting simulation and rendering system. In *Proc. SIGGRAPH Conference*, pages 459–472, Orlando, FL, July 1994. ACM.

[16] C.C. Weems, C. Brown, J.A. Webb, T. Poggio, and J.R. Kender. Parallel processing in the DARPA Strategic Computing vision program. *IEEE Expert*, 6(5):23–38, October 1991.

8 BIOGRAPHY AND ACKNOWLEDGMENTS

Wilhelm Burger received his M.S. degree in Computer Science from the University of Utah, Salt Lake City, in 1987 and a PhD. in Systems Science and Robotics in 1992 from Johannes Kepler University in Linz, where he currently works in the area of imaging and computer vision. His main interests include object recognition, autonomous navigation, and the application of machine learning techniques in vision. Dr. Burger has worked with the Honeywell Systems and Research Center in Minneapolis, MN, and has authored a book on "Qualitative Motion Understanding", published by Kluwer. He has served as a reviewer for several international journals and conferences and is currently president of the Austrian Association for Pattern Recognition.

Matthew J. Barth received his B.S. degree in Electrical Engineering / Computer Science in 1984 from the University of Colorado, Boulder. He completed his M.S. and Ph.D. degrees in Electrical and Computer Engineering (ECE) at the University of California, Santa Barbara (UCSB) in 1986 and 1990 respectively. Dr. Barth has worked with the Center for Robotic Systems in Microelectronics (CRSM) at UCSB and the Advanced Technologies Division of General Research Corporation in Santa Barbara. He spent several years as a visiting researcher at the universities of Tokyo and Osaka. Dr. Barth is currently an Adjunct Assistant Professor at the UC Riverside College of Engineering, and the group leader for transportation systems research at the Center for Environmental Research and Technology (CE-CERT). His areas of interest are transportation modeling and simulation, intelligent transportation systems, computer visualization, and real-time robotic perception / control.

During the research reported here, the first author was supported in part by the Austrian National Science Foundation (FWF) under grant P8496-PHY.

PART FIVE

Industrial Applications

20

Physical Models for Solving Off-Road Vehicle Motion Planning

Moëz Cherif

LIFIA & INRIA Rhône-Alpes

46, av. Félix Viallet, 38031 Grenoble Cedex 1, FRANCE

Tel: (33) 76574583   Fax: (33) 76574602

email: `Moez.Cherif@imag.fr`

Abstract

This paper deals with motion planning for a mobile robot moving on a hilly three dimensional terrain and subjected to strong physical interaction constraints. The main contribution of this paper is a planning method which takes into account the dynamics of the robot, the robot/terrain interactions, the kinematic constraints of the robot, and more classical geometric constraints. The basic idea of our method is to integrate geometric and physical models of the robot and of the terrain in a two-level motion planning process consisting in combining a *discrete search strategy* and a *continuous motion generation method*. It will be shown how each planning level operates and how they interact in order to generate a *safe* and *executable* motion for the all-terrain vehicle.

Keywords

motion planning, physical models, mobile robots, off-road vehicles, interaction constraints

1 INTRODUCTION

This paper addresses the problem of motion planning for a robot moving on a hilly three dimensional terrain, and subjected to dynamic constraints due to the interactions with its environment. During the last decade, most of works in robot motion planning have been focused on solving the problem of generating collision-free trajectories on planar areas considering non-holonomic kinematic constraints (see (Latombe, 1990)). Recently, few results have been obtained when additional dynamic constraints have to be processed, and when the robot has to move in a natural environment (for instance an off-road vehicle, or a planetary rover). Nevertheless, despite the ability of the proposed methods to solve some instances of such a planning problem (Shiller and Gwo, 1991)(Siméon and Dacre Wright, 1993), the automatic generation of *safe* and *executable* motions for a mobile robot subjected to strong physical constraints is far to be fully accomplished.

This comes from the fact that both the physical vehicle/terrain interactions and the dynamic constraints to satisfy cannot be processed using purely geometric and kinematic models. Indeed, such parameters play a major role in this context —because friction, sliding and skidding phenomena may strongly modify the behavior of the vehicle—. Moreover, this behavior results from the combination of various geometric and physical criteria: the mechanical architecture of the vehicle, the characteristics of the motion control law which is applied, the vehicle/terrain interactions, and the strategic orders given to the robot. This means that appropriate *physical models* have to be combined with more classic *geometric* and *kinematic models*, in order to integrate such complex constraints within the motion planning scheme (i.e. vehicle/terrain interactions and dynamic characteristics have to be accurately modelled and processed at the planning time).

2 THE APPROACH

2.1 The problem

Let $\mathcal{A}$ be the robot, $\mathcal{T}$ be the terrain, and Q_{start} and Q_{goal} be respectively the initial and the final configurations of $\mathcal{A}$. We denote in the sequel the workspace by $\mathcal{W}$, the configuration space of $\mathcal{A}$ by $\mathcal{CS_A}$, and its state space by $\mathcal{SS_A}$. The problem to solve is to find a safe and executable continuous motion $\Gamma(Q_{start}, Q_{goal})$ and the corresponding sequence of controls U allowing to move $\mathcal{A}$ from Q_{start} to Q_{goal} while respecting the constraints of the task. Γ is said to be safe if takes into account non-collision and contact relation constraints. These last constraints express the fact that contacts between several wheels of $\mathcal{A}$ and $\mathcal{T}$ have to be maintained, and cases of tip-over of $\mathcal{A}$ have to be avoided. Besides, Γ is considered to be executable if it verifies the constraints resulting from the non-holonomy and the dynamics of $\mathcal{A}$, and the constraints imposed by the set of forces and torques created by both the vehicle/terrain interactions and the control strategy to apply.

2.2 The Two-level Planning Method

The robot $\mathcal{A}$ considered in this paper is an articulated non-holonomic vehicle having a locomotion system composed of three axles having each one two motorized wheels (see Figure 1). $\mathcal{A}$ is equipped with a set of joint mechanisms on its axles allowing it to be constantly in contact with the irregular surface of $\mathcal{T}$. However, this leads $\mathcal{CS_A}$, and consequently $\mathcal{SS_A}$, to be of a high dimension since a full configuration Q of $\mathcal{A}$ is given by $6 + n_\delta$ parameters: *six* parameters $(x, y, z, \theta, \varphi, \psi)$ specifying the position/orientation (yaw,roll and pitch) of the main body in the reference frame of the workspace $\mathcal{W}$, and n_δ parameters specifying the values of the set of joint mechanisms. Besides, dealing with dynamics and physical vehicle/terrain interactions when solving for Γ may lead to heavy computational burden. This is due to the fact that we have to operate in the state space $\mathcal{SS_A}$ in order to cope with both complex differential equations depending on the execution constraints of the task.

The main idea of the approach we propose consists in solving the planning problem by combining and integrating the geometric and physical models of the task with a two-level technique in order to make the motion planning problem more tractable and to deal uniformly with the above mentioned constraints (Cherif et al, 1994a).

Figure 1 The six-wheeled robot on the terrain.

The Geometric Level

As mentioned earlier, only the configurations Q of $\mathcal{A}$ satisfying the safety constraints are considered. In that case, the parameters of Q are not independent and the inter-relations between them depend on the vehicle/terrain relation (i.e. distribution of the contact points according to the geometry of the terrain). A practical consequence of this property is that it allows to reduce the search space into a reduced subset $\mathcal{C}_{search}$ of $\mathcal{CS}_{\mathcal{A}}$ defined on (x, y, θ) —the horizontal position and the heading angle of the main body of $\mathcal{A}$ denoted from now by q. Afterwards, the solution is iteratively computed between q_{start} and q_{goal} corresponding to Q_{start} and Q_{goal}, by applying a discrete search technique (an A^* algorithm for instance) through an incrementally generated directed graph $\mathcal{G}$ representing the explored configurations of $\mathcal{C}_{search}$. Such an approach has already been used in (Barraquand and Latombe, 1989) and (Siméon and Dacre Wright, 1993) to find non-holonomic paths for a robot moving on planar and $3D$ areas, respectively. Two nodes $N(q)$ and $N(q_{next})$ are connected in $\mathcal{G}$ if $\mathcal{A}$ a simple non-holonomic path composed of a circular arc or a straight line segment may be generated from the current configuration q towards the next sub-goal q_{next}.

The Physical Level

Since the construction of $\mathcal{G}$ (i.e. generation of the nodes $N(q)$ and the non-holonomic paths) does not account neither the geometric shape of $\mathcal{T}$ nor the dynamics of the task, we process the second level in order to cope with such features. This consists in computing locally a continuous motion of $\mathcal{A}$ satisfying the execution constraints between q and q_{next}. q_{next} corresponds to the configuration of the best successor node of $N(q)$ when searching $\mathcal{G}$. This is solved by formulating the planning problem in $\mathcal{SS}_{\mathcal{A}}$ and using a physical model of the task (Cherif, 1994b).

Unlike methods operating in two stages: (1) processing a geometric path, and after-wards (2) generation of a full trajectory of the robot when considering the task execution constraints, our approach allows to introduce dynamics and vehicle/terrain interactions at the planning time. Indeed the exploration of $\mathcal{C}_{search}$ is mainly used to guide the search process, and to give us only potential intermediate configurations of $\mathcal{A}$ approximating the final solution (see Figure 2). The real trajectory of the robot is provided by the sequence of the local motions computed when the physical model of the task are processed. The main advantage of such an approach is to cope locally with the dynamic constraints of the task

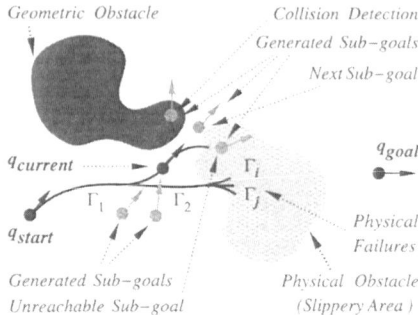

Figure 2 The general scheme of the planning approach.

as it has been already proposed in (Cherif et al, 1994a). At a given step of the algorithm, the potential successors of the current configuration q are explored in order to verify the safety constraints of the task (e.g. no-collision with the geometric obstacles). Afterwards, a motion $\Gamma(q, q_{next})$ is computed to move $\mathcal{A}$ from q towards its best safe neighbor q_{next}. The solution Γ is incrementally built until reaching the goal configuration q_{goal}. For instance, the motion solution Γ to be processed in Figure 2 is given by the previously computed motion Γ_1 between q_{start} and q, and the concatenation of the sub-motions $\Gamma(q, q_{next})$ and $\Gamma(q_{next}, q_{goal})$ which have to be processed. If the $\Gamma(q, q_{next})$ cannot be computed because of sliding or skidding of $\mathcal{A}$ on slippery areas (such as when computing Γ_i and Γ_j), the algorithm backtracks in order to select an other potential intermediate configuration to reach and to guide the search until q_{goal}. We present in §3 and §4 the physical modeling of the task and the physical level of the planning algorithm, respectively.

3 TASK REPRESENTATION USING PHYSICAL MODELS

3.1 Physical Modeling of the Vehicle $\mathcal{A}$

The dynamics of $\mathcal{A}$ is formulated using a mixed model combining the physics of particles and the mechanics of solids (see (Cherif et al, 1994a)). It is described by a network of interconnected rigid bodies Ω_i (corresponding to the components of $\mathcal{A}$ such as the wheels, the chassis or the axles) linked on specific points by visco-elastic relations (Cherif et al, 1994a). This enables to couple directly each element of the robot to those others that are in contact with it. It is different from the dynamic model based on a joint space formulation where every link is related to the one immediately before it, and every motion is sensed in relative and require successive transformations processing which can be time consuming.

Let $r_i(t)$ and $\alpha_i(t)$ be respectively the position and the orientation parameters at time t of a given Ω_i of $\mathcal{A}$. The motions of Ω_i are specified by the Euler/Newton equations: $F_i = m_i \ddot{r}_i(t)$ and $T_i = \dot{L}_i(t) = I_{\Omega_i} \ddot{\alpha}_i(t)$, where F_i and T_i are respectively the sums of forces and torques applied on Ω_i, $L_i(t)$ is the angular momentum about the center of mass G_{Ω_i}, and I_{Ω_i} is the inertia tensor of Ω_i about the frame axes. $\dot{L}_i(t)$ is also related to

Coriolis and centrifugal terms (Goldstein, 1983), but we will make the assumption that such terms are negligible. Then, F_i and T_i can be computed using the *Euler's* principle of superposition: $F_i = F_d + \sum_{force\,j} F_{i,j}$ and $T_i = U_i + \sum_{force\,j}(G_{\Omega_i}P_{i,j} \times F_{i,j})$. $F_{i,j}$ are the forces acting on Ω_i, $P_{i,j}$ are the points where the forces $F_{i,j}$ are applied, $G_{\Omega_i}P_{i,j}$ is the vector from G_{Ω_i} to $P_{i,j}$, and $\times$ is the outer product. The term F_d includes the gravity forces and additional viscous forces of the environment. The set of forces $F_{i,j}$ results from the the physical interactions of Ω_i with the other components of $\mathcal{A}$ —through the joints of $\mathcal{A}$— and with the involved components of the terrain —through the wheel/ground contact interactions. When Ω_i is a wheel, U_i corresponds to the torques generated by the control mechanisms (i.e. the "physical effector") applied on Ω_i, otherwise this term vanishes.

Each articulated mechanism associated with the joint δ_k of $\mathcal{A}$ is represented by a network $\Phi(\delta_k)$ combining a set of connectors c_r and a set of specific points selected on the rigid bodies Ω_i corresponding to the joint δ_k. The connectors are defined in terms of visco-elastic laws (i.e. combination of springs and dampers). For instance, we have represented $3D$ rotoïd joint mechanisms by two rigid bodies connected through two pairs of points respectively selected on them and belonging to the rotation axis (see Figure 3).

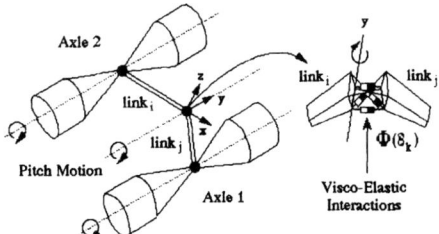

Figure 3 The physical model of the vehicle.

3.2 The Virtual Physical Model of the Terrain $\mathcal{T}$

Coping with robot/terrain interactions requires to build a virtual model $\Phi(\mathcal{T})$ of the terrain which is able to capture both the geometric and the physical properties of $\mathcal{T}$ and which allows the formulation of such interactions. For that purpose, we have implemented a model based upon the concept of *"deformable physical models"*, initially proposed for COMPUTER GRAPHICS (see (Luciani et al, 1991)(Terzopoulos et al, 1987)).

According to this concept, the terrain is represented by a set of interconnected particles $\Phi(P_i)$ having the following properties (Jimenez, Luciani and Laugier, 1991)(Luciani et al, 1991): (1) each particle is seen as a point mass m_i which obeys Newtonian dynamics —given by the equation $F_{P_i} = m_i \ddot{r}_{P_i}$ where r_{P_i} is the position of $\Phi(P_i)$ in $\mathcal{W}$— and which is surrounded by a spherical non-penetration "elastic" area; (2) the set of particles corresponds to the inertial and spatial occupancy characteristics of the modeled area of $\mathcal{T}$; (3) the particles are interconnected using interaction components referred to as the "connectors". Each connector corresponds to a type of interaction, and is modeled using appropriate physical laws allowing several types of behaviors (e.g. visco-elastic or elastic cohesion, dry friction interactions).

The discretization of the terrain in terms of such elementary physical components accounts several criteria such as the terrain surface shape, the average distribution of the contact points between the wheels and the ground, and the complexity of $\Phi(\mathcal{T})$ (i.e. the number of the processed particles). In the current implementation of the system, the particles distribution is determined by computing a set of spheres S_i whose profile approximates the surface of $\mathcal{T}$ given by the set of the initial geometric patches. This is done in such a way that each point of the terrain surface should be located on the surface of at least one S_i. Afterwards, a dynamic behavior is "given" to the computed set of spheres by placing a particle $\Phi(P_i)$ at the center of each S_i (Figure 4). The main advantage of such a representation is related to the fact that it allows us to maintain the geometric features of the motion planning problem (i.e. checking the geometric constraints as the non penetration in the ground), and to uniformly process the physical behavior of $\mathcal{T}$ and its interactions with $\mathcal{A}$.

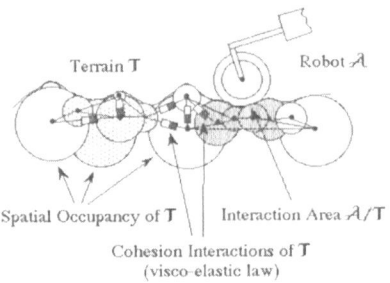

Figure 4 The physical model of the terrain.

4   THE PHYSICAL MOTION PLANNING LEVEL

The purpose of the physical planning level function) is to check for the existence of an executable motion allowing $\mathcal{A}$ to move between two successive configurations $q_{i,p}$ and q_i, i.e. a motion which satisfies the kinematic/dynamic constraints of the task. Task constraints consist of the kinematic/dynamic constraints of $\mathcal{A}$, the constraints imposed by the vehicle/terrain interactions, and the constraints coming from the applied control strategy.

4.1   Coping with Vehicle/Terrain Interactions

The interactions between $\mathcal{A}$ and $\mathcal{T}$ are based on the formulation of a set of dynamic laws depending on both the distribution of the contact points and the type of the surface-surface interactions. Describing $\mathcal{T}$ in terms of spheres requires to handle a smaller amount of information to represent its geometric shape (as shown in Figure 4). Furthermore, the combination of such simple primitives with an appropriate hierarchical description of the wheels allows us to compute easily the distribution of the contact points using a fast distance computation algorithm involving the structured sets of the considered spheres (Hopcroft, Schwartz and Sharir, 1983). Once a contact between the wheels of $\mathcal{A}$ and a sphere S_i corresponding to $\mathcal{T}$ is detected, the corresponding physical interaction is easily computed by activating the associated dynamic laws.

The surface-surface interactions are processed using two types of constructions: a visco-elastic law associated with the set of S_i involved in the contact, and surface-surface inter-actions. In order to solve the second point, we make use of a *finite state automaton* since complex phenomena like dry friction basically involves three different states: no contact, gripping, and sliding under friction constraints. The commutation from one particular state to another is determined by conditions involving gripping forces, sliding speed, and relative distances. Each state is characterized by a specific interaction law. For instance, a visco-elastic law between the interacting points of the wheels of $\mathcal{A}$ and $\mathcal{T}$ is associated with the gripping state, and a Coulomb equation is associated with the sliding state (Jimenez, Luciani and Laugier, 1991).

4.2 The Physical Planning Algorithm

The main advantage of the virtual model of $\mathcal{T}$ and the physical model of $\mathcal{A}$ is to exhibit consistent numerical (and graphical) behavior of the robot which can be useful in predicting the resulting motion and planning safe and executable trajectories. Thanks to this concept, the motion of $\mathcal{A}$ can be computed by uniformly integrating the differential equations of motion corresponding to the components of the physical models of the robot and the terrain, when controlling the wheels. In order to achieve such a process, we use a motion generation scheme proposed and described in (Cherif, 1994b).

The main difficulty is to find an appropriate way to generate a motion $\Gamma(q_{i,p}, q_i)$ which allows $\mathcal{A}$ to move from $q_{i,p}$ to the next subgoal q_i. Let δt be the time increment of the motion generation process, $s_{i,p}$ be the state of $\mathcal{A}$ corresponding to the configuration $q_{i,p}$, and let $s(n\delta t)$ be the state of $\mathcal{A}$ obtained after having applied n elementary motion steps when starting from $s_{i,p}$ (i.e. after having applied a sequence of n controls on the "physical effectors" of $\mathcal{A}$). Determining the required sequence of controls U to apply to $\mathcal{A}$ can be done by executing an iterative algorithm involving two complementary steps. The first step consists in hypothesizing a nominal sub-path $\mathcal{P}_i^k$ between the current configuration $q(n\delta t)$ and the next sub-goal represented by q_i, and the second step allow to track $\mathcal{P}_i^k$ while processing the physical vehicle/terrain interactions, as illustrated in Figure 5.

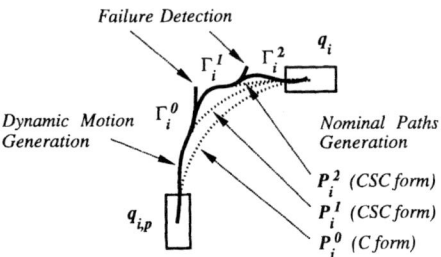

Figure 5 The local motion generation scheme.

The nominal path $\mathcal{P}_i^k$ is constructed using a technique derived from the Dubins' approach (Dubins, 1957). The obtained sub-path is a smooth curve made of straight line segments S and circular arcs C (of the form CSC). The tracking function operates on the dynamic model of $\mathcal{A}$ and the *virtual* representation of $\mathcal{T}$. It takes as input the velocity controls applied on each controlled wheel during a time increment δt. These controls are computed from the linear and steering speeds which are associated with the reference point of $\mathcal{A}$ when moving on $\mathcal{P}_i^k$. They are converted into a set of torques $U(t)$ to be applied to the wheels of $\mathcal{A}$. Since the motion generation step accounts physical phenomena like sliding or skidding, the configuration $q^\star(n\delta t)$ corresponding to the state $s^\star(n\delta t)$ of $\mathcal{A}$ obtained after having applied n successive controls may be different from the nominal configuration $q(n\delta t)$. The processed motion generation step will be considered as a failure when $q^\star(n\delta t)$ is too far from its nominal value $q(n\delta t)$. The previous algorithm is iterated until the neighborhood of $s_{i,p}$ is reached or until a failure is detected (see Figure 5). Figure 6 shows a local trajectory obtained when $\mathcal{A}$ is controlled to cross an irregular area of the terrain.

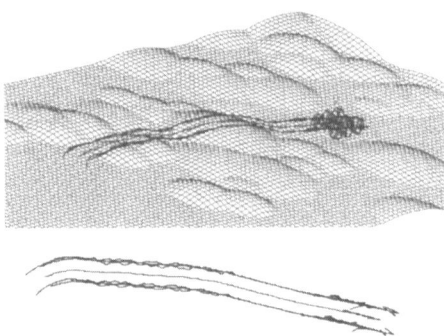

Figure 6 Local motion of the vehicle $\mathcal{A}$.

5 SIMULATION RESULTS AND CONCLUSION

We have shown in this paper how physical models can help in solving some classes of complex motion planning problems of an articulated all-terrain vehicle. A method to plan safe and executable motion for a rover moving in a natural environment and strongly subjected to physical interactions with the terrain has been also presented. This is based on the use of a "virtual" model of the terrain, and combines a continuous motion technique and a discrete search strategy in order to deal with several non-trivial features such as collision avoidance, kinematics and dynamics of the vehicle and its physical interactions with the terrain. The approach presented in the paper has been implemented on a Sun Sparc workstation. Several experiments for a non-holonomic six-wheeled rover $\mathcal{A}$ have been successfully performed in simulation. For instance, Figures 7 and 9 show the trajectories generated by the planner when $\mathcal{A}$ moves on an irregular area of the terrain. In figure 8 and 10, we show the resulting motions when the robot has to move on slippery areas.

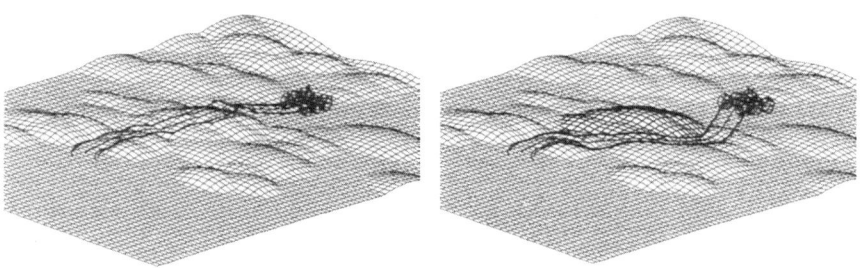

Figure 7 Moving on irregular areas.   Figure 8 Avoiding slippery areas.

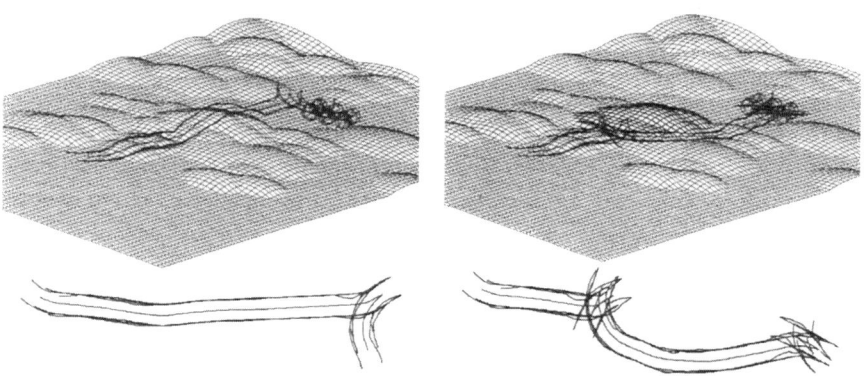

Figure 9 Maneuvering on irregular areas.   Figure 10 Maneuvering on slippery areas.

Acknowledgements

This work has been partly supported by the CNES (Centre National des Etudes Spatiales) through the RISP national project and the MRE (Ministère de la Recherche et de l'Espace) . It has been also partly supported by the Rhône-Alpes Region through the IMAG/INRIA Robotics project SHARP.

REFERENCES

Barraquand, J. and Latombe J.-C. (1989). On non-holonomic mobile robots and optimal maneuvering. *Revue d'intelligence Artificielle* **3**(2), 77–103.

Cherif, M., Laugier Ch., Milési-Bellier Ch. and Faverjon B. (1994a). Planning the motions of an all-terrain vehicle by using geometric and physical models. In: *Proc. of the IEEE*

Int. Conf. on Robotics and Automation. San Diego, CA (USA). pp. 2050–2056.

Cherif, M. (1994b). Dynamic motion planning of autonomous off-road vehicles. Research Report RR-2370. INRIA Rhône-Alpes. Grenoble (F).

Dubins, L.E. (1957). On curves of minimal length with a constraint on average curvature, and with prescribed initial and terminal positions and tangents. *American Journal of Mathematics* **79**, 497–516.

Goldstein, H. (1983). *Classical Mechanics.* Addison-Wesley.

Hopcroft, J.E., Schwartz J.T. and Sharir M. (1983). Efficient detection of intersections among spheres. *Int. Journal of Robotics Research.*

Jimenez, S., Luciani A. and Laugier Ch. (1991). Physical modeling as an help for planning the motions of a land vehicle. In: *Proc. of the IEEE/RSJ Int. Workshop on Intelligent Robots and Systems.* Osaka (J).

Latombe, J.-C. (1991). *Robot motion planning.* Kluwer Academic Press.

Luciani A. *et al* (1991). An unified view of multiple behaviour, flexibility, plasticity and fractures: balls, bubbles and agglomerates. In: *IFIP WG 5.10 on Modeling in Computer Graphics.*

Shiller, Z. and Gwo Y.R. (1991). Dynamic motion planning of autonomous vehicles. *IEEE Trans. Robotics and Automation.*

Siméon, T. and Dacre Wright B. (1993). A practical motion planner for all-terrain mobile robots. In: *Proc. of the IEEE/RSJ Int. Conf. on Intelligent Robots and Systems.* Yokohama (J).

Terzopoulos D. , Platt J., Barr A., Fleischer K., Elastically deformable models, *Computer Graphics*, 21(4), July 1987.

BIOGRAPHY

The author received the engineering degree from the University of Tunis, Tunisia, in 1990, and the M.S. degree in computer science from the Polytechnic Institute of Grenoble (INPG), France, in 1991. He is currently preparing a Ph.D. in computer science within the robotics group of LIFIA-IMAG Laboratory and INRIA Rhône-Alpes. His research involves mobile robots, motion planning, dynamic modeling and simulation, and virtual reality.

21

Integrating Applications into a Virtual Prototyping Environment

Uwe Jasnoch, Holger Kress, Joachim Rix
Fraunhofer-Institut für Graphische Datenverarbeitung
Wilhelminenstr. 7
64283 Darmstadt - Germany -
Tel. : +49 6151 155 (245 | 212 | 221)
Fax: +49 6151 155 299
Email : (jasnoch | kress | rix)@igd.fhg.de

Abstract

The paper will show and discuss different strategies of integrating applications into a Virtual Prototyping Environment. The ability to rapidly prototype a proposed design in a Virtual Environment is becoming a key contributor towards fulfilling the business requirements embodied in a short time-to-market, in cost-effective and high quality manufacturing, and in easy support and maintenance. The ability of integrating already existing applications into such an environment is fundamental. The paper first clarifies the different notations used for the different forms of integration. Afterwards a communication system as a basis for integration is proposed, which tackles most of the problems concerning the communication between heterogeneous applications. Finally, different strategies to integrate an already existing CAD system are shown and discussed.

Keywords

Virtual Prototyping, Integration Strategies, Communication

1 INTRODUCTION

Virtual Prototyping Environments gain more and more importance. In complex virtual prototyping environments, frameworks are often used as the software basis for these environments. As in CoConut , a framework is a computing environment which for itself is not very useful unless a number of applications are integrated. The phrase "tool/application integration" has

been widely used in the beginning. It has different meanings to different people, as stated in .
Some distinguish between loose integration and tight integration, black-box integration and
white-box integration, or integrated tools versus attached tools „. In the past, a better understanding of integration led to the distinction between three "kinds" of integration: white-box,
grey-box, and black-box integration .

These kinds have to be seen in the context of three basic areas: control, user interface, and
data. This leads to a three-dimensional quantity of integration as shown in figure . As a mini-

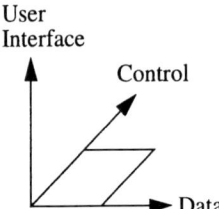

Figure 1: Integration Measurement

mum, a framework has to have some control over the application in order to start and stop the
application. For data integration, a certain knowledge about the data structure is necessary. The
dotted area of figure indicates a possible degree of integration, where the framework has the
knowledge of control and some knowledge of data. In this example, the application does not
use any user interface service. With the help of this figure, the different kinds of integration
become more clear: white-box integration means the maximum expansion on a certain axis.
Grey-box integration is an expansion between the maximum and the origin. Black-box integration means no expansion on a certain axis. But at least, to have a minimum control over the
application, some expansion on the control axis is necessary.

Nevertheless, white-box integration is only possible for applications, where the source-code
is modifiable or when applications are designed with the knowledge of the provided services in
mind. For applications, which are not designed to operate in a framework environment, the
phrase "tool/application encapsulation" has been established for the integration. From the data
integration point of view encapsulation means to have a wrapper for converting the data
between the framework schema and the application schema and the conversion of the access
functionality. Data integration itself, is the ability to share information throughout the environment . For an open design environment, data interchange between several applications is fundamental. This means in the real environment, on-line data interchange via the data
management service of the underlying framework. On the other hand, a flexible Communication System is needed, to enable the control integration of the new application.

This paper first discuss some major issues of the communication system as a basis for the
control integration of an application. Afterwards, the integration of an already existing CAD
system into CoConut is discussed as an example of a complex integration task. The paper ends
with a conclusion.

2 THE COMMUNICATION SYSTEM AS A BASIS FOR CONTROL

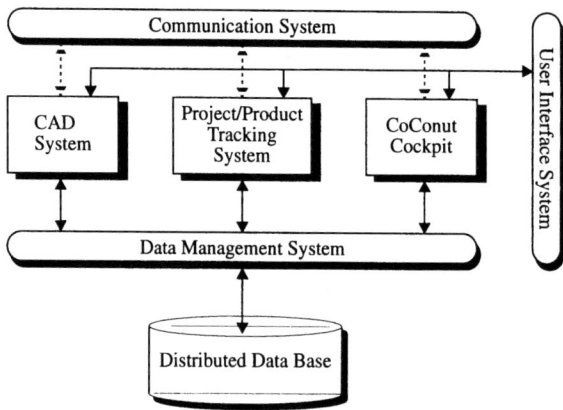

Figure 2: The CoConut Environment

INTEGRATION

The task of the communication system is to deliver messages between different application and/or framework components through the environment. Hereby, the different scopes (the user scope, the environment local scope, and the environment global scope) of the messages must be supported by the communication system, according to the purpose of the message. One major requirement is the independency between the applications. This means that an application must be able to send a message, without knowing the receiver in detail. The application only requests a certain task or gives information about an event, without the knowledge of the concrete partner. This independency is the basis of the integration of new applications in a heterogeneous environment like CoConut.

In a heterogeneous environment the application for solving a certain task could change for the different platforms. This does not mean changing only the code due to the different architecture, this also includes the usage of a completely different application. In our first prototype environment, the design task is handled on an IBM with the CAD system CATIA, while on a SUN this task is solved by the CAD system COSMOS. CoConut must be able to decide based on the integrated applications and the environment to start the correct one after receiving a message from an application indicating the request for designing an object. Taking this into account, the independency between applications regarding the communication is substantial.

One of the basic applications in CoConut is the cockpit. The cockpit starts the necessary applications and monitors the environment state for a user. Whenever an application needs an additional support by another application, it only sends a message and the cockpit knows how to react. If an application which solves the request does not run, the cockpit will start it. Figure presents an overview of the environment and the integrated application in this context. Here, there we do not distinguish between the dependencies of computer architectures and applications. These are hidden as much as possible for the end-user by the environment.

The basis for this controlling actions of the cockpit is the message service provided by the communication system. With the help of this service, the cockpit is informed about the requests and knows on the other hand about currently running applications. When a new application is integrated, there exists an assignment between the messages this application can react on and which messages this application could send. As a minimum, the messages for starting and stopping the application should be present. If these messages are not present, a wrapper must be provided delivering these messages. The benefit of this loose coupling between the application is that the cockpit could decide by its own which application could solve the request based on the current environment.

3 INTEGRATION STRATEGIES FOR A CAD SYSTEM

We have chosen the CAD system CATIA (Dassault Systemes, France) as an example for an integration of an existing application. The system is a characteristic representative for a grey-box integration strategy. As a commercial system the source code of the system is not available and therefore changes to the basic architecture of the system are impossible. Consequently, the integration strategy is focused on the provided functionality of the system's API and the services of the framework.

Three areas of integration will be considered: the integration on the user level, the integration aspects of the information model and of the data management.

Integration on the user level

One major goal of the framework is the consistent handling of heterogeneous applications in the Virtual Prototyping Environment. This is achieved by the encapsulation of the applications and by the control mechanisms of the framework. The consequence for the user is, that his application is not longer a pure single user system. In order to fulfil the integration issues the framework provides different services which have to be used. These services simplify the interaction and communication within the environment.

The most obvious service is the initial login-procedure entering the framework's cockpit. This procedure gives the user access to his personal working environment and makes his presence known to the other participants in the environment. After this he is able to start applications using the start-up service of the cockpit. To launch the CAD system, several ways of interaction are possible. Considering the idea of an integrated working environment, an interaction can only be done in a specific context. For a CAD system the context is the design of an assembly or a single part. Therefore the start-up procedure can be activated e.g. in an application which visualizes the product structure of the current product. The CAD system is started and the cockpit initiates the loading of the selected assembly or part into the CAD system. This example shows, how the usual dialogue using a load menu and selecting files is altered into a context-dependent selection of objects. The load procedure is done completely by the framework without further interaction of the user. This includes conversion between different formats and the extraction of the data out of the global into local data repositories.

For interactions during a design session the user interface of the CAD system is extended using the API functions. CATIA provides two interfaces for this: the IUA (Interactive User Access) language and the GII (Graphics Interactive Interface) interface. Both can be used to provide a specific user interface component for the communication within the environment, e.g. additional load operations, interim and end of session write operations, or communication services.

Integration aspects of the information model

The integration on the data level depends on the availability of a common data repository based on a standardized information model, which can be accessed by all involved persons. Therefore, CoConut is based on an underlying object-oriented model, which was developed by TC184/SC4 of the International Standardization Organization (ISO). This model is part of the ISO standard 10303 (STEP) . STEP is an acronym for „Standard for the Exchange and the Representation of Product Model Data" and characterizes the emerging standard for the exchange of product model data. STEP is the only standard which offers an information model covering various phases in the life-cycle of a product.

In the CoConut approach we use an information model which is especially developed for the area of the design of mechanical products with a three-dimensional shape representation. This model is standardized as document number 203 of ISO-10303 with the title „Application Protocol 203: Configuration Controlled Design" . In detail, the application protocol specifies the structure and the versions of complex mechanical products and their use in assemblies. A configuration management model gives the possibility to define different configurations of the same product. The design process is described as a chain of actions which are started by a start or change request. These requests are approved by authorized persons who are described by an organization model. The following shape representations are covered by the geometrical and topological model of the application protocol: wireframe representation, surface representation, facetted boundary-representation, advanced boundary-representation and solid models. Other information about security classification, contracts, documents or certification is also covered by the model. As the application protocol defines all necessary information used in the design process of mechanical products it can serve as the common information model for the CoConut environment, which ensures the integrity and consistency of the product data.

Each application in the environment handles only subsets of this information model. A project management application manipulates a subset which includes the information elements for e.g. project progress, approval status, and person and organization related aspects. The Product Structure Tracking System uses the information about the assemblies and parts of the product, and the CAD system deals with all information elements which are related to the shape of the product. This includes primarily the geometric and topological representation of the shape. Therefore the underlying information model has to allow application-dependent subsetting and the data management system must be able to handle these subsets in a consistent and uncontradictory way. For this purpose the data management system supports the data sharing with specific data converters for every application. The converters transform the application data from their native format to the neutral STEP format of the environment and vice versa.

Integration aspects of the data management

The data management system depends highly on the services of the object-oriented data base management system (OODBMS) of the environment. The OODBMS provides logging mechanisms, access to the data, and ensures the consistency of the data sets. In a heterogeneous environment different clients of the data base can run on various hardware platforms and enable a distributed access to the common data base. Figure shows a configuration in which the data sharing is completely handled by the OODBMS. No further communication links are necessary.

In case of grey-box integrated applications which are not able to directly interact with the OODBMS, other integration strategies must be provided. Figure gives an overview over the proposed solutions.

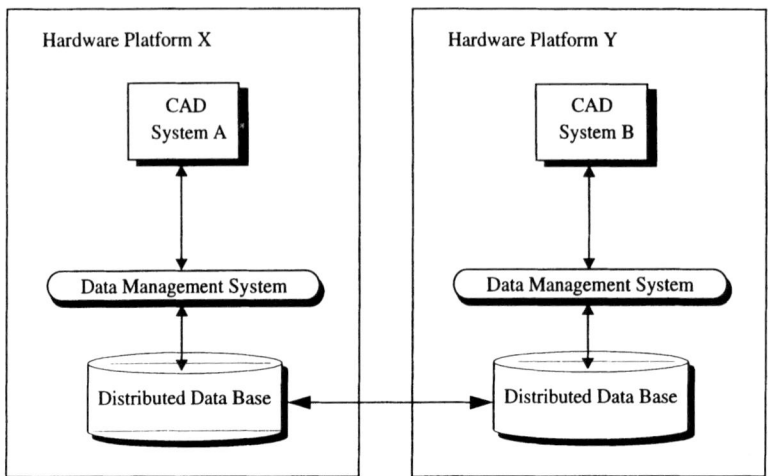

Figure 3: Data Management using a Distributed Data Base on heterogeneous Platforms

The API of the CAD system CATIA offers a set of functions called CATGEO subroutines used to add, modify or read basic data of the CAD system. This API is used to establish the connection to the data base. Three basic ways have been identified: a RPC interface, a physical file exchange, and the access to STEP working form repositories via NFS.

The RPC interface gives the functions of the data management system access to the functionality of the API subroutines. The approach follows the standard RPC implementation on heterogeneous platforms.

In the second case a sequential, ASCII file is used to exchange the CAD data among the different platforms. The STEP working form management system converts the native CAD data into instances of an intermediate, internal format. Accessing the instances of the internal format, the file formatter/reader generates the physical file according to the ISO specification 10303-11 . This file can be transferred via standard file transfer mechanisms. The receiving system generates with the file reader the internal format and transfers the data to the data management system.

The working form management system can store the instances of its internal format in an own data repository. This repository can be directly reached via NFS to pass the data to the data management system.

4 CONCLUSION

The paper described the communication system and the integration strategies for existing applications in CoConut, a system which supports the realization of parallel product development methods in the design phase. The described system contains several major features. One feature is the availability of a common, distributed data repository, which stores the relevant data of the design process according to the Application Protocol 203 *Configuration Controlled Design* of the ISO 10303 Standard STEP. The adaption of standards is a central objective in

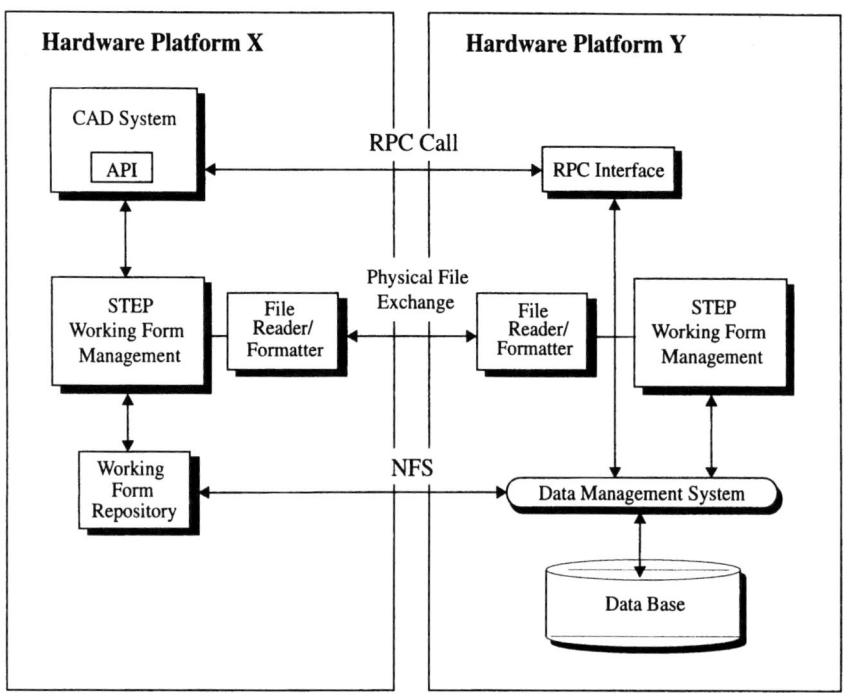

Figure 4: Data Management using a Data Base on a single Hardware Platform

CoConut and ensures the openness of CoConut for future developments.

The CoConut system represents an open, extendable framework for the integration of engineering applications. Future extensions to the system can address the underlying information model as well as the integration of applications which support specific tasks in the process chain of product development, such as kinematic or FEM analysis. The underlying data model can be extended by integrating additional STEP schemata. Thus, the whole data model will develop in the direction of an integrated product model.

The described CoConut environment integrates existing applications such as a CAD system in a heterogeneous computing environment. Nevertheless, the concept is open for the support of the whole engineering process which embraces the Virtual Prototyping Process.

5 REFERENCES

CFI Architecture Committee: „ *CAD Frameworks Users, Goals, and Objectives*", Version 0.92, 1990.

CFI Architecture Committee: „*Framework Architecture Reference*", Version 0.87, 1991.

European Computer Manufacturers Association: „*Reference Model for Frameworks of Soft-*

ware Engineering Environments", ECMA TR/55, 1991.

U. Jasnoch, H. Kress, K. Schroeder, M. Ungerer: *„CoConut: Computer Support for Concurrent Design Using STEP"*, in Proc. of the IEEE Third Workshop on Enabling Technologies: Infrastructure for Collaborative Enterprises, Morgantown, 1994.

T. Kathöfer et. al.: *„The Architecture of the Object Management System within the Cadlab Framework"*, in Rammig, Waxman (Eds.): Electronic Design Automation Frameworks, North Holland, 1991.

N. Kraft: *„Embedded Tool Encapsulation"*, in Rammig, Waxman (Eds.): Electronic Design Automation Frameworks, North Holland, 1991.

M. Lacriox, M. Vanhoedenaghe: *„Tool Integration in an Open Environment"*, Proc of the European Software Engineering Conference 1989, p 311-323, 1989

B. Prasad, R. S. Morenc, and R. M. Rangan: „Information Management for Concurrent Engineering: Research Issues", Concurrent Engineering: Research and Applications (1993) 1, 3-20, Academis Press, London 1993

ISO IS 10303-1: „Industrial automation systems - Product data representation and exchange - Part 1: Overview and fundamental principles"; International Organization for Standardization; Geneve (Switzerland); 1994

ISO DIS 10303-21: "Industrial automation systems - Product data representation and exchange - Part 21: Implementation methods: Clear text encoding of the exchange structure"; International Organization for Standardization; Geneve (Switzerland), 1993.

ISO DIS 10303-203: *„Industrial automation systems - Product data representation and exchange - Part 203: Application protocol: Configuration controlled design"*; International Organization for Standardization; Geneve (Switzerland); 1993

K.J. Cleetus: „Virtual Team Framework and Support Technology"; published in „Concurrent Engineering: Tools and Technologies for Mechanical System Design"; ed. by E.J.Haug; Springer 1993

6 BIOGRAPHY

Uwe Jasnoch received his university diploma in Computer Science from the Technical University of Darmstadt in 1989. Then, he was a software engineer with Philips Kommunikations Industrie AG for one year. Afterwards, Uwe Jasnoch was a researcher with the Interactive Graphics Systems Group at the Technical University of Darmstadt, where he was involved in several R&D projects. Since 1992, he has been a researcher with the Fraunhofer Institute for Computer Graphics in the Industrial Applications Department. His main research topics are data modeling, open environments, and consistency management.

Holger Kress is a researcher in the Industrial Applications Department of the Fraunhofer Institute for Computer Graphics since 1991. He is currently involved in research projects in the area of product modeling, groupware, and CAD frameworks. He received a masters degree in mechanical engineering from the Technical University of Darmstadt in 1991. His research interests include concurrent engineering, product modeling, and computer supported cooperative work.

Dr. Joachim Rix is head of the department for Industrial Applications of the Fraunhofer Institute for Computer Graphics (IGD) in Darmstadt, Germany. From 1991 to 1992 he was Associate Manager of the Fraunhofer Computer Graphics Research Group (today; Fraunhofer Center for Research in Computer Graphics, Inc. (CRCG) in Providence, RI). Mr. Rix received his Diploma and Ph.D. in Computer Science from the University of Darmstadt. His topics of interest are in Computer Graphics, CSCW, CAD, and Product Modelling. This includes the integration and use of computer graphics with its presentation and interaction techniques in industrial applications, like CAD, CAM, Concurrent Engineering , and Groupwork Computing. Mr. Rix is member of the standards Committee of ISO/IEC JTC1/SC24 „Computer Graphics" and was Rapporteur of the study group „PREMO" (Presentation Environments for Multimedia Objects). Since 1985 he is member of the national and international committees (DIN NAM 96.4, ISO TC184/SC4) developping STEP (Standard for the Exchange and Representation of Product Model Data). Since 1994 he holds the position of a deputy convenor of its WG 3 „Product Modeling". Since 1981 Joachim Rix is member of the Eugrographics Association.

22

Virtual Prototyping – Design and realistic presentation of industrial products

Jürgen Gausemeier, Andreas Sabin
Heinz Nixdorf Institut - University of Paderborn
33098 Paderborn, Germany, sabin@hni.uni-paderborn.de
Wolfram Lewe
Gepade Polstermöbel, Pamme GmbH & Co.
33129 Delbrück, Germany, Tel.: ++49/5250/515-0

Abstract

In this paper we present a concept for the general application of an integrated Engineering–system (CAE–System) in an upholstery–enterprise. The expositions describe a system–independent solution and an exemplary implementation by using standard–software–tools. This paper focusses on the design–stage of the product development process. The benefits for enterprises by designing upholsteries with an integrated computer–based system are shown, for example drastically reduction of conventional manufacturing of specimen and reduction of product development time.

Keywords

CAE–System, 3D–Modeling, Virtual Prototyping, Product development

1   INTRODUCTION

Problem

Enterprises are acting in surroundings, which are characterized by radical changes over the past years. The fulfillment of customer needs is now the fundamental basis for the competitiveness of enterprises. However their products shall be very reasonable, of first rate and also developed, manufactured, and delivered in short time.

One start to achieve these goals and to enhance the competitiveness is the reduction of the product development process, in the following called CAE[1]–process. Enterprises like furniture–manufacturers, car–makers, clothing industry, aircraft–manufacturers, and shipyards, which all produce consumer goods with high importance of excellent product design have to transform the following activities to shorten the time of the CAE–process:

- Use of computer–aided design–tools to develop virtual (computer–based) prototypes, which fulfill the customers needs and also achive a drastical reduction of the conventional production of specimen.
- General support of the product development process with CAE–applications based on an integrated product data model.
- Provision and utilization of standardized parts to reduce and control the variety of modules and parts.
- Integrated project– and process management tools to plan, control and supervise development projects in one or more enterprises.

Goal

Goal of this paper is to present a concept for a drastically acceleration of the CAE–process. The focus of this paper is the application of a computer–aided design–tool and the general information flow based on a product data model. Using suitable design–tools enables the development of products which correspond to the customers needs in short time. Additionally they can be placed and configurated directly with the customer. The product development time can be drastically reduced and the so far indispensable production of specimen can be avoided as far as possible.

Procedure

The following chapter presents the concept to integrate virtual prototyping into the CAE-process. Chapter 3 describes a successfully realization of our concept for the design of upholsteries and the benefits of the realisation for manucfactures of upholsteries. The final chapter 4 gives an overview of possible further developments on virtual prototyping of consumer goods.

2 CONCEPT FOR VIRTUAL PROTOTYPING IN THE PRODUCT DEVELOPMENT PROCESS

Virtual Prototyping as a component of the Integrated Engineering –System

The fundamental for the presented concept and the realized case study forms the integrated engineering-system (CAE-system), presented in figure 1 and explained in (Gausemeier, Frank, Genderka, 1993) and (Gausemeier, Frank, Genderka, 1994).

The CAE–system provides non–application–specific tasks as system-configuration, user-interface, user-management, general CAE-applications (e.g. process-, cost-, quality-management) and data-management (Rammig, Steinmüller, 1992). Additionally specific CAE–applications are integrated into the framework of the CAE–system. These CAE-applications solve special tasks in the product development process, for example FEM (finite element method), technical

---

1. CAE – Computer Aided Engineering

design modeling, drawing, etc. A suitable design tool, which uses the full functionality of the CAE-system has to be integrated into the CAE-system as the central component of virtual prototyping.

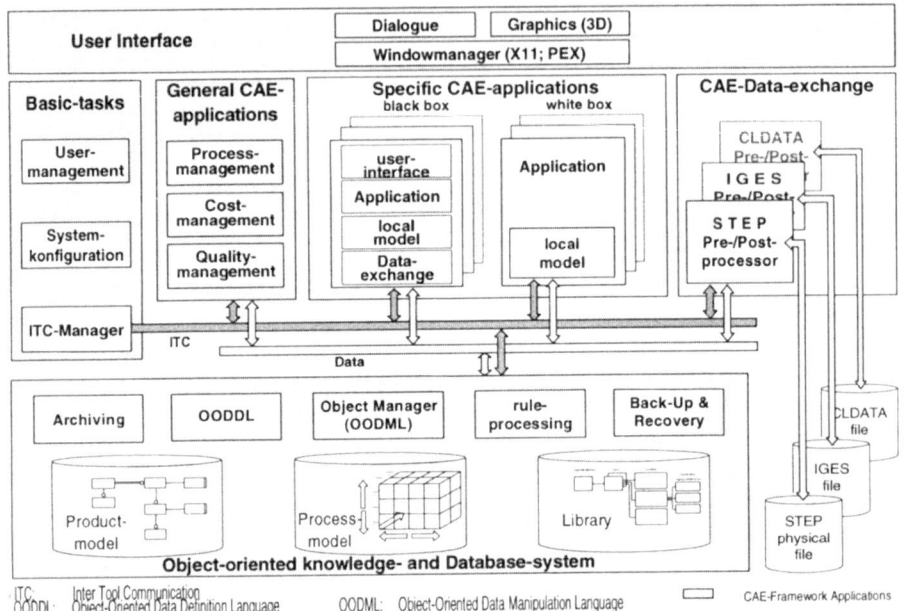

Figure 1   Structure of an integrated engineering–system (CAE–system).

The CAE–system provides non–application–specific tasks as system-configuration, user-interface, user-management, general CAE-applications (e.g. process-, cost-, quality-management) and data-management (Rammig, Steinmüller, 1992). Additionally specific CAE–applications are integrated into the framework of the CAE–system. These CAE-applications solve special tasks in the product development process, for example FEM (finite element method), technical design modeling, drawing, etc. A suitable design tool, which uses the full functionality of the CAE-system has to be integrated into the CAE-system as the central component of virtual prototyping.

Using the "CAE-data exchange" functionality, CAE-applications can communicate with other CAE-applications or external systems by using data exchange protocols, e.g. STEP[1] or IGES[2]. A further characteristic feature of the CAE-system is the object-oriented knowledge- and database-system, which contains product model data, standardized elements, and the underlying process model for the entire product life cycle.

---

1. STEP – Standard for the Exchange of Product Model Data
2. IGES – Initial Graphics Exchange Standard

Demands on Virtual Prototyping in the Product Development Process

In the first step, requirements on a design–specific CAE–application, which consider the goal „Accelerate the CAE–process by using computer–aided design–tools", now have to be defined:

- The CAE–application „design" has to be realized as a specific application of the CAE–system and has to use the full funcionality of the CAE–system.
- Easy redesign of the model by defining parameters and constraints to enable fast and easy design–variations.
- Configuration–management – by using standardized elements, different models can be easily be generated and changed. Libraries of standardized elements are a prerequisite to reduce and control the variety of modules and parts.
- The designed product has to be portrayed in all possible perspective views by reflecting the particular light conditions (orthochromatic, in perspective, realistic shadow–, and surface–presentation).
- Photorealistic visual presentations enable sales and marketing departments to make strategic decisions without conventional production of specimen.
- The intuitive working method of the designer has to be supported by the application. The designer should not be inhibited in his creativity by using the application.
- The customer shall recieve a paper documentation as well as a presentation of the product on the screen. That means, he will get a rapid visualization of his ideas. Changes can be realized directly at the system. Furthermore the customer can take along colored prints of the product model.
- The data of the outline, which are generated in the design process have to be applied in further process steps, e.g. creating of flattenings, cutting optimization, or drawings. This refers to the use of a standardized product data model and the accessible precision of the data in the design process.

Draft of a Design–Tool for Virtual Prototyping

The above mentioned requirements lead to the conceptual solution which is presented in figure 2. Core of the concept is a 3D–oriented modeler, based on the NURBS–technology[1]. According to the high amount of sculptured surfaces especially in design–applications, the use of the NURBS–technology a prerequisite for the generation of realistic looking models (Mortenson, 1985). All parts and their related components, which were designed by the 3D–oriented modeler are filed in a parts library. The 3D–modeler defines the shape of the objects. Aspects like for instance coloring, surface characteristics, lighting conditions are not considered in this first step of the Virtual Prototpying process.

A further possibility to make the shape of objects available to the system is the later digitizing of physically existing objects. Suitable techniques are for example optical scanning or digitizing with 3D–digitizing–pens by defining of surface–points of the existing objects. The accuracy of the model thereby will increase with the number of grasped points.

---

1. NURBS – Non–uniform Rational B–Splines

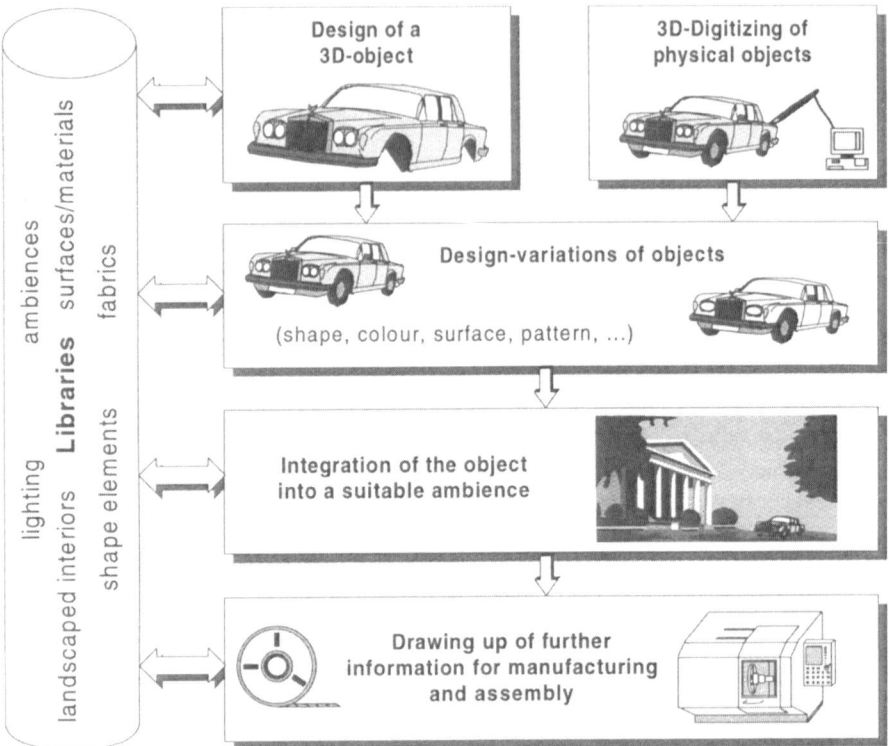

Figure 2   Tasks at Virtual Prototyping of consumer goods.

In the next step information on surface characteristics are attached to the generated components and products. Surface characteristics can be divided into the aspects:

- Materials, e.g. textiles, leather, metal, plastics, lacquer, timber, porcelain.
- Color, with typical characteristics like tone, reflexion, brilliance.
- Structure, e.g. fold, relief, roughness, pattern, texture.

These kind of information are also available in libraries. They are mapped on the surface of the particular components and product models (Texture mapping). In this early design stage a realistic looking model of the later product can already be generarted by assigning surface–characteristics to the objects. Shape variations are possible in every stage of the design process, because the object is modeled with a 3D–modeler which offers the possibility to define parameters and constraints. Visual modifications can be made by changing surface characteristics. Furthermore, the object can be viewed from any direction, because a 3–dimensional body and not a 2–dimensional drawing is modelled with the system.

Up to now the result of the design activities is the complete modelling of an object and the assignment of it´s surface characteristics. The next step is to place these objects into an adequate ambience, which can be fetched from an ambience–library, generated from different ambience–modules (e.g. buildings, landscapes, interiors, ...) which themselves are stored in libraries or at least it can be totally new generated. Furthermore, scanned photografies, e.g. landscapes can be used for special applications. By placing the modelled objects in ambiences the effects of shape and color of the objects can be excellently visualized and a nearly realistic reflection can be presented.

Sources of light have to be defined after positioning the object in an ambience to generate shadings, reflections, and so on. Characteristics for sources of light, which can be positioned in the ambience in any number and at any place are:

- Color.
- Intensity (brightness).
- Radiation characteristics (spotlight, diffuse scattered light, indirect lighting, etc.).

After specifying the characteristics of the sources of light and positioning them in the ambience, techniques for visualizing are applied. Examples for visualization techniques are Shading, Ray–tracing, Anti–aliasing (Muhar, 1992), (Rooney, Steadman, 1990). These techniques consider and convert the effects of the sources of light, for example shadings, hues or reflections.

The result of this process is a nearly photorealistic presentation of the object in a suitable ambience. These photorealistic presentations can be used by the enterprise to obtain decisive factors for strategic planning of products and product families, and further to use as sales– and promotion support for the sales department on the operative level.

According to the fact that the objects are modelled with 3D–oriented modelers, the generated product model data can be processed in the next task „Drawing up of further information for manufacturing and assembly". Examples are the generating of flattenings by using flattening–tools, e.g. for fabrics or sheet metals; cutting–optimization by using interlocking–tools to reduce the need of materials; generating of templates; modelling of the technical design and interior design (e.g. timber frame of an upholstery, ventilator and motor of a hair dryer); FEM–applications and so on. Corresponding CAE–applications are called up by the CAE–framework relating to the underlying work–flow–process.

Integrating a suitable design–tool into the integrated CAE–system drastically accelerates or even enables the Virtual Prototyping process of consumer goods with high requirements on excellent design. The concept for Virtual Prototyping presented in this paper enables the generation of nearly photorealistic presentations of objects. A drastically reduction of the (so far necessarily) conventional production of specimen can be achieved. By using 3D–modeling techniques and libraries for products, modules, surface characteristics or ambiences any changes of the objects will be possible. An adjustment of an object to the customers requirements so will be easy and fast converted.

3   VIRTUAL PROTOTYPING IN THE DESIGN-PROCESS OF UPHOLSTERIES

The concept presented in chapter 2 was realized in cooperation with Gepade, a German upholsterer and market leader of stylistic upholsteries (Lewe, 1994).

It was the goal of this project to demonstrate the principle feasibility of our concept by using in the first step standard–software–tools and to show the benefits of an integrated CAE–system with design–specific applications for the enterprise.

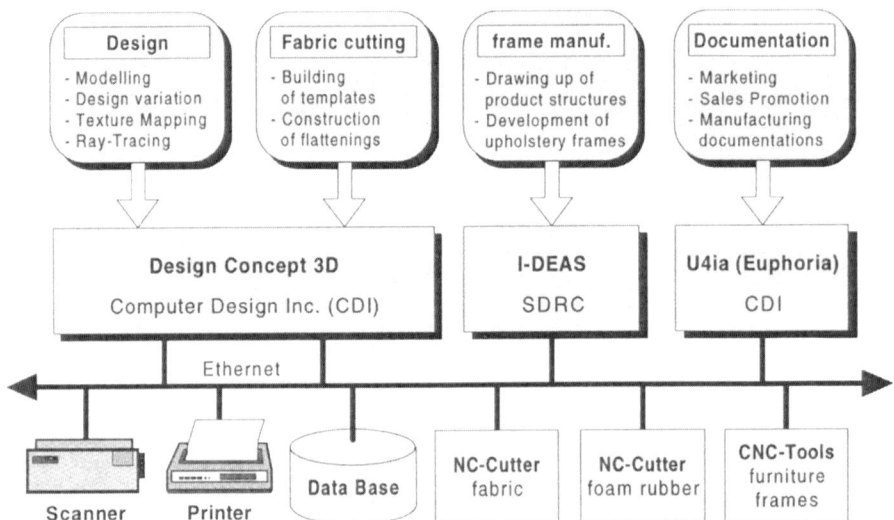

Figure 3   Configuration of the CAE–system for Virtual Prototyping of upholsteries (Lewe, 1994).

The requirements of the concept for Virtual Prototyping and company and industry specific factors lead to a selection and implementation of tools, which are presented in figure 3.

As the CAE–application for the specified design tasks the tool *Design Concept 3D* by *Computer Design Inc.* was applied. *Design Concept 3D* supplies functionalities for generating 3D–surface–models, parametrics, definition of constraints, and definition and application of surface characteristics. Furthermore the tool supplies functionalities for the creation of flattenings, generation of templates, and further functionalities especially for upholsterers, for example definition of cuttings, seams, fabric overlaps. As required, *Design Concept 3D* provides libraries for products, modules, fabrics, sources of light, ambiences and so on. Additionally photorealistic presentations of objects, placed in ambiences, can be generated. Applications of *Design Concept 3D* for Virtual Prototyping of upholsteries are:

- Generating the design of upholsteries and their components.
- Design–variations of shape, surface characteristics, lightings, etc.
- Generating of photorealistic presentations.
- Flattenings and cuttings of fabric.
- CAD/CAM–integration of NC–cutter (cutting of fabrics and foam rubber).

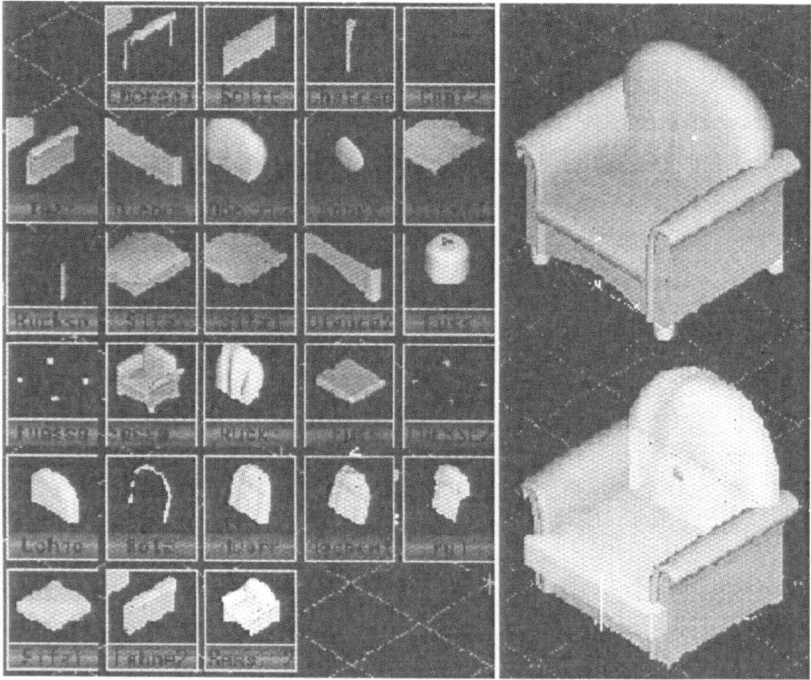

Figure 4   Design–variations by using standardized elements.

Figure 4 and the following figures 5 and 6 present functionalities of the choosen design–application. Figure 4 shows the realization of design variations by using standardized elements, for example armrests, blind arches, seats, back supports and so on, which are fetched from specific libraries. After assigning surface characteristics to the upholstery, the object can be placed in a landscaped interior and sources of light can be positioned, presented in figure 5. Whenever the requirements are fulfilled by the arrangement, manufacturing and assembly information can be generated. Figure 6 shows an example for the fabric–flattening of an upholstery.

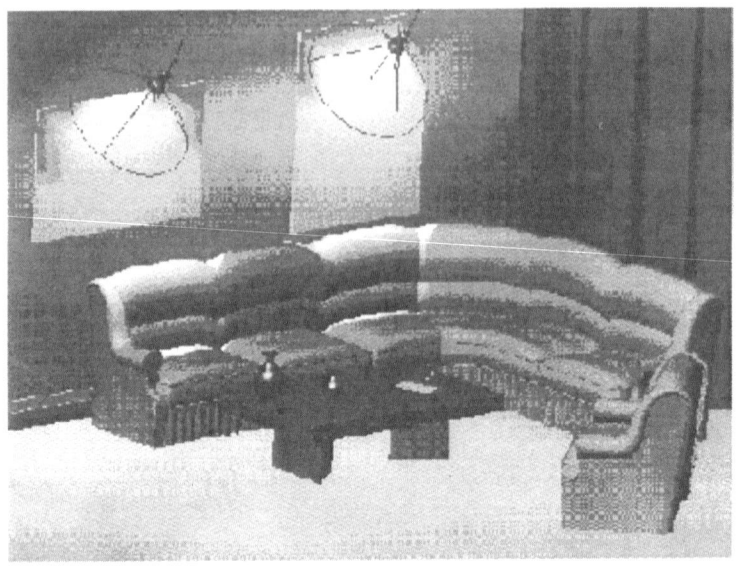

Figure 5   Creation of a landscaped interior and adjustment of lighting conditions.

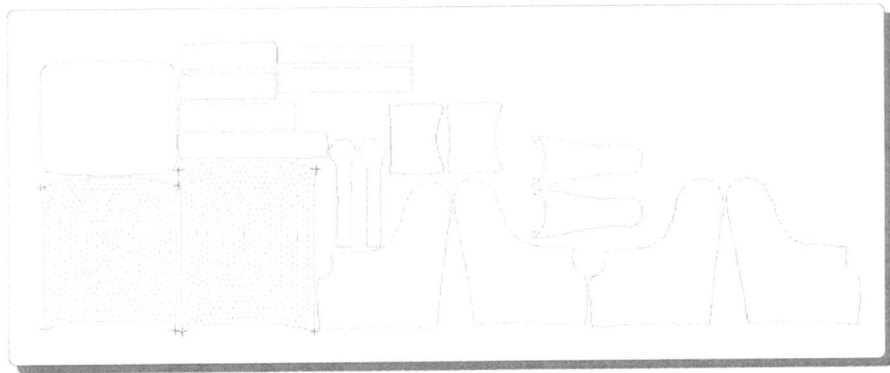

Figure 6   Flattening of the fabric of an upholstery.

In the next step the timber frame of the upholstery has to be developed under consideration of the fixed upholstery–design. Requirements on the frame–design are on the one hand the use of standardized frames or frame–components as far as possible and on the other hand the use of the outer shape of the upholstery as a general set–up for the frame. One guideline of the project was to use standard software–tools as far as possible. Because *Design Concept 3D* does not offer all the required functions for 3D–mechanical design, a further software–tool is choosen for the design of upholstery–frames. Employed is *I–DEAS Master Modeler* from *SDRC*, a 3D–solid–modeler. The coupling between *Design Concept 3D* and *I–DEAS Master Modeler* and the assigned tasks to the tools is presented in figure 7.

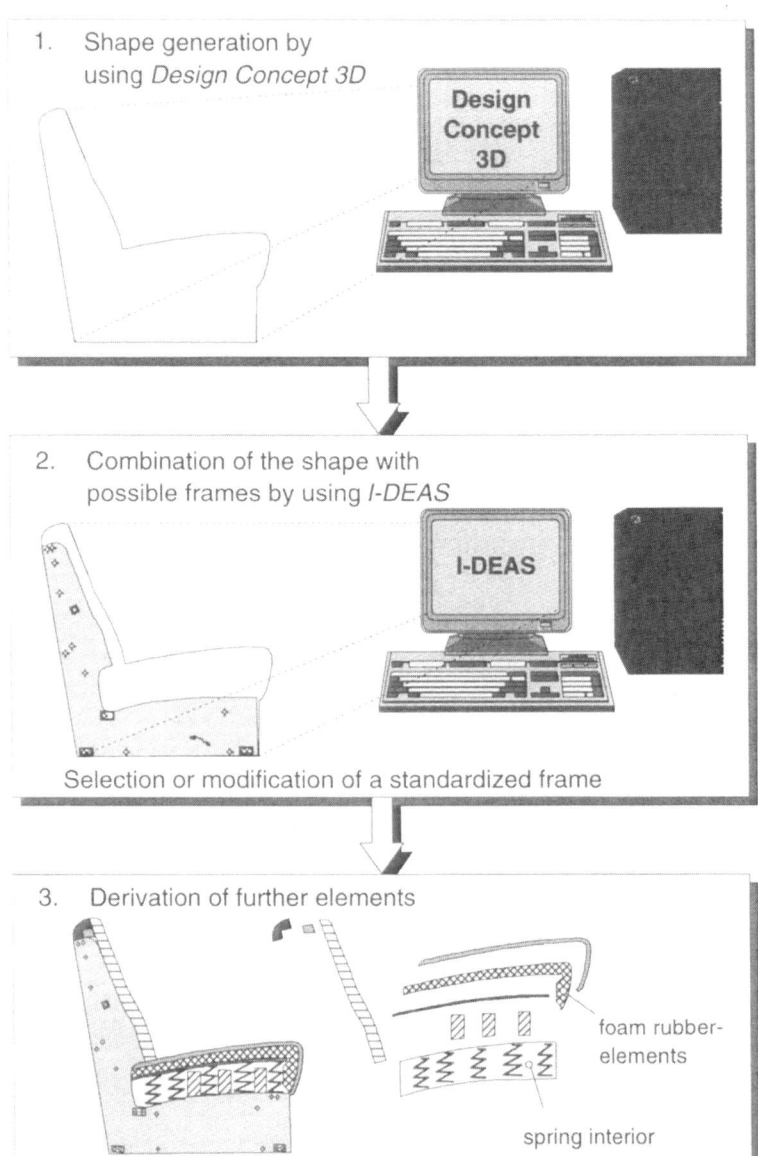

1. Shape generation by using *Design Concept 3D*

Design Concept 3D

2. Combination of the shape with possible frames by using *I-DEAS*

I-DEAS

Selection or modification of a standardized frame

3. Derivation of further elements

foam rubber-elements

spring interior

Figure 7 Coupling and allocation of tasks between Design Concept 3D and I–DEAS Master Modeler.

The *I–DEAS Master Modeler* has to fullfil the following tasks in this project:

- 3D–Frame–design by using as far as possible standardized components.
- Frame–variations by using parametric– and constraint–functionalities.
- Product data management.
- CAD/CAM–integration to manufacturing of frames (CNC–machine tools).

Figure 8 presents a timber–frame for an upholstery which was modeled with *I–DEAS Master Modeler*:

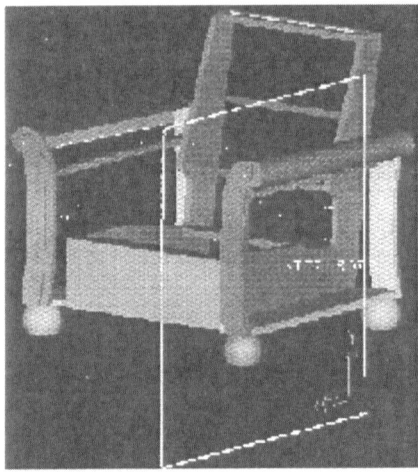

Figure 8   Upholstery frame modeled with I–DEAS Master Modeler.

For the graphically output of the photorealistic presentations of the modeled objects, generated with *Design Concept 3D* is used *U4ia (Euphoria)*, a software–tool also provided by *Computer Design Inc.* In this project it is employed for generating sales– and promotion documentations. A high–performance color–printer provides paper–printouts, edited by *U4ia*.

A further requirement on the concept for Virtual Prototyping was the ability to work with fabrics, which are used by the upholsterer to assess realistic optical effects (e.g. color, alignment, pattern–structures, ...) of different fabrics on upholsteries. This functionality is absolutely necessary for the supply of fabrics in the right size.

Gepade does not produce fabrics itself and computer–based fabric–patterns of the used fabrics are not available. Hence a large size color–scanner is employed to generate computer–based patterns of the desired fabrics. Furthermore the color–scanner can be used to scan photografies, for example portraits, landscapes, frontispieces or clippings of magazines and newspapers to improve the realistic impression of ambiences.

Benefits of the implementation for an upholsterer

The implementation of the above presented concept "Virtual Prototyping for the design of upholsteries as a component of the integrated CAE–system" opens up a lot of benefits for an upholsterer. These benefits can be classified as general benefits for the enterprise and as department–specific benefits (Lewe, 1994).

General benefits:

- drastically acceleration of processes
  - generation of a high number of design–variations in very short time;
  - utilization of the product model data in the complete product development process;
  - reduction of manual data exchange processes and coordination–processes.
- drastically reduction of the conventional manufacturing of specimen.
- flexible consideration of customer requirements and convertion in product–design.
- improvement of the utilization of standardized elements in every stage.
- enhancement of product– and process–quality.

Department–specific benefits:

- Design
  - easy, fast and interactive possibilities for computer–aided variations of shape and surface–characteristics of designed objects;
  - nearly unlimited design–variations by using standardized elements for components, fabrics, surface–characteristics and so on;
  - direct visuell control of the realized tasks.
- Development / technical design
  - manufacturing of specimen only for promising variants after Virtual Prototyping;
  - derivation of suitable timber–frames by using the design and standardized frame–components;
  - direct computer–aided generation of manufacturing and assembly–information, e.g. templates, flattenings, cuttings, NC–programs.
- Sales department / marketing
  - generation of photorealistic presentations of upholsteries for sales–negotiations;
  - reduction of misunderstandings between customer and sellers;
  - higher flexibility by changed customer requirements or customer tastes;
  - planning and assessing of the success of new product families without expensive manufacturing of specimen (drastically reduction of time and costs).

4 DEVELOPMENTS OF FUTURE VIRTUAL PROTOTYPING-SYSTEMS

This chapter describes developments of tools and process–innovations, which are suitable to improve and increase the benefits of Virtual Prototyping. It has to be distinguished between developments, which reference to the general Virtual Prototyping–concept was presented in chapter 2 and developments which are particularly interesting for Virtual Prototyping of upholsteries, refered to chapter 3.

Application of Virtual Reality–techniques (VR–systems)

Virtual reality is a new kind of computer technology, which makes the user believe he or she is immersed in a computer–generated world, the so–called "Cyberspace" (Foley et al, 1992), (Wexelblat, 1993). The development of VR–systems is currently in the beginning stage. Due to capital intensive hard– and software, industrial applications are only practical in special areas, for example the space industry or the military sector. Because of the high range of applicabilities, the VR–technology will also be applied in future research and development of standard industrial products (Gausemeier, Ebbesmeyer, Grafe, 1994). When adjusted VR–systems are fully technical developed and economically justifiable, an enterprise can use these techniques to support the Virtual Prototyping–process and for the presentation of its products to the customers.

In the upholstery–example it will be possible that a customer supplied with suitable VR–equipment immerses himself into a landscaped interior, created on his requirements. In "his" landscaped interior the customer can move totally free, shift upholsteries or other furnitures to desired places, change the arrangement by using other upholsteries from libraries, change fabrics, colors and so on. If other catalogues are available, the customer can check the composition of different types of furniture e.g. upholsteries, chairs, tables, cupboards with carpets, curtains, wallpapers, paintings, and other design elements like flowers or art objects.

Up to now 3D–modeled and photorealistic objects are presented in a 2–dimensional medium, for example a screen or on paper documents. By using VR–techniques, these 3D–modeled objects can also be presented to the user on a 3–dimensional medium. The user can model and view 3D–objects in a 3D–medium. The clarity and graphic nature of the scene will tremendously increase and the design quality will improve significantly.

Electronic sales catalogues

A further revolutionary change will emerge by electronic sales catalogues, which will remove the current paper catalogues. The electronic sales catalogue will not contain examples of available products with accessory charts of permissible combinations of e.g. furnitures, fabrics, colors, or optional extras. An electronic catalogue contains standard elements and eventually combination rules, which will be arranged according to the customers requirements. The salesperson analyses together with the customer the customers needs and arranges afterwards components, fabrics, materials, colors, and so on, which are available from libraries of the sales catalogue. Thus all available combinations, which are supplied by the enterprise can be presented. By using large size projectors the modeled objects can be presented life–sized, which will furthermore improve the clarity of the object.

Data transfer systems

The so far presented concept focusses on Virtual Prototyping for one enterprise as an approach to increase the competitiveness of the enterprise. The cooperation of different enterprises enables a further increase of competitiveness and offers new possibilities on the market.

A manufacturer can for example endow his direct customers, the specialized dealers, with the above explained electronic sales catalogues. Using data exchange technologies between manufacturer (e.g. upholsterer) and specialized dealers, the data about prices, delivery dates or terms of delivery can be transfered immediately. The manufacturer can include new orders directly in his production planning system and the customer will get exact delivery–information.

Standards to enable data exchange for commercial information are for example "EDI"[1] (Banerjee, Golhar, 1994) or "EDIFACT"[2] (Scholz-Reiter, 1991), which are yet available or in the development process.

Layout planning systems

By integrating further manufacturers into the data transfer system, for example manufacturers of carpets, curtains or cupboards, entire systems can be planned, e.g. living rooms, kitchens or offices. Using layout planning systems enables the arrangement of landscaped interiors, which are adjusted in color, style, and size.

Customers will profit from thus layout planning systems, because their desired landscaped interior is planned according to their requirements and they can get immediately a presentation and information about the latter appearance in reality but also about prices and terms of delivery. Manufacturers and specialized dealers can increase their planning–accurancy combined with a decrease of delivery periods.

5 SUMMARY

The involvement of a design–application into the integrated CAE–system ist the key component of the presented concept for Virtual Prototyping in the product development process. To gain the competitive advantage, a very close teamwork of task–specific CAE–application, embeded in the CAE–system is essential. This is reached by an integrated product data model which covers the whole product life cycle.

The presented concept for Virtual Prototyping was realized for the design of upholsteries by using standard software packages. The expected benefits of Virtual Prototyping for enterprises could be proved and confirmed.

Finally developments are presented, which have to be integrated into the general concept and into the example "design of upholsteries". These developments are Virtual Reality (VR), electronic sales catalogues, data transfer systems between one or more manufacturers, suppliers and customers and layout planning systems.

---

1. EDI – Electronic Data Interchange
2. EDIFACT – Electronic Data Interchange for Administration, Commerce and Transport, IS 9735

Figure 9   Example for a virtual landscaped interior, modeled with the presented implementation.

6 REFERENCES

Banerjee, S. and Golhar, D.Y. (1994) Electronic Data Interchange. *Information & Management*, volume 26, issue 2, 65–74.

Foley, J.D., Dam, A.v., Feiner, S.K. and Hughes, J.F. (1992) Computer Graphics – *Principles and Practice*. 2nd edition, Addison–Wesley Publishing Company.

Gausemeier, J.; Ebbesmeyer, P. and Grafe, M. (1994) Virtuelles Modellunternehmen für Forschung, Lehre und Technologietransfer, Proceedings *Virtual Reality–Forum '94*, Stuttgart.

Gausemeier, J.; Frank, T. and Genderka, M. (1993) Entwicklungstendenzen integrierter Ingenieursysteme. Proceedings *EDM–Congress 1993*, (ed. Ploenzke AG).

Gausemeier, J.; Frank, T. and Genderka, M. (1994) Erfolgspotentiale integrierter Ingenieursysteme (CAE). Proceedings *CAD '94*, Paderborn.

Lewe, W. (1994) Entwicklung eines Konzeptes für ein integriertes Ingenieursystem (CAE) in einem Unternehmen der Polstermöbelindustrie. Technical Paper Nr. 6 of the HEINZ NIXDORF INSTITUT, Paderborn.

Mortenson, M.E. (1985) Geometric Modeling, John Wiley&Sons, New York.

Muhar, A. (1992) EDV-Anwendungen. Ulmer, Stuttgart.

Rooney, J. and Steadman, P. (1990) CAD – *Grundlagen von Computer Aided Design*, Munich, Berlin.

Rammig, F.J. and Steinmüller, B. (1992) Frameworks und Entwurfsumgebungen. *Informatik Spektrum*, volume 15, Heft 1.

Scholz–Reiter, B. (1991) CIM–Schnittstellen. 2nd edition, Oldenbourg, Munich Vienna.

Wexelblat, A. (1993) Virtual Reality – *Applications and Explorations*, Academic Press Professional, Cambridge.

7 BIOGRAPHY

Prof. Dr.-Ing. Jürgen Gausemeier, born 1948, is Professor for Computer Integrated Manufacturing at the Heinz Nixdorf Institut of the University of Paderborn. He completed his doctorate at the Institut für Werkzeugmaschinen und Fertigungstechnik of the Technical University of Berlin with Prof. Spur in 1977. While working in industry for twelve years, Dr. Gausemeier was head of the Research & Development Department for CAD/CAM-systems and finally head of the Production Department process management systems with a major Swiss Corporation.

Dipl.-Wirt.-Ing. Andreas Sabin, born 1965, studied mechanical engineering and business administration at the University of Paderborn and works since 1993 at the Heinz Nixdorf Institut in the department of Computer Integrated Manufacturing. His major research topics are Virtual Prototyping and Product Development Processes.

Dipl.-Wirt.-Ing. Wolfram Lewe, born 1967, studied mechanical engineering and business administration at the University of Paderborn and works since 1994 with Gepade-Polstermöbel, Pamme GmbH & Co. in the information technology department.

23

Virtual Prototyping Using Graphical Simulation and Advanced Programming Techniques

K.-D. Thoben, U. Berger, J. Bauer, A. Schmidt
BIBA - Bremen Institute for Industrial Technology and
Applied Work Science at the University of Bremen
P. O. Box 33 05 60, D-28335 Bremen, Germany
email: as@biba.uni-bremen.de

Abstract

This paper presents a practical approach towards Virtual Prototyping. The fields of application are One-of-a-Kind products and industrial prototypes. The first example is taken from the problem area of typically high complex capital goods being virtually modelled and simulated prior to the real setting-up. The strategy of complimentarily developing both, the prototype's mechanics and its functionality and the issues of concurrent and distributed team-work are characterised. Enabling technologies for the iterative setting-up of the real prototype are described with reference to practical experience. The second example is the virtual and the touchable prototyping of sheet metal parts, incorporating the FEM simulation of the process as well as the closed prototyping chain for a fast verification of the results. An enabling technology is the triangulation of parametric CAD surface models, a technique to represent random solid bodies as a polyedrical mesh, from which "real" and "virtual" prototypers will equally profit. The possibility to include surface data acquired by 3D measurement systems is also discussed. As a conclusion, the modelling of process chains is seen as an adequate approach to achieve better clarity in production as well as in the product development, describing actions and streams in between them. Their addition leads to a reference model for rapid product development and One-of-a-Kind Production, providing an abstract view of the various process chains in a neutral and logical manner. This is a necessity to reach the aim of production in the end of this century: "make just what is needed when it is needed".

Keywords:

Virtual Prototyping, graphical simulation, industrial robots, car disassembling, Neural Network, off-line programming, optical 3D-measurement, CAD surface model, sheet metal forming, Rapid Prototyping, Concurrent Engineering, triangulation, Stereolithography STL-file, Reverse Engineering, rendering

1.   INTRODUCTION

The competitiveness of industrial countries depends on their ability to anticipate customer needs, respond quickly to the needs they can articulate and develop new, unique products. According to actual market data, the "time-to-market" phase requires ever larger investments of time and money for product development. With relatively high production costs, getting to market first is the key to profitability in the short term.

Rapidly developing and manufacturing according to customer wishes, so-called *on-demand manufacturing*, is the challenge. The time has gone when products can be developed first and then offered to a market or a potential customer. The implications for manufacturing systems design are profound. Therefore, there is a need to develop the capability, both technically and organisationally, to rapidly transform new ideas into prototypes (first-of-a-kind). Additionally related manufacturing equipment has to be developed, maintained and enlarged. As a consequence, advances in information and communication technologies and related organisational approaches have to make the required breakthroughs possible.

Besides other approaches, the computer-based visualisation of an early version of a product, parallel to the possibility of analysing, testing and manipulating this non-physical representation in a computer-generated virtual environment, plays an important role in achieving a shorter "time-to-market". As the basic principle of Virtual Prototyping is the early consideration and integration of entire life-cycle related requirements and the resulting early prediction and evaluation of the product behaviour, various approaches are possible. These areas of application range from simple visualisation to attract a customer by a simple but early virtual version of the product requested (i.e. product visualisations during tendering) to most complex FEM applications analysing the strength, functionality, and manufacturing and product behaviour during operation.

As, in principle, the idea of Virtual Prototyping is not really new, new hardware and software techniques and technologies can make Virtual Prototyping an excellent extension to existing approaches like Conjoint Measurement, DfX (Design for Manufacturing, Design for Assembly, etc.), Feature-Based Design, NC simulation, etc. and Virtual Prototyping can integrate and make best use of these single-minded integration approaches. In addition to this, further benefits of Virtual Prototyping can be achieved by a dynamic and iterative integration of the real and the virtual world and by an integration of the various views of different partners during a globally distributed product development.

In the following, experiences gained in different projects applying various approaches of Virtual Prototyping and related enabling technologies for speeding up product development of capital and consumer goods will be described. In a next step these approaches and enabling technologies will be characterised and classified to define a systematic way for improving product development and production through Virtual Prototyping.

2.   PROTOTYPING OF FREE-FORMED SHEET METAL PARTS

Regarding the product development cycle itself, the basic difference lies between the still widely used sequential way of defining a new product, and the more advanced methods with a consequent parallelism and a well-organised iterative approach. These methods are often referred to as *Concurrent Engineering*, *Simultaneous Engineering* and *Agile Engineering*, and they suggest the following rules to product developers:

- Gather any information needed in the starting phase. Begin with a large interdisciplinary expert team; decrease number of people while refining the solution.

- Make changes early, when this is still inexpensive.

- Make decisions early. Test out a large variety of solutions and select the best at the beginning instead of repairing mistakes resulting in large expenses at the end.

- Use communication tools (CSCW).

Basically, Concurrent Engineering can be performed with conventional means like drawing boards up to a limited complexity. But for planning complex products like aircraft, ships and manufacturing plants, more advanced design and communication tools - CAD, CAM, CAE, RP and EDM Systems - are highly recommended.

At first will be given an insight into potentials and demands observed with computational geometry manipulation and RP Techniques. Then, these consequences will be extended towards Virtual Prototyping.

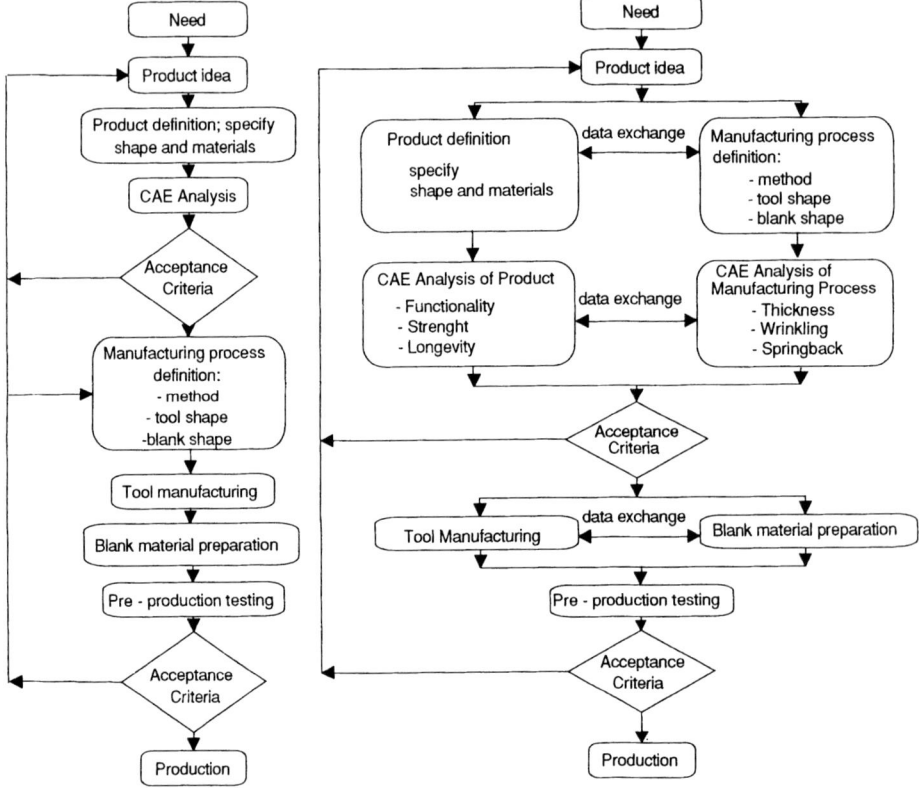

Figure 1 Schematic process chain: left with conventional method, right with integrated process chain

With Rapid Prototyping Technologies, models with good surface quality and accuracy can be built today. However it is not yet possible to produce a sheet metal forming tool directly by Rapid Prototyping. The potential of the new manufacturing techniques can only successfully be used by *a sequence of process steps* incorporating RP technologies. This sequence consists of:

- *Techniques of triangulating trimmed surfaces in parametric space.*
  The standard data exchange format in Rapid Prototyping is STL. It incorporates a primitive polyhedrical representation of three dimensional bodies. This has the advantage of simplicity and wide application range, but the disadvantages are limited accuracy, large file sizes and comparatively difficult data preparation in the case of converting surface CAD-models to STL - format.

- *Reverse Engineering*
  Quick modelmaking like RPT supply a powerful "downstream" path, from the abstract geometry data towards the physical representation of a part. "Reverse Engineering" tools are supposed to do just the opposite: with 3D measurement systems, a geometrical shape aquisition is performed first. Then, an abstract description of the object is reconstructed with the aid of dedicated software tools. The result of this is a valid CAD model of the part, even if it was comletely unknown before.

- *Rapid Prototyping Technology (in this case Stereolithography) is used to produce a mould.*
  With so-called Rapid Prototyping Techniques (synonyms: Layer Manufacturing, Solid Freeform Manufacturing, ...), parts with nearly unlimited geometries can be built automatically, with fully automated path preparation. On the other side, the material properties and the accuracy of the built parts are within fixed limits.

- *Casting technique using a high strength concrete material.*
  With methods like plaster moulding, vacuum casting and investment casting, moulds, tools and functional prototypes can be derived from the RP part directly. The casting materials palette can cover a wide range of material properties, e.g. high strength and high wear resistance in the case of moulds.

The main idea behind a new process chain is to build forming tools by using easily castable amorphous material and integrate the Stereolithography process for fabrication of the casting shell (negative geometry). To attain these goals, an information transfer process was developed from a CAD surface model of the formed part by means of Rapid Prototyping Technologies towards an advanced sheet metal forming process [3, 4].

The cast material consists of a mineral binding material with addition of ultra fine particles, which are typically 100 times smaller than the cement particles. These small particles, combined with a dispersing agent make up an extremely dense packing which results in a very high strength as well as the ability anchor to particles and fibres in the matrix [5]. It is a cheap material which is easy to handle, with no environmental risks connected to its preparation and disposal [6].

- *Application of a forming operation which is specially suitable to produce small batches.*
  Tool manufacturing is very expensive and ineconomical for small volume production. For sheet metal forming, methods with "effecting means" are available, without the necessity of the male and the female die. These are rubber pad forming, fluid cell forming and pressure forming.

The rubber-pad-forming technology is established for flexible and cheap forming operations. It uses a special type of rubber-die pressing in which the rubber die is totally enclosed, thereby exerting fluid pressure on the workpiece. The forming pressure is up to 1000 bar.

The main advantage of this principle is the possibility to form a part with only one die, where normally an upper and lower die are necessary.

The steps for manufacturing were:

- Production of blanks with a punching/nibbling machine after the CAD data had been postprocessed.
- Applying the rubber pad forming process to the blanks.

Figure 1 shows the conventional method and the integrated method for process chains for prototype parts.

2.1. Integrated Process Chain for Sheet Metal Parts: a Case Study

As an example, a prototype protection cap was realised for a position lamp (shipbuilding industry). The material was stainless steel (wear and corrosion resistant) and the required thickness of the sheet metal was 0,7 mm. (Figure 2)

Both a computer-aided optimisation of the forming process (finite-element methods) as well as close co-operation with tool manufacturing experts are necessary; otherwise some undesired results in the manufacturing process could occur, such as wrinkles; spring-back effects, cracks or distortions.

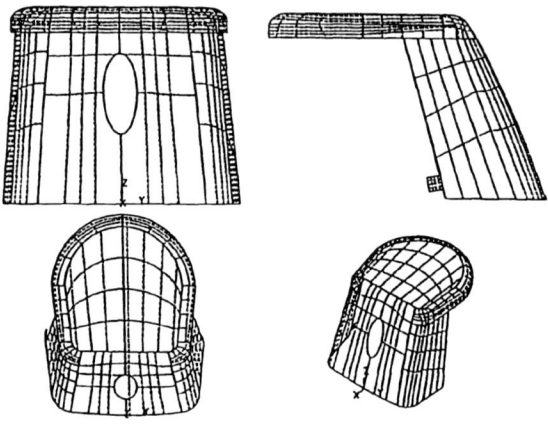

Figure 2 CAD surface model of protection cap

Derived from the product's CAD data, counterparts for the tool construction were designed. After triangulation and slicing on computer, these casting shells were fabricated on a Stereolithography machine. Then, the casting was performed with a special kind of ultra-fine concrete.

Parallel to the tool manufacturing, the preparation of the blank material was started. Based on the workpiece and tool data, the geometrical design of the sheet metal was done on a 2D-CAD/CAM system.

Now the metal part production started: the blanks were manufactured on a CNC-controlled punching- and nibbling - machine, and the forming operation was performed on a rubber pad forming press.

2.2. Triangulation of Trimmed Surfaces in Parametric Space

An essential key to automated manufacturing processes is the manufacturing data preparation, after the CAD model of a product or a tool is available. On the field of data-pre-processing for Stereolithography, a software package was written that is capable to triangulate parametric surfaces to a polyedrical representation (Stereolithography STL-file). It is able to find faults in STL-files and repairs them automatically, can handle layer information and create STL out of it again, and has visualisation and graphical user interface. For "closing the loop" from CAD model to physical part and back again, it will be of interest to integrate 3D optical sensors with the aim of doing Reverse Engineering, too.

The faceting of parametric surfaces is of great interest for many CAD/CAM applications. After the faceting, a surface model can be displayed with a nearly natural appearance in real-time. The faceting technique is also applied to other numerical calculations for surface models.

The authors of [1] were particularly motivated by the task of writing an interface between CAD systems and a Stereolithography apparatus (SLA). The current interface to an SLA requires that a part to be machined must be triangulated. Since a triangulation for an SLA must correspond to a real object, no cracks and improper intersections are allowed in the model. Further, triangles must meet along with common edges, and triangle normals must point away from the object. These requirements are easily met by a solid model. In practice, however, the sophisticated models found in the automotive and aerospace industry consist of free-form surfaces. This makes the triangulation of the models much more difficult, especially if trimmed surfaces are handled.

Unfortunately, most of the known algorithms are not designed to process a trimmed surface. An extension of an existing triangulation algorithm performs triangulation for trimmed surfaces completely in parametric space. Strategies for avoiding cracks between patches and surfaces are also implemented.

To design freeform surfaces with nearly random shape, three mathematical methods are applied: first, the whole surface is dissected to smaller pieces (called "patches"); each of these is described by a polynom with limited degree. Second, the polynoms are not described explicitly as s = f(x,y,z) in euclidean space, but as functions s = f(u,v) with u,v being parameter values in a "curved" coordinate system - easily describable as "run length" of the surface in the two different directions u and v. Third, undesired portions of the so-defined freeform surface can be cut off by "trim curves". These are defined in the same parameter coordinate system as the surface patches themselves.

So, a "piecewise polynomial surface is a grid of parametric polynomial patches which can be considered as a tensor product of two parametric variables, u and v". The basic principle of the following triangulation procedure is, to perform it in the parameter coordinate system u, v instead of the euclidean space x, y, z. The advantage is, that thet a considerable number of algorithms now

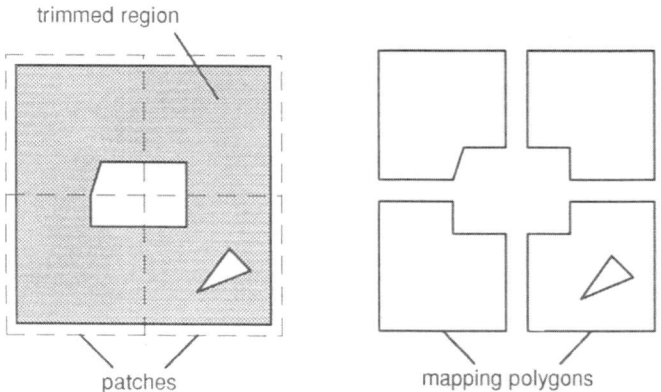

Figure 3 Generation of mapping polygons by splitting of trimmed regions.

By splitting the surface into its patches again, the parameter space is subdivided into a set of polygons, called mapping polygons. Each of these polygons thus corresponds to a patch on the surface. To generate the mapping polygons in parametric space, the splitting and refinement procedures can be efficiently performed using the Euler operator [8]. After splitting, the boundary of a mapping polygon consists of splitting lines and trimming curves (Figure 3). By checking the neighbourhood information for pairs of half-edges, one can easily detect cracks.

Avoiding cracks between surfaces is more difficult. This can be done by calculating the shortest distance between two line segments (edges), since the trimming curves have already been approximated by line segments. If two trimming curves match, they are geometrically merged assigning the same coordinates to the matching points on the curves, to prevent cracks and improper intersections between surfaces.

To ensure that the model resulting from the merging is manufacturable by the SLA, it is required that the model satisfies the manifold criteria [2] as follows:

- Every edge is shared exactly by two faces.
- Every vertex is surrounded by a single cycle of edges and faces.
- Faces may not intersect each other except at common edges and vertices.

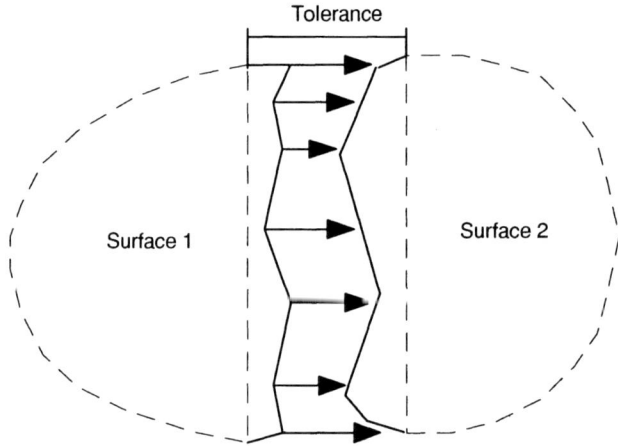

Figure 4 Merging of boundary curves of two matching surfaces

On the basis of these criteria, each edge on the boundaries can be merged only once. The merging procedure expands by following matching edges in both directions until the end points of edges are reached (there, naturally, no further match can occur). Finally, two pairs of end points of merging edges are assigned to the same co-ordinates in Euclidean space.

Merging boundaries of neighbouring surfaces cause some artefacts. However, these modifications are all within usually small, specified tolerances. On the other hand, the modifications are inevitable to make a surface model manufacturable for SLA machines at all (Figure 4).

Although the algorithm has been implemented as an interface for SLA, it is expected that it would also be suitable for the rendering procedures needed for realistic image generation as needed in Virtual Reality applications.

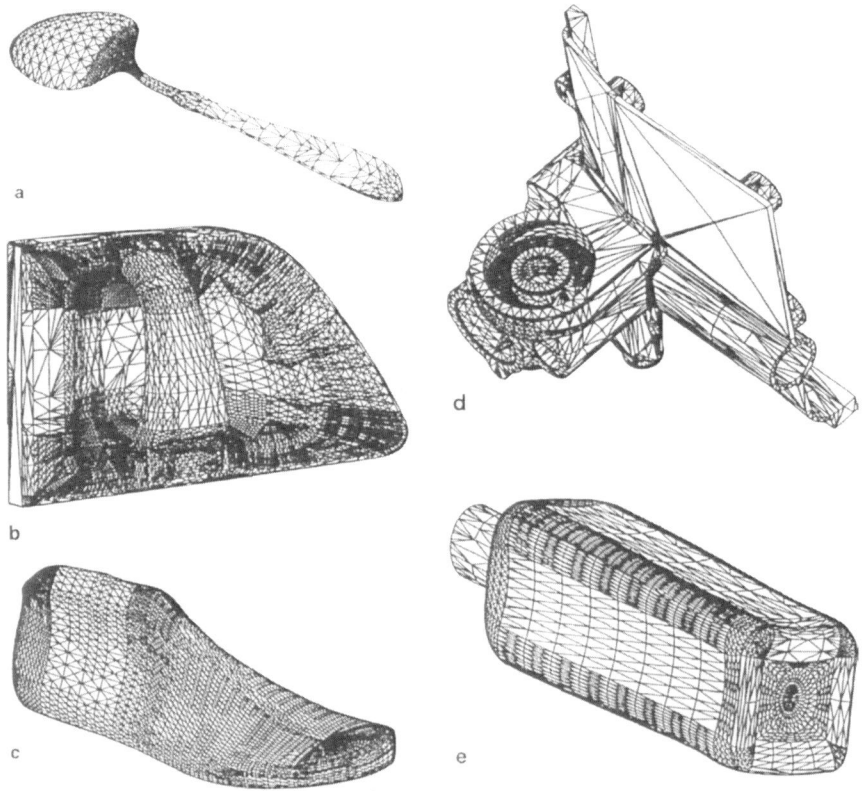

Figure 5 Triangulation models created by the described algorithms; (a) spoon, (b) door handle, (c) shoetree[1], (d) engine part[2], (e) liquor bottle[3]

2.3. Integrating sensor data in a polygonal shape representation

For "Reverse Engineering" on the basis of 3D sensor data, some very similar problems occur as described before. One major problem is again to meet the manifold criteria. Presently the STL interface software will be extended to make it capable to create a mesh out of the sensor data and sweep it to a solid model. The regions that are "behind" the measured points, representing the parts of the object that cannot be seen by the sensor, are represented by solid "sticks", stretching out behind the triangle mesh to infinity. Then, such "solid views" are intersected, generating an STL representation of the whole object once all views are performed. The advantage of this procedure is supposed to be that it will not need a CAD system to do a shaded visualisation or direct replication via Rapid Prototyping.

[1]Courtesy of KF Desma, Achim, FRG
[2]Courtesy of Porsche AG, Weissach, FRG
[3]Courtesy of Cisigraph GmbH, FRG

2.4. Results

The described approach chain can be consequently supported by Concurrent Engineering Methods and Layer Manufacturing Technologies. Tremendous time reduction was realised. The experiments show that the prototype could be manufactured with this process. Currently, coating the dies by wear-resistant surface materials (nickel, chromium) is under research.

Software tools using a triangular representation of freeformed bodies proved to supply a good means for supporting early phases of production and product development, where only rough data is available. A software package initially dedicated to support process planning of Rapid Prototyping facilities shows to be a basis that can easily be expanded towards follow-up processes like casting moulds, FEM simulation, 3D sensor data integration, and many more.

The described process chain allows for cheap and fast production of an experimental sheet metal forming tool. Thus the test phase of the tool can start at an early stage of the product development. Because one starts from a CAD model and uses a Rapid Prototyping Technology, the tool can be changed quickly and cheaply. Several tests can be made without violating time frame and budget. A final steel tool can be produced starting from the same data. The possibility of tests using an experimental tool results in a greater safety concerning functionality of the tool and quality of the part. Designer and toolmaker both can create an optimal product and tool design easier.

3.   VIRTUAL PROTOTYPING OF A ROBOT CELL FOR DISASSEMBLING

The prototype development of a robot cell for disassembling of car components is an example that represents One-of-a-Kind capital products. It demonstrates that the physical complexity of the One-of-a-Kind product often corresponds to its functional complexity. Because the functional concept the mechanical structure have a strong impact on each other, both sides have to be tackled at the same time. For the development of solutions that have a high functional and geometric complexity the concepts of Virtual Prototyping offer effective support.

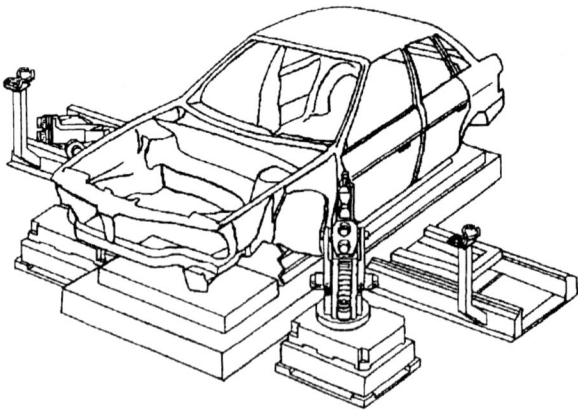

Figure 6 Principle of the NEUROBOT robot cell for car disassembling

Often experts from different companies are involved in the development. Then, such projects are, in fact, distributed over different sites. This means that the project has to be carried out by distributed specialist groups working concurrently in close contact with each other. Consequently, the technical and organisational concepts of Virtual Prototyping and Concurrent Engineering have to be merged by enabling technologies to support the development of such One-of-a-Kind capital products like this disassembling cell.

The NEUROBOT[4] prototype cell is designed to be a self-consistent, error tolerant and autonomous intelligent robot and multi-sensor system based on Neural Network and Adaptive Control concepts. It will be able to perform disassembling operations in non-structured environments and in the absence of design and manufacturing descriptions of the car components (Figure 6). Besides the fact that the recycling of old cars is a problem of increasing relevance there are a number of technical challenges that motivate the work on NEUROBOT. The project will demonstrate the economic and technological potential for the automation of disassembling tasks of a large variety of car components.

The mechanic and sensor structures of the cell have to have a high degree of problem adjustability and kinematic freedom to reach the highest possible flexibility for the effective disassembling of as many as possible components with a large variety. The system has to acquire autonomically the information necessary for the disassembling process. The required task sequences need to be planned for each individual disassembling problem by the system itself based on the information acquired before. The programs for the robot cell must be generated automatically considering reachability and collision avoidance.

3.1. Complementary Virtual Prototyping of the Mechanic and Functional Components

While many fields of advanced technology are touched within this project and because of their strong cross relations, the mechanical design, the development of the sensor and control systems as well as their programs is carried out concurrently by different expert teams. For these purposes a computer environment with powerful simulation functionality for the development of the Virtual Prototype is established.

1. The structure of the car disassembling robot system is described and the interfaces for the information exchange between the system components itself and between the virtual and the real environment are defined (Figure 7).
2. The development tools for the Neural Networks in the NEUROBOT controller and sensor systems are chosen.
3. The simulation environment for the development of the virtual prototype is established by joining IGRIP with the controller development tools.

Because of the complexity of the disassembly task various different approaches for the generic car model, the generic disassembling plan, the network architecture, the learning rules, etc. are assessed and the performance of each approach is evaluated. The performance of the whole system will be tested frequently to avoid cost-intensive changes at a late time. Thus, the development phase is an iterative process, distributed between a number of concurrently working project partners.

As the Virtual Prototyping environment has interfaces to the real world (e.g. translators for different robot control languages or sensor ports) the real system components can be tested

[4]The development is carried out in the context of the ESPRIT project no. 8338 "NEUROBOT - Neural Network Based Robots for Disassembling and Recycling of Automotive Products" funded by the Commission of the European Communities.

with the virtual system components while these components haven't been set up or while these components are at another project partner's development site.

The function of the system in different situations can be tested with the real sensor system connected to the virtual model. Or in turn, the real robot can be tested in connection with the virtual controller, and so forth. This allows the test of the entire system at any stage of the work so that the soundness of the development can be assessed frequently.

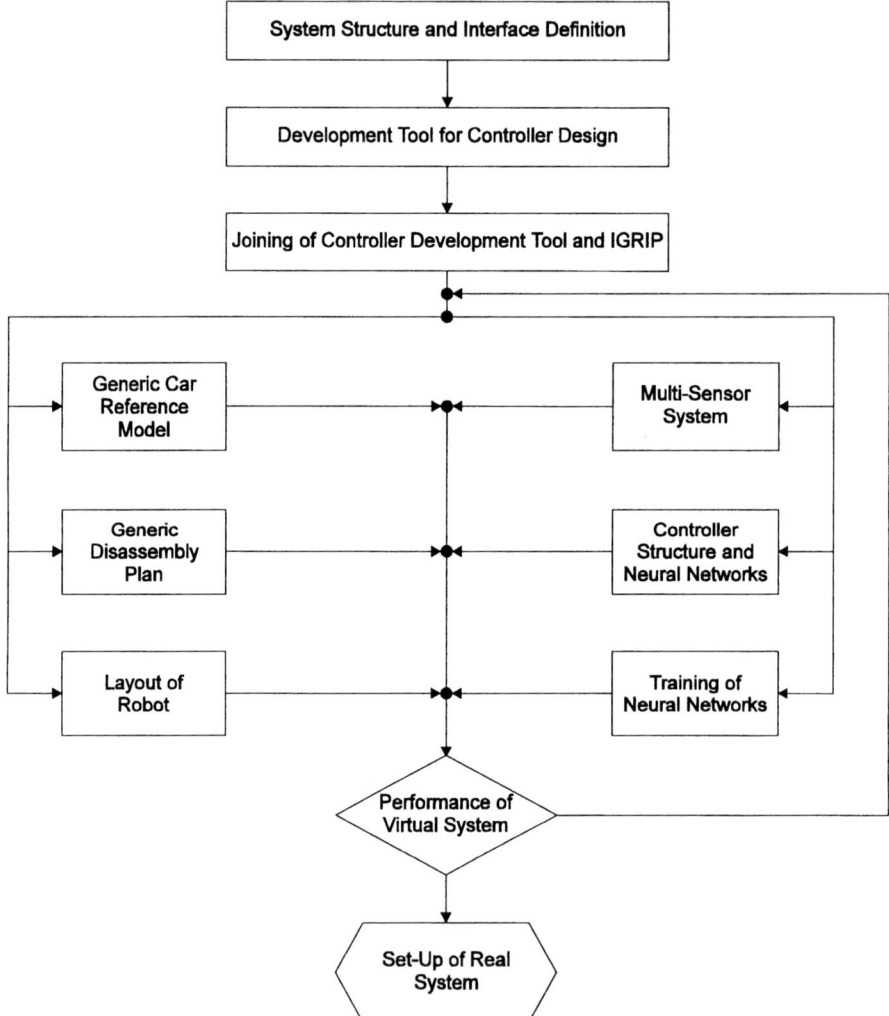

Figure 7 Iterative structure for the Complementary Virtual Prototyping of NEUROBOT.

3.2. Characteristics of the Complementary Virtual Prototyping

The development of One-of-a-Kind capital goods using approaches of concurrent, distributed complimentary Virtual Prototyping has the following characteristic features:
- changes, new additions or different options are easily included,
- results and cross impacts of different approaches become apparent immediately,
- functional concept and mechanical structure are developed complimentarily,
- interfaces to the real world from many points within the virtual prototype,
- real components are tested together with virtual components,
- opportunity for the assessment of the entire system at any stage of the setting-up,

Thereby, the described method has the following advantages over the traditional way of development:
- opportunity for distributed and concurrent development,
- conceptual weaknesses are rectified in the virtual prototype,
- the soundness of the development can be assessed frequently,
- errors are easier to identify,
- functionality and mechanical structure are consistent,
- test bed representing the entire system is available during the whole setting-up
- expenditures for changes in the virtual prototype are far less than for changes in the real prototype
- work is more efficient,
- prototype is developed faster,
- prototype with a higher quality.

3.3. Linking Graphic Simulation with Development Tool for Intelligent Controllers

The development of this prototype robot cell is carried out in a virtual simulation environment. This environment consists of the powerful graphic simulation and robot off-line-programming tool IGRIP of DENEB Robotics and development tools for Neural Networks and Artificial Intelligence applications. The user interface comprises of many tools from a CAD and layout environment as well as simulation, translation and communication tools (Figures 8 and 9).

Figure 8 Screen from the IGRIP off-line
programming tool

Figure 9 Robot cell and its simulation

The off-line-model of the disassembling cell will be linked via socket communication with the Neural Network development tool. Through this communication link, simulated sensor signals, graphic and geometric information and control data are exchanged. This facilitates the generation of realistic training data for the Neural Networks. The connection to the robot 32-bit controller is established via IGRIP up- and download. The measurement PC and IGRIP can exchange data via TCP/IP and NFS.

3.4. Linking Graphic Simulation with 3D-Sensor and Industrial Robot

With an optical 3D-sensoric system on an industrial robot linked with a CAD/CAM environment it is possible

1. to carry out the work piece handling and the manufacturing sequence in a virtual production scenario to avoid program faults, collision, exceeding the work range or payload limit during the real robot action [7] and

2. the product tolerances can be compensated through an in-step co-ordination of work piece geometry, tools and peripherals, process programs and CAD data to meet high quality standards [8].

In iteratively performing these two functions for small steps of the total manufacturing process the benefits of the predictive planning of Virtual Prototyping can be used without making compromises in the process reliability and the product quality. The feedback incorporating 3D-sensor, robot and off-line programming is a possibility to meet the required high degree of flexibility in the production processes of the One-of-a-Kind products (Figure 10).

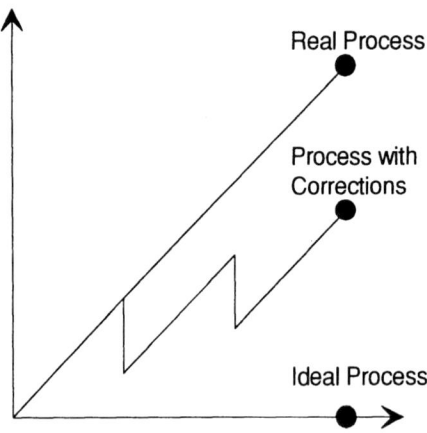

Figure 10 Tolerances of real, ideal and corrected processes

So One-of-a-Kind production can be automated, supplying repeatable results. By this, the tight time limits of prototype manufacturing can be held, and One-of-a-Kind products can be "right the first time", preserving from expensive and time-consuming corrections of defects.

3.4.1. Measurement System

Figure 11 Steel construction part being measured by optical fringe projection system and coded light approach.

The 3D sensor utilises an active optical 3D-measurement process: the *coded light approach*. This is an absolute measurement method that requires only a small number of images to obtain a full depth-image within seconds. This is done by projecting sequences of line-patterns (Figure 11).

For the feed-back of geometric information acquired as 3D-clusters it is necessary to extract special points. First, collision spaces in the cell are examined. When there is no more risk of damage, the 3D-measurement tool and the CCD-cameras can be moved to the "points of interest". For many applications it is sufficient to search for corner points.

3.5. Functionality of the Laboratory Robot Cell

- cell layout design with IGRIP graphic robot simulation system,
- simulation of flame cutting, arc welding and assembly tasks for the CLOOS dual robot cell (13 axes),
- simulation of 3D-measurement with optical sensory system,
- translation of simulation programs and download of system control codes,
- acquisition of 3D-clusters with coded light approach,
- extraction of special points from a 3D-cluster,
- calibration of basic optical characteristics of 3D-measurement system,
- upload of system control codes and 3D-clusters,
- download of corrected system control codes.

3.6. Results

In practice, the main needs that motivate the development of the described system are:

1. For the effective use of off-line programming, the layout data of the robot cell has to be updated by accounting the tolerance-caused offsets.
2. To supply for a high process reliability the robot's kinematic variations caused by payloads as well as the work piece tolerances have to be calibrated. The handling of flexible parts is also facilitated.
3. Faults and accumulated tolerances have to be detected in an early stage of the manufacturing process to minimise the costs and the delay for corrective actions. In addition the quality can be assessed in situ and can be documented.
4. When sequentially producing first a virtual prototype and then the real prototype the plan divergence and the process tolerances accumulate. This can be minimised by stepwise alternating between planning in the virtual environment and verification and calibration of the real environment.

4.   CONCLUDING REMARKS

The ability to get a product out into market quickly is becoming the dominant requirement for manufacturing companies. This has to be realised with respect to an increasing demand for small quantities - in the extreme case of One-of-a-Kind products and a decreasing product life- and innovation cycle. The motivation to be flexible and able to "make just what is needed when it is needed" is becoming stronger and stronger. The term "Rapid Response Manufacturing Systems" is used as proposed by Kohls and Moehring [9].)

"Concurrent Engineering (CE)" and "Integrated Product Development (IPD)" are generic terms for various approaches to allow designers and manufacturing engineers to communicate early in the design phase of a product. Multi-functional teaming is a means for increasing communication (and co-ordination), and has become the common approach towards improving responsiveness. The ability to do design iterations early in the product development cycle enhances design, improves quality and reduces costs [10]. The earlier that product improvements occur in the design process, the greater the savings will be. Beside various information management and groupware-tools, prototyping is a crucial concept to facilitate this early communication of ideas. Prototyping approaches range from soft prototypes, i.e. computer models of product or process design (Virtual Product Development, Virtual Manufacturing), to actual 3D physical engineering "hard" prototypes developed from CAD model data by a variety of Rapid Prototyping Technologies.

The modelling of process chains is an adequate approach to achieve better clarity of the prod- uct development. Process chains describe "what" has to be done and the information technological transactions between different activities. As well as for the description of the "As-Is" and the "Should-Be" situation, process chains have to be extended towards "How" it has to be done. The advantages of using this approach are the clarity and the possibility to choose and combine the most adequate activities aiming at a rapid response in product development.

Development of a generic reference model for rapid product development and One-of-a-Kind Production will provide an abstract view of the various process chains in a neutral and logical manner. In this sense the functions and relations of a reference model should be seen as a generic structure to derive the system architecture of product and part-specific future rapid response manufacturing process chains.

5. REFERENCES

[1] Hirsch, B.E., Sheng, X.: Triangulation of trimmed surfaces in parametric, space computer-aided-design, Vol. 24, No. 8, August 1992.

[2] Mäntylä, M: An Introduction to Solid Modelling, Computer Science Press. USA, 1988.

[3] Berger, U., Thoben, K.-D., Müller, H.: Rapid Prototyping Technologies for Advanced Sheet Metal Forming. SME Rapid Prototyping & Manufacturing '93 Conference, Dearborn, Michigan, USA, 1993.

[4] Müller, H., Berger, U., Thoben, K.-D.: Successful Application of Rapid Prototyping Technologies for Advanced Sheet Metal Forming and Investment Casting. Intelligent Manufacturing Systems, International Conference on Rapid Product Development, Stuttgart, Germany, 1994.

[5] König, W.: Fertigungsverfahren, Band 5: Blechumformung, Düsseldorf: VDI, 1990.

[6] N. N.: Company brochure of DENSIT©. Ålborg, Denmark.

[7] Berger, U., Schmidt, A., Wolf, H.: Optical system for robot based One-of-a-Kind manufacturing. EOS/SPIE European Symposium on Optics for Productivity and Manufacturing, Frankfurt/M., Germany, 1994.

[8] Barone, P. A.: "Techniques for Developing a Multi-Robot Cell for Production Processing of Complex Airframe Assemblies", 1989 SME Intern. Conf. and Exposition, pp. MS89-399-1 to 28, Detroit, USA: SME, 1989.

[9] Kohls, J.B., Moehring, S.M.: Achieving Rapid Prototyping - Issues and Technologies to Facilitate Design and Prototyping, Third International Conference on Rapid Prototyping. Dayton, USA, 1992.

[10] Wozny, J.M.: "Research Trends in the U.S. for next generation CAD-Systems", CAD´92: Neue Konzepte zur Realisierung anwendungsorientierter CAD-Systeme. Berlin, Germany: Springer, 1992.

6. BIOGRAPHY

Dr.-Ing. Klaus-Dieter THOBEN studied mechanical engineering at the TU Braunschweig. After finishing his studies, he was employed at the University of Bremen in the Faculty of Production Engineering. There, he received his doctor of engineering degree in CAD applications in 1989. In the same year he joined BIBA, where he led the CAD/CAM laboratory. Since 1991 he is head of the Department for Computer-Aided Design, Planning and Manufacturing. His special interests are computer-aided techniques and applications in product development, CIM and simultaneous engineering.

Ulrich BERGER holds a degree of mechanical engineering from the University of Stuttgart and expects to finish his Ph.D. at the University of Bremen in May 1995. He is the technical manager of the BIBA institute. The professional experience of Mr. Berger contents responsible functions in the industrial automation and manufacturing industry.

Jürgen Friedrich BAUER graduated in Mechanical Engineering at University of Stuttgart in 1993. After a study project on Rapid Prototyping materials, he went to Daimler-Benz Research Centre in Ulm then, were he worked on 3D measurement systems for digitizing purposes in the context of the Intelligent Manufacturing Systems (IMS) initiative. Since 1994, he is scientist at BIBA Institute were he coordinates research activities in the field of Rapid Product Development and does technical project management in several EU funded RTD projects (Brite EuRam BE 5814 "Manufacturing Technologies for Sheet Aluminium Forming" (Hyprform), Brite EuRam BE 5278 "Optimisation of Rapid Prototyping Techniques for Automotive Industry").

Achim SCHMIDT graduated in Mechanical Engineering at University of Stuttgart. From 1983 to 1993, he was research assistant at Fraunhofer Institutes (IPA, IBP) and the University Computer Centre at Stuttgart for data processing, software development, graphic simulation and robot control. Since 1993 he is research scientist at BIBA in the fields of robotics, off-line simulation, virtual environments, autonomous systems and optical sensors. Since 1994, Mr. Schmidt is involved in the ESPRIT project No 8338: NEUROBOT - Neural Network-Based Industrial Robots for Disassembly and Recycling of Automotive Products.

24

Prototype Validation in Virtual Environments – Vision and first Implementation –

K. Böhm[*], *H. Wirth*[†], *W. John*[†]

[*]*ZGDV – Zentrum für Graphische Datenverarbeitung e.V.*

[†]*Technical University of Darmstadt, Department of Computer Science, Interactive Graphics Systems Group*

Wilhelminenstrasse 7
D–64283 Darmstadt, Germany
tel. +49 6151 155 243
fax. +49 6151 155 299
email: {boehm, wirth, john}@igd.fhg.de

Abstract

This paper describes the differences we perceive in the process of virtual prototyping as opposed to classical prototyping. An environment for virtual prototyping as it could be in the future is introduced. In this introduction, tasks to be fulfilled in this environment and the tools needed are identified. Furthermore, a first implementation of a virtual prototyping system is described and evaluated.

Keywords

Virtual Prototyping, 3D Interaction, Virtual Reality, Distributed System, Simulation

1   VIRTUAL PROTOTYPING

The term Virtual Prototyping does not define a new technology per se. Instead, it is a combination of existing computer science disciplines. Among them are CAD, CAM, databases, simulation and animation. The development of a new product involves three main steps. The first step is the generation and modelling of a prototype, the second step a test and validation whether this prototype corresponds with the initial specification. Finally, the result of a validation can be a redesign, request for some minor changes, requirements for a more detailed model, or some other optimizations. After that, the prototyping cycle starts again. So validation is not the final step in the development process, but a means to enhance product quality in the early stages of the production cycle.

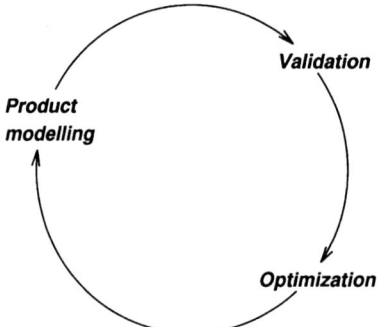

Figure 1 Product development cycle.

From our point of view, the product development cycle is mainly controlled by the following people: the designer, the test engineer, and the manager or marketing department. The basic idea of virtual prototyping is to avoid the construction of real physical prototypes in the design, test and optimization phases in the cycle (as shown in figure 1). The main goal of virtual prototyping is to decrease the costs in the development process. By reducing the costs, with the same efforts in time and money, such an approach could lead to a better product.

An object is being modelled with a geometric shape and a functional behavior description. Additional properties such as coordinate and geometric tolerances, or material properties are attached to it in order to test and evaluate the prototype with simulation. The result of the product design is an object in a computer generated virtual environment. With an active reference specification, the developers and managers are able to verify their ideas immediately with the help of simulation and animation. For nontechnical persons involved in the production cycle, e.g. a manager, an adequate presentation and interaction interface has to be provided. Presenting in a virtual space has the potential to perform presentation and decision sessions with different people not necessarily located on the same building, town, or country. Another advantage is that the production cycle becomes more transparent to all people involved. For further information see also (Dai and Göbel, 1994).

2 VALIDATION IN VIRTUAL ENVIRONMENTS

2.1 Vision

On the basis of the topics laid out in the previous section, one can envision the whole prototyping process taking place in a virtual environment. From the creation of the first 3D-models using a CAD package to the assignment of physical properties to different parts of the system and from the simulation of the fully described system to the redesign of faulty parts, today, many of the steps of prototyping are far from being integrated into a single set of tools. At the moment, many steps of the development are already carried out with the help of computers, e.g. using CAD programs to define the geometrical shape of objects. Other steps, like testing a prototype, are carried out with custom-made hardware. One can imagine several advantages in integrating all steps of the product development cycle in an environment on a computer:

- In a computer controlled environment, keeping and comparing several versions of the developed system is no problem.
- Computer controlled environments allow different, simultaneous views on the simulated system which would be impossible in reality, like for example an observer watching the mechanics of a steering system while the test engineer "drives" a simulated car.
- Additional information, such as documentation, results of simulation runs, etc. can be kept in the environment.

Another advantage, when following the integrated approach, could be the possibility to hold meetings of e.g. design engineers in a virtual environment. This will become more and more feasible with the installation of fast WANs. In a virtual space, engineers or management people could meet without the need of changing their physical location (see figure 2, for further information see also (Tijerino et al., 1994)). The use of such a virtual space could comprise nearly all of the above mentioned tasks of a virtual prototyping system:

- Engineers could share a virtual space where several people work on the same design or present new developments to all participants.
- Documentation for different parts could be edited by different people while remaining in a common, consistent environment.
- Evaluations conducted by a test engineer can be observed by other persons involved, who do not have to meet at one physical location.
- The test engineer can present and discuss the results of the testing to the development engineer in the test environment.

Both environments, the single and the shared one, need a consistent, intuitive and easy to use user interface, though. In order to be able to build a new type of simulator, one has to identify

Figure 2 Vision of a round-table in a shared virtual environment (from (Tijerino et al., 1994)).

the different tasks the user has to be able to fulfill with the help of the user interface of the simulator. Typical applications for virtual prototyping are mechanical or electro-mechanical systems, which normally are controlled by devices like levers, knobs, buttons, etc. Some systems require simultaneous input of different values. Most input devices give some sort of feedback to the user, either by their state or position, or by giving some sort of force feedback. The other type of feedback is the general system output or feedback, which is independent of input devices.

2.2 Using interactive virtual environments for validation

For the visualization of 3D-geometry we already have more or less adequate techniques e.g. fast Gouraud-shaded rendering supported by the hardware. However, the methods of interacting in virtual environments are still in an experimental state. Especially testing and validating prototypes requires a great amount of intuitive and natural interactions. Therefore, in this section, we want to focus on the subject of interaction in simulations in a virtual environment for validation tasks. The subject of interaction in a 3D-virtual CAD environment is not covered here, for further information see (Bleser, 1993).

To develop a new user interface metaphor for software based simulations which is not based on traditional 2D-window based systems, one has to take different problems into account: First, users should not be forced to learn or to accustom themselves to 2D-window based user interfaces and their usage metaphors. Furthermore, most real world input devices have little in common with a mouse or a keyboard, so their usage has to be modelled by some metaphor. Second, 2D-window based user interfaces do not allow synchronous input with more than one hand, which does not suffice for many simulated systems. Third, the user should be presented with a look-alike of the real interface they expect. A solution to this problem could be a virtual reality user interface. The scene is presented to the user in pseudo-3D, and the user can interact with the devices displayed in the scene.

Our solution to the problems as described in the previous paragraphs is the concept of "virtual input devices"*. The naming of this concept is based on the fact that these input devices do not really exist. They only exist as 3D-depictions of real input devices in the scene rendered on the screen – as a virtual environment for the user. The users can interact with these devices in a way they are quite accustomed to – the user can, for example, touch and press buttons. The devices on the screen are controlled by a 3D-model of a hand which in turn is controlled by the user with the help of a data glove and a position/orientation sensor. There are other approaches which are similar (see for example (Nomura et al., 1992) and (Codella et al., 1992)), but none of them combines the simulation/VR aspect with development and validation facilities.

Several benefits can be derived from such a system: Little or no custom hardware needs to be built for simulators, input devices like dataglove are usable for different types of simulations, as the definition of virtual input devices is mostly independent from the actual input device used and one can assume that the interface requires less user training than a mouse-based one.

The definition of a virtual input device has to cover different aspects. First, there has to be a geometry with material attributes. Second, the constraints of movement for the virtual input devices have to be known. Third, a set of parameters has to be defined which are the values delivered by the virtual input device representing its different states. Finally, interaction techniques must be defined for the handling of the virtual input devices, e.g. gestures for grabbing objects and methods for pressing buttons (see figure 3).

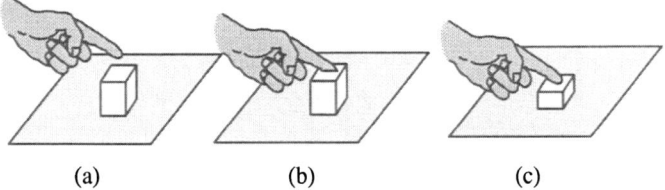

(a)   (b)   (c)

Figure 3 Button pressing sequence.

---

* The term "virtual input device" has been coined before in (He and Kaufman, 1993). Unlike the definition used in this paper, which defines virtual input devices as those abstracted from the real system and displayed in the virtual environment, the virtual input devices in (He and Kaufman, 1993) are abstractions of input devices of the VR toolkit (like flying mouse, data glove etc.)

For effective work with virtual input devices, the user needs the appropriate input device (in hardware) to allow him or her the interactions described above. We think that using a data glove for this task is most appropriate at this moment for the following reasons: Input devices like 3D-mice have a "usage metaphor" built into the hardware which does not allow the development of input-hardware independent interfaces. Clicking on a pseudo 3D-scene with a 2D-mouse is even less intuitive. In an environment where a head mounted display is used to enable an immersion in the virtual environment, a 2D-mouse is not usable at all. The use of a data glove allows multi-dexterous input, as another glove for the second hand or other hardware like a position/orientation sensor can be added easily without overloading the user. Position/orientation sensors allow input of other analog values, for example as a replacement for a gas pedal. A data glove can theoretically be equipped with facilities to provide force feedback in order to enhance the realism and usability.

3   MUSE AS THE VIRTUAL PROTOTYPING PLATFORM

Our first realization of a virtual environment for prototype validation is made in the context of the *MuSE** project. This section describes the approach *MuSE* takes towards virtual prototyping. *MuSE* is designed as an integrated software environment for both design engineers and test engineers. Figure 4 shows the architecture of the complete *MuSE* environment, it can be divided into three parts.

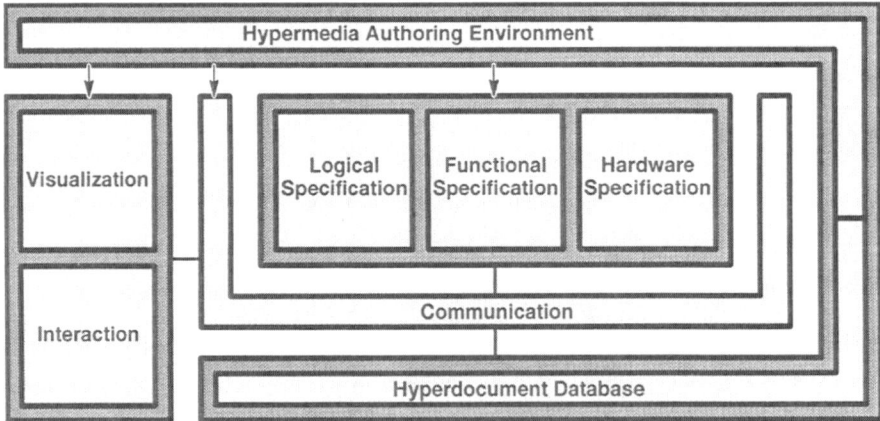

Figure 4 The *MuSE* architecture

MuSE is the acronym of the german project title *Mu*ltimediale *S*ystem *E*ntwicklung (Multimedia–based Systems Engineering). The project is sponsored by the "Deutsche Forschungsgemeinschaft" (German research foundation) under grant number He 1170/5–2

- The entry point to the *MuSE* system for both test engineers and developers is a hyper network, based on the hypermedia database front-end SEPIA (detailed description can be found in (Streitz et al., 1992)). Different types of users are offered different electronic spaces where they have certain rights to access data, execute simulations, etc. The documents and the specifications of the different modules of the system are stored as nodes in the hyper network. Running a simulation, creating annotations or creating variants of the system are operations on hyper nodes. The *MuSE* environment stores all information which results from the design process in the underlying database.

- *MuSE* supports the system engineer with tools for specifying the system's behavior. Typically, a technical system consists of various parts or modules. To allow rapid prototyping, each module may be specified in an incomplete state. *MuSE* uses high level specification languages for the description of the physical properties of the system. The description of the system is divided into three levels for which we use different languages with different properties. The reference specification is written in State Event Logic (see (Große, 1994)). It is used for off-line simulation (for example to ask questions about ways which may lead to erroneous system states) and offers a user interface for that task (see (Große, 1993)). The reference specification in *MuSE* is implemented in Prolog. As the reference specification is not suitable for real-time simulation, the second level of the specification is implemented in a functional language. The executable functional language SAMPλE (pronounced "sample", with a detailed description found in (Jäger et al., 1988)) is used to specify the simulation functions. The structure of the system is mapped onto a process net, with several processes running in parallel. To be able to do this, SAMPλE was extended by a concept of concurrent processes similar to CSPs (see (Hoare, 1985)). Electronic system components can be modelled in a third level using the VHDL language, which allows the description of parallelism even inside single circuits (details in (Deegener and Huss, 1993)). Both SAMPλE and VHDL can be used at the same time to describe and co-simulate the system.

- The system's behavior is examined by simulating it. There are two types of simulations. One is called an on-line simulation, the other an off-line simulation. In an on-line simulation the user controls the system directly for example via our virtual environment interface. The user's actions may be recorded in a database in this mode. With an off-line simulation, the user's actions are taken from the database and fed into the simulation component of *MuSE*.

4 IMPLEMENTATION

At the moment, a prototype of the software environment has been realized. For the realization of the VR front-end the user interface toolkit GIVEN (Sokolewicz et al., 1993) was used.

GIVEN already provides the following features: Input device independence, individual object behavior, gesture dialog including advanced gesture recognition, and interaction modelling including collision checking. Thus, for the realization of the VR interface we were able to concentrate on describing the object specific behavior of the visible objects. An example of this object specific behavior is the reaction of the steering wheel when the users grabs and tries to turn it. The combination of GIVEN and *MuSE* is illustrated in figure 5.

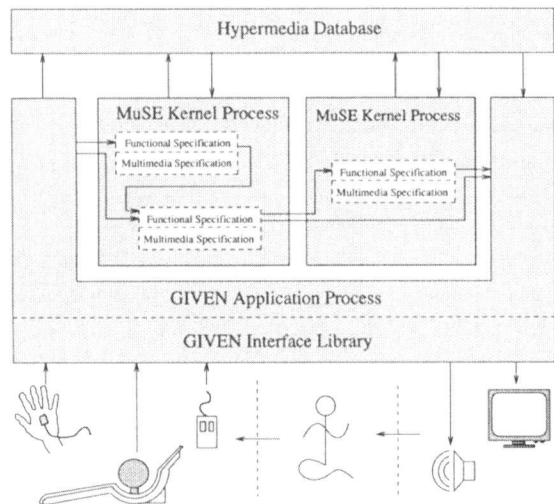

Figure 5 Combination of the *MuSE* Kernel and GIVEN.

An important issue is the realization of the virtual input devices. Their description as well as the runtime control is not part of the simulation module but an integrated user interface component. This means, we do not send all input events coming from the devices to the simulation tool, instead we only send the states of the virtual input devices to the simulator. For example, the control of a button press (shown in figure 3) requires real-time collision detection between the cursor (virtual hand) and the button object as well as the dialog control. Both necessary features are supported by the underlying toolkit GIVEN.

4.1 Hardware, software and peripherals

The system is running on a Silicon Graphics VGX 320 Workstation. We are using the CyberGlove from Virtual Technologies which is equipped with a Polhemus FASTRACK sensor as well as the VPL Research DataGlove™ Model 2 system. At the moment, the simulated gas and brake pedals are controlled by MIDI pedals which are connected to the workstation via serial interface. The system is programmed in C and C++ under UNIX. For the rendering of the scene we are using Silicon Graphics GL. The output device is a standard monitor.

Our test scene contains 5268 polygons. With Gouraud-shaded rendering and on-line collision detection and gesture recognition, we have a frame rate of 6 frames per second.

5   EXPERIENCE MADE WITH THE FIRST IMPLEMENTATION

The quality of the user interface is the deciding factor as to whether the system is acceptable or not. Therefore, this section gives an evaluation of the first virtual environment implementation for the system validation. It is a summary of our personal experience during developing, testing and demonstrating the system. In addition we received a great amount of valuable remarks from several leading researchers after demonstrating the system to them.

5.1   Environment

As an example of a realistic system serves the rear wheel steering system of a small all-purpose truck.

For the VR user interface, we modelled the interior of the driver's cabin as well as the chassis of the small all-purpose truck. The test engineer is offered a view like they would be sitting in the driver's cabin, looking forward (see figure 6). Situated on the dashboard are a steering wheel, several buttons and different status lamps. The control devices are realized with the concept of virtual input devices (see chapter 2.2).

In the so called driving mode, the user is able to control the car by pressing the buttons and by grabbing and turning the steering wheel. The user is able to perform actions using the input devices dataglove or spaceball. At the moment, the gas and brake pedals are controlled with two MIDI pedals. In the navigation mode, the user is able to move around in the scene in order to have a look at different parts of the truck.

5.2   Results

The users as well as the observers understand quite well the paradigm of how to interact with the system. The visualized information (e.g. the steering wheel) is self–explaining concerning the required user action. The user grabs and turns the steering wheel and pushes the buttons. This showed us that the concepts were well developed and the implementation, at least for users accustomed to automobiles, was realistic. However, during the use of the system we were able to identify a number of places where the system needs to be improved. The following points need to be addressed in order to make the handling easier for the users and even more intuitive:

- **Depth perception:** Since we are using a standard monitor for the visual output the 3D-feeling for the interaction is not sufficient. Problems occur especially when the user wants to press a button. Stereoscopic output with head mounted displays (HMD), Boom or shutter glasses give a better depth perception. For standard monitors, special interaction techniques, e.g. 3D-snap, can help to overcome this problem.

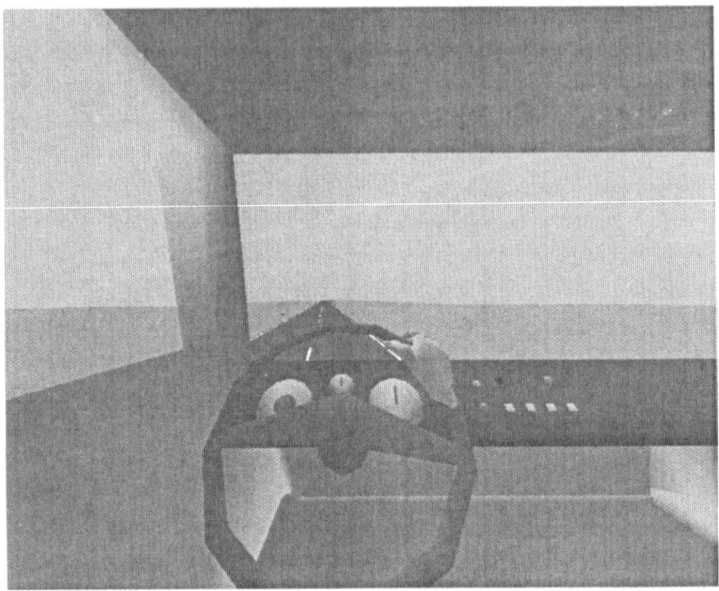

Figure 6 *MuSE* cab interior: grabbing the steering wheel.

- **Perspective:** Another problem which occurs here because of the use of standard monitors is the relatively unrealistic perspective (see figure 6). Sitting in a car the user expects to see the front and parts of the side-windows. But a flat screen can not represent the normal view angle of humans. Therefore we changed the perspective in order to receive this impression. The disadvantage is the *unnatural* presentation of the object in the scene. Furthermore, a viewpoint change based on head movements of the user would be very intuitive, especially in our kind of application – sitting in a car – which is very common to almost all adults.
- **Frame rate:** Being connected on-line to the simulation module our system generates currently only a few frames per second, which is not sufficient for a satisfying work. We can identify several bottlenecks which need to be improved:
  1. Rendering polygonal description of the scenes.
  2. Simulations including the communication between VR-System and simulator.
  3. Collision checking for interaction between objects.
- **Simultaneous Input:** Testing and validating a car requires simultaneous input such as steering and changing gears. A hardware gas-pedal has been constructed as an addition to the system, since accelerating and steering have to be done simultaneously. This is actually a contradiction to the concept of virtual input devices. Solutions to the simultaneous-input problem, such as using a dataglove on each hand, need to be

explored. An additional input channel – speech – is currently integrated in the system. Although it is not common to talk to a car interior, together with the right metaphors it could simplify the interaction.

- **Functionality:** The first implementation of the virtual environment and the simulation do not cover all required functionalities such as changing gears. The next version of the system has to support more functionality.

- **Gesture-based input:** Using the dataglove as the input device offers the potential of gesture-based interactions. We are using gestures for:
  - Grabbing and releasing the steering wheel, a grab gesture simplifies the grab-procedure.
  - Navigating (walking) around the truck for detailed examination.

  Gesture input does not represent the real world interaction, however interaction tasks like navigation can hardly be controlled in a really realistic way (walking). We think that a subset of less than 5–6 gestures will easily be understood by the user and simplifies the interaction.

- **Force Feedback:** The interaction with the virtual objects would be much simpler with force feedback, but we do not see the availability of general purpose input devices with force feedback in the next future. Therefore we focus on visual and audio feedback for the interactions.

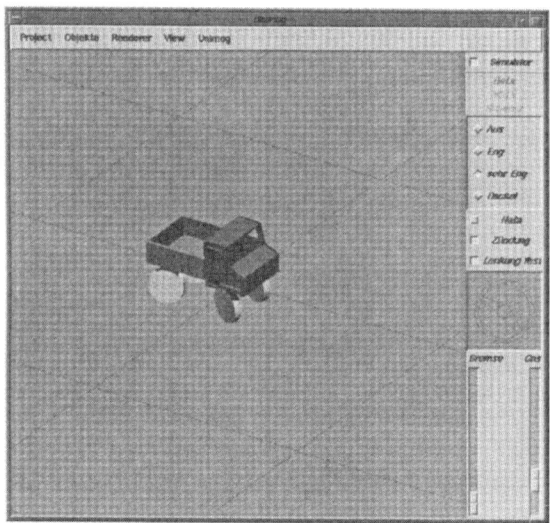

Figure 7 2D-*MuSE* User Interface

5.3   Conclusion

Although we have identified a lot of points which need to be tackled in order to make our virtual environment really applicable for the validation process, we believe that this is the right way towards the vision described in chapter 2. Alternative user interfaces (see figure 7) to the simulations module have also been developed in this context. This interface is essential for the *MuSE* software developer in order to perform immediate testing of the system part they are working on. The capabilities of the whole *MuSE* environment are available to them e.g. on-line simulation, database access. However, a realistic environment for a test engineer can not be achieved based on this user interface.

6   REFERENCES

Bleser, J. (1993) *Analyse der Benutzeranforderungen an CAD Systeme im Hinblick auf 3D-Interaktion, sowie Entwicklung neuer Interaktionstechniken für diesen Bereich innerhalb des Systems GIVEN*. Diplomarbeit, Technical University Darmstadt.

Codella, C., Jalili, R., Koved, L., Lewis, J.B., Ling, D.T., Lipscomb, J.S., Rabenhorst, D.A., Wang, C.P., Norton, A., Sweeney, P. and Turk, G. (1992) Interactive Simulation in a Multi-Person Virtual World, in *Proceedings of the ACM CHI '92*.

Dai, F. and Göbel, M. (1994) Virtual Prototyping – Concepts and Realization. *Computer Graphik Topics 1/94*, 9–10.

Deegener, M. and Huss, S.A. (1994) Ein Verfahren zur Kopplung standardisierter sequentieller und nebenläufiger Beschreibungssprachen für die Simulation komplexer Systeme, in *Proc of the GI/ITG Workshop CAD-Umgebungen und Methoden des Entwurfs von Schaltkreisen und Systemen*, Dresden.

Große, G. (1993) State-Event Logic, in *Forschungsbericht im Projekt MuSE muse–93–1*, Technische Hochschule Darmstadt, Fachbereich Informatik, Darmstadt.

Große, G. (1994) Propositional State Event Logic, in *Proceedings of the European Workshop on Logics in AI (Jelia '94)*, Springer LNAI.

He, T. and Kaufman, A. (1993) Virtual Input Devices for 3D-Systems. *Visual Computer 93–080*.

Hoare, C.A.R. (1985) *Communicating Sequential Processes*. Prentice Hall.

Jäger, M., Gloger, M. and Kaes, S. (1988) SAMPλE – A Functional Language, in *Lectures in Computer Science*, volume 328, Springer.

Nomura, J., Ohata, H., Imamura, K. and Schultz, R.J. (1992) Virtual Space Decision Support System and Its Application to Consumer Showrooms, in *Visual Computing* (ed. T.L. Kunii), CG International Series, Springer-Verlag, Berlin & Heidelberg.

Streitz, N.A., Haake, J., Hannemann, J., Lemke, A., Schütt, H., Schuler, W. and Thüring, M. (1992) SEPIA: A cooperative Hypermedia Authoring Environment, in *Proceedings of the 4th ACM Conference on Hypertext (ECHT '92)* (ed. D.Lucarella, J. Nanard, M. Nanard and P. Paolini), ACM Press, New York.

Sokolewicz, M., Wirth, H., Böhm, K. and John, W. (1993) Using the GIVEN++ Toolkit for System Development in *MuSE*. *Eurographics Workshop on Virtual Environments*, Barcelona, Spain.

Tijerino, Y.A., Mochizuki, K. and Kishino, F. (1994) Interactive 3-D Computer Graphics Driven Through Verbal Instructions – Previous and Current Activities at ATR –. *Computers & Graphics, Journal*, Special Issue: Advanced Interactions, Pergamon Press Ltd.

7   BIOGRAPHIES

Hanno Wirth studied computer science at the Technical University of Darmstadt, from which he graduated in February 1992. During his studies, he spent a term at the Bosch research and development facilities in Stuttgart–Schwieberdingen designing and implementing a graphical user interface for a mobile test and validation tool for computer controlled rear wheel steering systems for cars. Since receiving his degree, he has been working as member of the scientific staff at the Interactive Graphics Systems Group of the Department of Computer Science of the Technical University of Darmstadt. His research interests include virtual reality, simulation of user interfaces and human computer interaction.

Klaus Böhm studied computer science at the Technical University of Darmstadt, and graduated in March 1991 with the degree of Diplom Informatiker. He is currently working at the Zentrum für Graphische Datenverarbeitung (Computer Graphics Center) in Darmstadt as a member of the scientific staff. His research areas include virtual reality, 3D interaction and multimedia interaction techniques.

Werner John studied computer science at the Technical University of Darmstadt, and graduated in June 1991 with the degree of Diplom Informatiker. He is currently working at the Fraunhofer Institut for Computer Graphics in Darmstadt as a member of the scientific staff. His research areas include high performance 3D graphics, visualization techniques and object oriented user interface architectures.

25

Applying Virtual Reality to Electronic Prototyping - Concept and First Results

Fan Dai, Wolfgang Felger, Martin Göbel
Fraunhofer Institute for Computer Graphics (IGD)
Wilhelminenstraße 7, 64283 Darmstadt, Germany

Abstract

Electronic (or digital) prototyping means to use product data models instead of physical prototypes for analyses and evaluations of product design. In many cases, e.g., for conceptual design and product presentation, it is essential to provide the users the impression of using a real prototype. For these purposes, we proposed a concept called virtual prototyping. The idea is to integrate techniques of virtual reality with computer aided design and simulation, to provide an environment, with which a CAD model can be presented and manipulated like a real object. In this paper, our overall concept is described. Some typical examples are presented with discussions about their implementation. Finally, experiences based on these pilot implementations are summarized and future directions of R&D are discussed.

Keywords

virtual prototyping, virtual reality, conceptual design, modeling and simulation, object manipulation, realistic presentation

1 INTRODUCTION

Electronic (or digital) prototyping is an alternative to physical prototyping technologies. Based on CAD model data of a designed product, different analyses can be done to prove and improve design results. CAE systems such as FEM, kinematics and dynamics simulation programs provide very detailed information about a product's function. Of course, these systems are designed for and primarily used by specialized engineers.

In the early phase of product development, e.g. for conceptual design, detailed analysis are not required. More important are features like form, color, placement and some functional features like motion constraints, accessibility etc. Additionally, prototypes are often used in supporting discussions between people with different technical backgrounds. Designer, managers, or engineers discussing about a product design, are used to point on a model directly and say: it could be a peak there. They sometimes want to modify the model immediately and use the new model for further discussions. For such purposes, electronic prototyping using CAD/CAE systems based on conventional mouse-menu driven user interfaces is not suitable and sufficient.

Recent developments in computer graphics and computer simulation have now provided more sophisticated tools for electronic prototyping. Specifically, virtual reality (VR) techniques offer the possibility to experience virtual worlds with very high realism. This includes visual and audio presentation and techniques towards intuitive object manipulation. Combining VR with modeling and simulation methods opens new possibilities of prototyping based on CAD models. In the future, it will be possible to take the data model of a product as a virtual prototype instead of a real object to model and analyze geometry, functionality, and the manufacturing of designed products interactively. We call the resulting technology *Virtual Prototyping* [0].

Virtual Prototyping is a method to be used to evaluate different design alternatives very quickly. In contrast to physical prototypes, a virtual prototype is made very fast, can be manipulated and modified directly and the data is reusable. Applying virtual prototyping will rapidly reduce the number of real prototypes and speed up the product development process. This implies a tremendous reduction of development costs.

To implement the idea of Virtual Prototyping, considerable research and development is necessary. Untill now, only a few experiments are known (e.g.: [0]). IGD, the Fraunhofer Institute for Computer Graphics in Darmstadt (Germany), has performed research and development on several related topics and has recently started the integration of techniques and software to develop an environment for Virtual Prototyping. The first results of this effort have been completed, which can be experienced live in the *Virtual Reality Demonstration Centre* at IGD [0, 0]. In this paper, our overall concept of virtual prototyping is described. Some typical examples are presented with discussions about their implementation. Finally, experiences based on these pilot implementations are summarized and future directions of R&D are discussed.

2   VIRTUAL REALITY AS NEW MAN-MACHINE INTERFACE

A virtual prototype must reflect the characteristics of physical prototypes. These are in particular: *optical appearance, spatial presence and dynamical behavior*. Virtual prototypes can be viewed from different angles, scaled to fit in small spaces (e.g. models of buildings and ships) and manipulated with hands. Artistical as well as technical features like form, space, and measurements can be judged. Even characteristics such as stiffness, kinematics, and dynamics appear as spatial changes. Qualitative judgments about the products function can be made by manipulating the prototype and viewing these spatial changes. Therefore, a virtual prototyping environment has to present these spatial information and allow to interact with the designed objects directly in 3D. This can be achieved by using VR as new man-machine interface.

Virtual Reality is characterized by a realistic environmental simulation and the stimulation of human senses to give them a realistic impression of the virtual world. Up to now, considerable progresses are achieved in this area [0, 0, 0, 0]. Typical features of existing VR systems are:

- real-time visualization and auralization
- immersive presentation (HMD, CAVE)

- direct 3D interaction, like navigation, object manipulation, etc. (gloves, tracking systems) [0]

To apply VR to electronic prototyping, thus to implement virtual prototypes, we have to integrate existing VR techniques into the virtual prototyping system and to improve them to fit the requirements of virtual prototyping.

3 COMPONENTS OF A VIRTUAL PROTOTYPING ENVIRONMENT

Our concept of virtual prototyping is based on the integration of modeling, simulation and VR techniques. Figure 1 illustrates the proposed architecture of a virtual prototyping environment [0]. The main components are:

- VR-interface
- presentation unit
- interaction unit
- embedded tools
- VR data model

The VR data model is the description of designed objects as virtual prototypes. It contains information about the objects suitable for the presentation and manipulation in a VR environment.

The presentation unit makes all relevant data visible. This visualization is based on the VR data model, which contains the actual geometry and lighting information generated by the modeling, preparation, and simulation tools. The interaction unit interprets user actions and either changes the viewing parameters or generates logical events for simulation and manipulation. Furthermore, it transmits immediate feedbacks from simulation to the interaction devices.

Integrated simulation tools provide information about product functionalities on-line and allow the virtual prototype respond to a user's input. Modeling tools allow interactive changes of the virtual prototype.

The VR interfaces, together with the presentation and interaction units provide users with the impression of manipulating a real prototype. There are different VR devices available. Depending on the applications, suitable combinations of these VR devices must be chosen.

It should be noted that there is still much research and development work necessary to implement an interface with multi-sensoric realistic feedback.

Additionally, to use existing software and to integrate virtual prototyping into the product development process, following components are required:

- interface between the VR-system and the external tools, respectively
- interface to the product data model base

A virtual prototyping environment does not replace CAD systems. It works upon CAD data. Simulation systems provide additional functional information to a designed object.

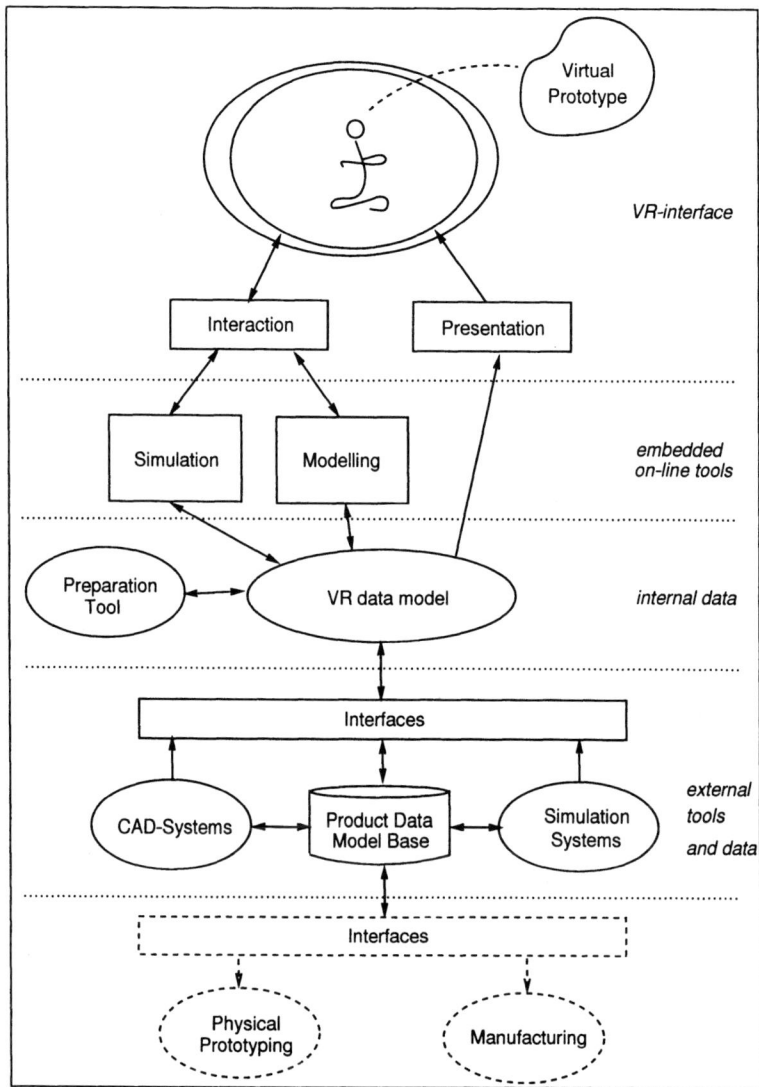

Figure 1 Virtual Prototyping Environment Architecture

In the future, the product data model base will contain the global, integrated product model data. Presently, this data base has to include individual model data from different CAD system vendors with their special data formats. CAD and simulation systems use special data formats which must be converted into the data format of the VR system.

Because CAD systems do not support the specification of materials, textures and other presentation attributes, a preparation tool is needed. Moreover, functional features like product structure, links and some physical properties have to be added.

4 SAMPLE APPLICATIONS

Our current work is focused on the integration of various existing results from the area VR, visualization and simulation to investigate these techniques. Some preliminary implementations of virtual prototyping demonstrates our concept through specific applications. It has provided an illustration of VR-supported design and simulation. Following, four typical application areas are described briefly.

Visual evaluation

Visual evaluation is one of the first application areas of virtual reality. This is also a kind of virtual prototyping, where audio-visual effects are of primary importance. Typically, a walk through is performed in order to get an impression about the spatial relationship of a 3D model. This is especially important in the field of architecture and interior design. In the past, IGD did numerous, high quality presentations together with industrial clients from the furniture industry, city planning, and construction offices.

Furthermore, first examples in the mechanical engineering domain were performed. The data model of the hospital area inside a huge ship has been visually evaluated (see Fig. 2, data provided by the German shipbuilding company Blohm&Voss). Due to highly complex pipeline systems inside a ship, unpermitted pipeline intersections can be constructed very easily. Such intersections have been identified very quickly, with the help of virtual reality walk through techniques.

In all projects we had to deal with "real world data", which means a geometric data base constructed by commercial modeling systems (e.g. AutoCAD, Applicon, ProEngineer). Presently, such systems do not provide sufficient information for virtual prototyping. Always necessary, is a preprocessing of the data base, before it could be passed to the VR system. Important issues are real-time requirements, data consistency, and data enhancement (e.g. specification of additional attributes, like texture, light, etc. for realistic rendering) [0].

Conceptual geometric design

For geometric design, the virtual prototype has to be manipulated in a way, that its form can be changed easily and directly. This implies an underlying data model which allows highly interactive form changes with real-time feedback. The manipulation method must be intuitive to enable a broad user acceptance. The integrated modeling tool does not have to be very powerful, but easy to use.

Fig. 3 presents a prototype system which features interaction techniques for real-time object manipulation. The interaction metaphor is the idea of a potter process. Using a glove a user can point onto the surface of a rotating cube. Real-time collision detection is applied in order to change the form of the cube according to the hand movement. This results in an on-line modification of the geometry. This first example is constructed with triangle meshes. With the experience gathered by this system, future work will concentrate on the integration of free-form-surfaces, like NURBS, and physically-based models (volume, elasticity, etc.).

Conceptual mechanical design

For mechanical design, kinematics is a typical problem often discussed in the conceptual design phase. Additional functional features are, for example, accessibility and assemblibility.

An assembly system has been implemented for research purpose (see Fig. 4) at IGD. With a glove, a user can control a full space cross-hair. Moving the cursor into an object and making a fist grabs this object. It will then follow the movements of the users hand. Open the fist, will release the object at its current location. With this the user is capable to assemble the objects. A precise collision detection with penetration avoidance is applied. When two objects are aligned properly to each other a snapping mechanism will assemble the parts automatically. Due to the lack of force feedback, acoustic signals give additional feedback to the user [0]. Dedicated sounds signalize the object grabbing and release, an object collision, as well as the object snapping.

There are generally two possibilities to evaluate mechanical systems in a VR environment [0] . One possibility is to simulate the mechanisms and assembly off-line, and to show the results in VR. More sophisticated virtual prototyping requires integrated on-line simulation. An example is presented in Fig. 5. During an immersive walk-through the cybernaut can interact with objects of the virtual environment, and these objects expose a realistic (i.e. physically correct) behavior. The lamp in Fig. 5 is represented by an open kinematic chain. It is mounted on the table and can be moved by grabbing parts of it. According to the constraints of the segment joints and the hand movement the on-line simulation results in a correct behavior of the lamp.

Training and prototype evaluation

A broad application field is the domain of training and prototype evaluation. The introduction of a new product on the market can be expedited, when last changes are easily possible and the training of involved people, like maintenance workers can be started before a real product is on hand.

For example, the change of a car engine can be done in a dry test and the generation of its appropriate maintenance manual description can be performed at a very early production level.

The evaluation of a prototype by potential customers can benefit from the VR technology too. No physical prototype mock-up is necessary and so the potential customers of the product (e.g. a new car) can contribute during the whole development process and not only at its final step, when larger changes are not possible anymore.

In contrast to the issues described in the other three application areas, the VR usage here is more oriented towards product user-relationship rather than the product developer-relationship.

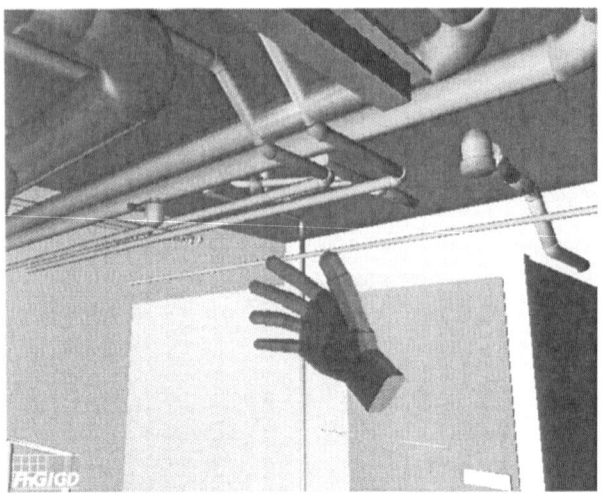

Figure 2 Visual Evaluation of a Ship Unit

Figure 3 Example of Geometric Design

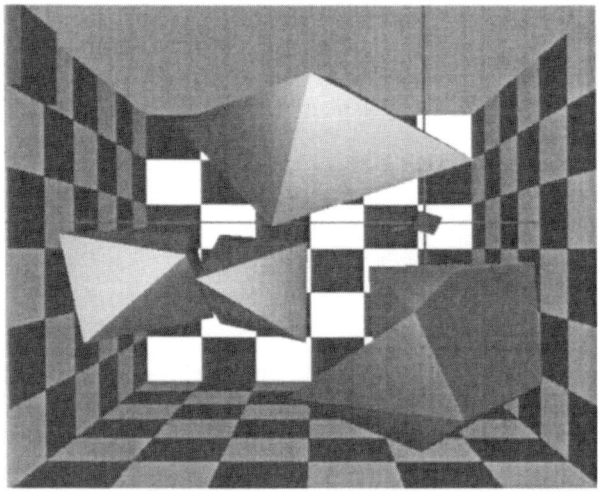

Figure 4 Example for Assembly in VR

Figure 5 On-line Kinematics Simulation

5 SUMMARY

The main problems of implementing a virtual prototype, using existing software and techniques are:

- the data models of graphics software, specifically VR systems, do not contain attributes for the description of product structure and physical properties
- CAD systems do not provide sufficient information for the presentation and simulation
- the VR interaction methods are not yet realistic enough
- there is no suitable tool for real-time geometrical analysis and physically-based simulations available

Based on the results of our investigations, we will continue to develop more sophisticated embedded simulation tools and to improve the VR interaction methods, especially for direct manipulation of virtual functional prototypes. Evaluation of the existing environment with additional design and engineering applications will be performed in order to develop requirements for additional system components. Furthermore, the integration of virtual prototyping into product development process requires the use of a common product data model and the combination of CSCW (computer supported cooperative work) and VR.

REFERENCES

Astheimer, P.: *What you see is what you hear - Acoustics applied to Virtual Worlds*, IEEE Symposium on Virtual Reality, San Jose, USA, October 1993

Astheimer, P., Dai, F.: *Dynamic Objects in Virtual Worlds - Integrating Simulations in a Virtual Reality Toolkit -* . European Simulation Symposium '93. Delft, Oct. 25. - 28. 1993.

Astheimer, P., Felger, W., Müller, S.: *Virtual Design: A Generic VR System for Industrial Applications.* in: Computers & Graphics, Vol. 17, No. 6, pp. 671-677, 1993

Astheimer, P., Felger, W., Göbel, M., Müller, S., Ziegler, R.: *Industrielle Anwendungen der Virtuellen Realität - Beispiele, Erfahrungen, Probleme & Zukunftsperspektiven.* in: Warnecke, H.J., Bullinger, H.-J. (Hrsg.): Virtual Reality '94, Springer-Verlag, 1994, pp. 261-280

Carrabine, L.: *Plugging into the Computer to Sense*, Computer-Aided Engineering, June 1990, pp. 16-26

Chapin, W.L., Lacery, T.A., Leifer, L.: *DesignSpace: A Manual Interaction Environment for Computer Aided Design.* Demonstration on CHI'94, Boston, Massachusetts USA, April 24-28, 1994

Dai, F.: *Integrated Planning of Robotic Applications in a Graphic-Interactive Environment.* Robotics and Autonomous Systems 8 (1991) 311-322. North-Holland, 1991.

Dai, F., Göbel, M.: *Virtual Prototyping - an Approach Using VR-Techniques* ASME International Computers in Engineering Conference 1994, Minneapolis, Minnesota, September 11. - 14., 1994.

Encarnacao, J.L., Astheimer, P., Felger, W., Frühauf, T., Göbel, M., Müller, S.: *Graphics and Visualization: The Essential Features for the Classification of Systems.* in: IFIP

5.2/5.10, Proc. ICCG 93, Bombay (India), Feb. 22-26, 1993

Felger, Wolfgang: *How interactive visualization can benefit from multidimensional input devices.* SPIE/IS&T Symposium on Electronic Imaging Science and Technology, Conference 1668: Visual Data Interpretation, San Jose (California), USA, February 9-14, 1992

Felger, W.: *Konzept und Realisierung eines Demonstrationszentrums für Anwendungen der Virtuellen Realität.* in: Proc. 3. GI-Workshop, Sichtsysteme, Wuppertal, 18./19. Nov. 1993

Fishwick P.A., Luker, P.A.: *Qualitative Simulation, Modelling and Analysis,* Series Advances in Simulation No. 5, Springer Verlag (1991)

Göbel, M. (ed): *Virtual Reality* (Special Issue), Computers and Graphics, Pergamon Press, Vol.17,6, November 1993

Hagen, H., Tomiyama, T. (eds.): *Intelligent CAD Systems I,* Springer Verlag (1987)

Kota, S., Chiou, S.J.: *Conceptual Design of Mechanisms Based on Computational Synthesis and Simulation of Kinematic Building Blocks.* Research in Engineering Design, 4, 1992. pp. 75-87.

Krüger, M.W.: *Artificial Reality,* Adison Wesly, 1991

Müller, S., Unbescheiden, M., Göbel, M.: *Genesis - Eine interaktive Forschungsumgebung zur Parallelisierung des Radiosity-Verfahrens für die virtuelle Welt,* Proceedings VR '93, Stuttgart, Februar 1993

NSF Report: *Research Directions In Virtual Environments.* Computer Graphics, Vol. 26, No. 3, August 1992.

Pahl, G., Beitz, W.: *Engineering Design.* The Design Council. Springer-Verlag, London 1984.

Wall, M.B., Ulrich, K.T., Flowers, W.C.: *Evaluating Prototyping Technologies for Product Design.* Research in Engineering Design (1992) 3, pp.163-177.

6 BIOGRAPHY

Fan Dai is a research stuff member and project manager of IGD. He was born in Changde, China 1962. He studied electrical engineering in Saarbrücken, Germany, where he received the Dipl.-Ing. degree. In 1991 he received his Dr.-Ing. degree (PhD) from the Technical University of Darmstadt.

Fan Dai was a researcher with the Interactive Graphics Systems group at the Technical University of Darmstadt from 1985 to 1990. Since 1990, he works with the Fraunhofer Institute for Computer Graphics (IGD) in Darmstadt. His recent work concentrate on dynamic virtual worlds and applications of virtual reality, especially virtual prototyping.

Fan Dai is a member of the IEEE Computer Society. He is also a member of the Association of Chinese Computer Scientists in Germany (GCI, group member of the Chinese Computer Federation), which he chaired from 1990 to 1992.

Wolfgang Felger was born in 1960 in Stockstadt am Rhein (Germany). He received his diploma in computer science from the Technical University of Darmstadt (Germany) in 1987. Since then he has been a staff member at the Fraunhofer Institute for Computer Graphics (IGD), where he received his doctoral degree (PhD) 1995. His research interests focus on scientific visualization and virtual reality. Currently he is technical manager of the virtual reality demonstration centre at Fraunhofer-IGD, with a focus on commercial applications and presentations using virtual reality techniques.

Felger made numerous contributions to international conferences and refereed journals. Furthermore he served as member of the programme committee in virtual reality conferences, organized by EUROGRAPHICS and IEEE.

Martin Göbel was born in 1955, he studied Computer Science at the Technical University of Darmstadt where he received the diploma degree in 1982. From 1982 until 1986 he was research assistant with the Interactive Graphics Systems group at the Technical University of Darmstadt. 1987 he joined the Fraunhofer Gesellschaft. Martin Göbel received the PhD (Dr.-Ing.) in 1990. He is author and editor of 5 books on Graphics Standards and Visualization.

Martin Göbel is head of the Visualization & Simulation Department of IGD. Currently, he is the project manager of the Fraunhofer Demonstration Centre for Virtual Reality in Germany. He chairs a special interest group on VR in Germany and has set up the first Eurographics Workshop on Virtual Environments in September '93 in Barcelona.

Martin Göbel is member of the IEEE Compter Society, EUROGRAPHICs and the German Computer Society (GI). He is actively partipating in EUROGRAPHICS working groups on SCIENTIFIC VISUALIZATION and VIRTUAL ENVIRONMENTS, and the German GI-group on IMAGING & VISUALIZATION.

PART SIX

Workshop Summaries

26

IFIP Workshop on Virtual Environments

José Teixeira
Centro de Computacao Grafica
Coimbra
Portugal

1   WORKSHOP OVERVIEW

The IFIP WG 5.10 Workshop on Virtual Environments was organised by CCG/ZGDV (Computer Graphics Center - Coimbra - Portugal) and was held at the Comissao de Coordenacao da Regiao Centro, in Coimbra, Portugal, on October 24-25, 1994. The workshop chairman was Prof. Dr. Jose Carlos Teixeira, CCG/University of Coimbra, and the Programme Committee joined scientists of high repute from nine countries. From the submitted contributions, the Program Committee selected twelve to be presented during the workshop. The workshop had 56 attendees from six countries (Brazil, Germany, Portugal, Spain, UK and USA). The main topics that had been proposed for the workshop were: Human Factors, Software Architectures, Modelling Tools and Techniques, Input and Output Devices, Interaction Techniques, Distributed Systems, Cooperative and Multi-User Systems and Applications.

The workshop devoted the first day to invited contributions and practical demonstrations and the second day to original papers and a final invited presentation. The invited speakers were: Mathias Wloka (Brown University, USA), Pierre duPoint (Division Limited, UK), Kirk Woolford (Academy of Media Arts, Germany) and Dr. Martin Goebel (IGD, Germany). Mathias Wloka described interaction techniques and discussed approaches to improve user participation in virtual environments. Virtual reality and promising applications for virtual environments were addressed by Pierre du Point from the application and commercial point of view. Among other applications, concurrent product design, automotive concept design, visualisation of architectural models and maintenance planing where presented. Kirk Woolford discussed tactile dialogue in networked environments and the development problems that have to be solved to enable touch feeling. The participants could feel the position of other visitors through physical stimuli; afterwards, it was discussed how this new approach could be used in industrial applications. The last invited presentation by Dr. Martin Goebel presented Virtual Environments as a new opportunity to improve industrial applications, and some approaches to solve technological problems. Examples of the use of Virtual Environments in industry were also discussed, as well as the technology improvements which should be expected. The presented papers were organised in four sessions:

- Simulation and Interaction in Virtual Environments
- Concepts and Tools
- Virtual Environments for Production
- Distributed Environments.

In the Simulation and Interaction in Virtual Environments session the three papers presented dealt with simulation support for virtual environments, motion planning for a mobile robot and the use of solid modelling in virtual environments. The use of distributed systems and different types of constraints were discussed and the results of the developments were presented.

In the Concepts and Tools session virtual reality was framed in the historical development of the reality representation techniques, the sensorial immersion problem was discussed and a new approach to real-time rendering was presented.

The importance of Virtual Environments for Production was discussed based on three perspectives: virtual prototyping, concurrent engineering and virtual manufacturing. The impact and opportunities of virtual prototyping environments for the industry and an architecture for supporting real-time collaboration in concurrent engineering within a distributed virtual environment were presented and discussed.

In the last workshop session, characteristics of large distributed virtual environments were discussed, as well as the use of 3D displays to perform application tasks, and the problem of high degree of realism and speed in virtual environments.

A virtual environment is an interactive computational space where the visitors interact directly with the models in an intuitive way. The interface and the support functionalities enable a direct communication between the user and the environment and increase the work productivity of the user. A virtual environment allows an user control of the space, time and viewing perspective. Therefore, some application areas (e.g. CAD, molecular modelling, urban planing, medical surgery, games, pedagogic software and tele-presence) are increasingly using virtual environments. We found the discussion of basic virtual reality technologies very important to better understand their application, as well as the problems and possibilities of virtual prototyping.

The participants classified the workshop as very interesting, since it combined the state of the art invited talks with hands-on demonstrations and research papers. The discussions during the workshop showed very well the high level of the speakers and attendees and the actuality of the work presented.

2   PROGRAMME COMMITTEE

P. Bono (FhG-CGRG, USA), P. Brunet (Univ. P. Catalunya, E), S. Bryson (NASA, USA), R. Earnshaw (Univ. Leeds, UK), J. Encarnaçao (GRIS, D), B. Falcidieno (IMA, I), A. Figueiredo (Univ. Coimbra, P), M. Gigante (RMIT, Australia), M. Goebel (FhG-IGD, D), G. Grinstein (Univ. Massachusetts, USA), M. Gross (ZGDV, D), R. Hubbold (Univ. Manchester, UK), T. Kunii (Univ. Aizu, Japan), L. Magalhaes (Unicamp FEE-DCA, Brasil), J. Rix (FhG-IGD, D), J.C. Teixeira (CCG/ZGDV, P), A. Wexelblat (MIT, USA)

27

Virtual Prototyping: Expectations and Realizations of a New Working Paradigm

Results of the First IFIP Workshop on Virtual Prototyping

Dr. Stefan Haas
Fraunhofer CRCG
167 Angell Street
Providence, RI 02906
Tel: (401) 453 6363
E-Mail: shaas@crcg.edu

1 WORKSHOP OVERVIEW

The IFIP WG 5.10 Workshop on Virtual Prototyping is the first international workshop organized by Fraunhofer CRCG's Virtual Prototyping group of Stefan Haas. The workshop was held in the Days Hotel from Sept. 21 - 23, 1994, in Providence, Rhode Island, USA. In response to the call for participation by the program chairs, Dr. Peter Bono, Fraunhofer CRCG, and Dr. Joachim Rix, Fraunhofer IGD, thirty-five participants from twenty-four institutions of the US, Germany, Italy, and the U.K. came together for this event. This was the first meeting dedicated to Virtual Prototyping, that merges CAD, virtual reality, scientific visualization, cooperative working, framework environments, product design, and manufacturing.

Beginning with invited talks by David Mizell, Boeing, and Prof. Henry Fuchs of the University of North Carolina, current research work and industrial applications in different areas of Virtual Prototyping were presented by fourteen papers in three sessions. Each of these sessions was followed by two parallel discussion sessions opening many exchange and communication possibilities in this new and rapidly growing area. Additionally, several participants used the opportunity to present short statements and introduce themselves.

The paper contributions were grouped in three sections:

- Environments,
- Product Models, and
- Technologies.

In the Environments session, four papers related to architectures and frameworks for virtual prototyping were presented. System integrators from American and German research and industry reported their requirements for open framework architectures, including distributed and mobile systems, concurrent engineering, cooperative work, and integrated product models.

In the Product Model session, more detailed aspects of the integrated product model were presented. Papers focused on feature modeling and feature recognition, as well as the role and the usage of the product model standard STEP.

Many applications using virtual prototyping components were presented in the third session about Technologies. Most of the applications showed the successful integration of virtual environments. The application examples showed how an immersed user can benefit from direct interaction with the virtual prototype.

The six discussion sessions were heavily used to discuss the participants' experience, expectations and views towards virtual prototyping. Main topics were technological questions about the impacts and barriers of the new technology of virtual prototyping.

Another frequently asked question was about the definition of virtual prototyping. As one result, this workshop turned out to be the first forum to discuss questions like this in detail. Therefore, an e-mail listserver (`vp@crcg.edu`) and a virtual library world-wide-web (WWW) home page (`http://vp.crcg.edu`) will be set up to continue this discussion forum electronically. For subscription to the mailing list send the `subscribe` message to the above address. The workshop program, the list of participants, and further proceedings can be obtained from `shaas@crcg.edu`.

Future activities of the whole group include setting up an informal consortium and organizing the next workshop to continue this successful meeting. For the next workshop, the time around ASME's 50th annual meeting focusing on the development of the digital design is being considered. Meanwhile, the consortium, in combination with the list server and the WWW page, will keep the discussion alive as well as share demo data and programs among this new community to ease the use of new technologies and to encourage evolving standards.

2 WORKSHOP RESULTS

2.1 What is Virtual Prototyping?

In recent years, new technologies like CAD/CAM, distributed work, client server architectures, and virtual reality as well as the advances in 3D computer graphics have had strong impacts on the way of working in manufacturing automation. More and more steps from the very first design steps to the manufacturing as well as life cycle control and product management can be computer supported.

One major step with strong influences on product prices, quality, and availability is the product prototyping. Traditionally, prototypes have to reflect all attributes of the later product prior to starting its production. Depending on the requirements of the product, some or even many prototypes have to be built; a time and resource-intensive process. Especially when testing complex and interfering features, successive prototypes have to be built anew rather than being modified.

The increasing usage of digital information storage and manipulation of product information raises many questions, such as:

- Can the drawbacks of physical prototyping be overcome by using digital processing techniques ?
- Which new working techniques are possible, if several applications and users can interact simultaneously with the virtual prototype?
- How can we step ahead out of the existing work environment?

These general questions extend the basic scope of computer-supported creation of product specification data, e.g. its shape and properties, to a virtual product having identical (virtual) properties and being able to appear and respond exactly like the real product.

This is challenging information technology to deliver physically correct results within the product tolerances. Even though the basic arithmetic accuracy of computers is already extremely high, other requirements like real-time processing, handling of large data sets, and immersing human user interfaces have to be solved efficiently to make the virtual product (almost) real.

The IFIP Workshop on Virtual Prototyping was the very first discussion and presentation forum dedicated to this still rather vague definition. Several papers discussed ongoing R&D work and showed the benefits of first virtual prototyping applications. Each paper section was followed by parallel discussion sections reflecting the need for this forum. Discussion is Especially important as virtual prototyping has not yet defined itself and its position in the CAD/CAM, computer graphics, and manufacturing world.

2.2 New Perspectives for Design and Manufacturing

Virtual prototyping is more than the integration of digital geometry data. Throughout the workshop and all discussions the participants left no doubt about this. They expected virtual prototyping to be more, to go beyond existing working techniques demanding new methodologies for the CAD/CAM environment.

Thus, virtual prototyping is redesigning existing concurrent engineering strategies towards interaction between people, tools, and product data. This overcomes the restrictions of existing product development tools and environments in which parallel product development technologies are mostly independent and unaware of each other.

One paradigm found in many virtual prototyping approaches is visualization-based interaction with the product model, such as in immersive VR mock-up environments or between different CAD tools. This yields interactive and simulation-based design changes rather than iterative refinement loops typical of classical concurrent engineering.

Many workshop participants considered this paradigm as a major keystone of their companies agility, one of the most important aspects for productivity, flexibility and quality of their work.

Another important aspect characterizing virtual prototyping is the ability to model complex product informations in an exchangeable, preferably standardized way. This is necessary to utilize the versatility of virtual prototyping tools together with existing ones, which do not support a seamless integration.

STEP with its target towards modeling scenarios, called application protocols, including data, methods, access, and the necessary migration steps, was considered to play an important role for the integration task. STEP not only provides a standard way of expressing CAD entities, but also defines the environment in which they exist and the context semantics build by these structures.

High level application protocols define the relationship of all basically required informations for a given application scenario. This can include parametric design and feature models, all of which have great impact on virtual prototyping. Interactive design changes can be supported efficiently, one of the major requirements for the success of virtual prototyping.

Several demonstrations showed how these requirements can be utilized with existing state-of-the-art computer technology. The mostly virtual-reality-based applications showed assembly testing and ergonomics studies, exploiting the high degree of flexibility of parameterizable models, e.g., for cockpits and consoles. Other applications focused on the integrated production environment and its ability to generate individualized products and generate manufacturing information while the customer is waiting. Another group of applications focused on the communication between existing applications by means of CSCW (Computer Supported Cooperative Work).

2.3   Discussions on Virtual Prototyping

In between the paper sections parallel discussion sections gave the workshop participants time to talk about the new subject of virtual prototyping. Therefore, the discussions covered different technical aspects and tried to enlighten:

- Barriers of virtual prototyping
- Opportunities of virtual prototyping
- What has to be done next

The following were generally seen as the most significant barriers to virtual prototyping:

- The State-of-the-Art – virtual prototyping systems are still of limited practical value due to output constraints stemming from both overall virtual prototyping system design and the functionality of individual system components.
- Education – very few decision makers in business organizations have anything beyond a cursory understanding of virtual prototyping. It is a technology in need of articulate internal champions.
- Human Factors – virtual prototyping is still difficult for the potential user to become accustomed to, and represents a substantial change in the way an engineer will work. Therefore, the introduction of virtual prototyping will need to overcome occupational entropy.
- Cost/Benefit/Quality – before virtual prototyping can be adopted by an organization, its advocates must be prepared to present a clear, concise argument on its behalf which outlines the cost, but also details the benefits, including gains, according to a recognized metric.
- Lack of Competition – if there were competitive vendors offering virtual prototyping systems, there would be more attention paid to the technology.
- Migration – non-digital and existing data should be usable in the virtual prototyping system. On some systems, this has to be done even without shutting down the system
- Data Exchange – has to be bi-directional without losses between all involved CAD systems

The opportunities and driving factors for virtual prototyping were seen in the following areas:

- Human Factors – take the person as a person rather than a machine user
- Maintainability – how can large systems be maintained, especially if they are one-of-a-kind?

- Safety Issues – try to predict safety issues, as that can arise from bad console assembly arrangements when designing cockpits. As virtual prototypes are cheap and accessible for everyone, do security tests in parallel
- Compliances/OSHA/EPA
- Cost Reduction – try to reuse digital data to avoid data generation costs and errors
- Time reduction – speed up the design of new products or new product versions
- Quality – reliability is increased by pre-tested prototypes
- Innovation Ability – respond quickly to new customer requirements, e.g., for airplane or car customization
- Interactive Design – integrate the customer in the design decision process. Be aware of improving the understanding, e.g., of alternatives, for the customer rather than irritating with too much flexibility.

3 WORKSHOP SUMMARY

The response of all participants at this three day workshop showed that there is much need for an information and discussion forum in the field of virtual prototyping. Besides the scheduled paper presentation, an additional short paper section gave further people the possibility to report about their activities, which was accepted by several participants. Likewise, discussions were only initiated by the discussion sections, but most of them continued afterwards.

The results of this workshop were mainly in finding out about ongoing activities and discussing how to continue working on this topic. The following statements comprise these spirit and should be used to continue the current activities:

- First results show the impacts of the new working paradigm, but we have to keep on going.
- Tools have to improve performance
- Digital product models have to support all product data
- Digital tools must be extended to achieve better support for manufacturing processes, especially when working together
- Immersive interactions have to free human users rather than tying them
- Education – Very few decision makers in business organizations have anything beyond a cursory understanding of virtual prototyping. It is a technology in need of articulate internal champions.
- Migration from existing systems, either digital (plotting etc.) or non-digital and for up-and-running systems
- Integration of simulation and feedback from physical systems
- The information exchange and discussion has to be continued, e.g. by the second IFIP Workshop on Virtual Prototyping (to be held Nov./Dec. 1995 in Arlington, TX)

INDEX OF CONTRIBUTORS

KEYWORD INDEX

Printed by Books on Demand, Germany